LS 23898

D1256248

PHYSICS OF FORMATION OF FeII LINES OUTSIDE LTE

ASTROPHYSICS AND
SPACE SCIENCE LIBRARY

A SERIES OF BOOKS ON THE RECENT DEVELOPMENTS
OF SPACE SCIENCE AND OF GENERAL GEOPHYSICS AND ASTROPHYSICS
PUBLISHED IN CONNECTION WITH THE JOURNAL
SPACE SCIENCE REVIEWS

VOLUME 138

PROCEEDINGS

PHYSICS OF FORMATION OF FeII LINES OUTSIDE LTE

PROCEEDINGS OF THE 94TH COLLOQUIUM OF THE
INTERNATIONAL ASTRONOMICAL UNION
HELD IN ANACAPRI, CAPRI ISLAND, ITALY,
4–8 JULY 1986

Edited by

ROBERTO VIOTTI

Istituto di Astrofisica Spaziale (CNR), Frascati, Italy

ALBERTO VITTONE

Osservatorio Astronomico di Capodimonte, Napoli, Italy

and

MICHAEL FRIEDJUNG

Institut d'Astrophysique (CNRS), Paris, France

D. REIDEL PUBLISHING COMPANY

A MEMBER OF THE KLUWER  ACADEMIC PUBLISHERS GROUP

DORDRECHT / BOSTON / LANCASTER / TOKYO

Library of Congress Cataloging in Publication Data

International Astronomical Union. Colloquium (94th : 1986 : Anacapri, Italy)
   Physics of formation of FeII lines outside LTE.

   (Astrophysics and space science library; v. 138)
   Includes index.
   1. Astronomical spectroscopy—Congresses. 2. Iron—Spectra—
Congresses. 3. Astrophysics—Congresses. I. Viotti, Roberto, 1939–
II. Vittone, Alberto. III. Friedjung, M., 1940–   . IV. Title.
V. Series.
QB465.I54 1986   523.01′584   87–26384
ISBN 90–277–2626–4

Published by D. Reidel Publishing Company,
P.O. Box 17, 3300 AA Dordrecht, Holland.

Sold and distributed in the U.S.A. and Canada
by Kluwer Academic Publishers,
101 Philip Drive, Assinippi Park, Norwell, MA 02061, U.S.A.

In all other countries, sold and distributed
by Kluwer Academic Publishers Group,
P.O. Box 322, 3300 AH Dordrecht, Holland.

TABLE OF CONTENTS

SESSION 4 - FUTURE PLANS AND CONCLUSIONS

PREFACE

The same kind of physics is frequently common to very different fields of Astrophysics, so experts in each of these fields have often much to learn from each others. It was therefore logical that the International Astronomical Union should sponsor a colloquium about an ion which produces many spectral lines that can be used as a diagnostic for many sorts of objects, and which may sometimes have a major influence on physical processes occurring in astrophysical sources.

The lines of singly ionized iron (FeII) are present in absorption and emission in the spectra of objects such as the Sun, cool stars, circumstellar envelopes of hot stars, novae, diffuse nebulae including the supernova remnants, and active galactic nuclei. These lines are very often formed far from LTE, and their interpretation is not easy in view of the complex Grotrian diagram for FeII, and the gaps in the knowledge of various physical parameters. In addition, the density of very strong FeII lines becomes very large in the ultraviolet, and the lines can play a major role in the line blanketing. They need therefore to be taken into account in any energy balance argument.

This volume presents the proceedings of what we think to have been only the second meeting on this subject (*), and the first sponsored by the International Astronomical Union. Specialists in many different fields came to the Capri Colloquium on the 'Physics of Formation of FeII Lines Outside LTE' and gave their contribution to the discussion on many aspects of the problem. Different sessions of the colloquium were devoted to: (1) Basic atomic data. (2) Observation of FeII in different astrophysical objects from the solar photosphere to stellar chromospheres and winds, from novae to active galactic nuclei. (3) Theory of line formation and models, including data analysis techniques and spectral synthesis. (4) Prospects of future research, including new laboratory work, ground and space observations.

There were 29 invited and contributed papers and 13 posters. Each talk was followed by lively discussions which unfortunately could not be included in the proceedings owing to the constraints on the size of the book.

(*) A previous meeting on FeII was organized by M. V. Penston at the European Space Agency IUE Station of Villafranca (Madrid, Spain) on 3-5 October 1979, and was reported in Nature, Vol.282, p.557 (1979).

Both recent progress and outstanding problems were highlighted. Work is still very much humpered by the lack of good atomic data. Though oscillator strengths are beginning to be well known, much work is still required on collision strengths and particularly on ionization parameters. Nobody seems to be working on the latter, while the difficulties of interpreting the FeII spectra of active galactic nuclei could be only due to badly known physical quantities. In addition, theory and observations are often far apart; the theoretical calculations of line intensities and semi-empirical methods of analysis of observed line intensities, need to be brought closer together. Thus much work still remains to be done before the 'FeII Problem' can be considered to be solved.

This Colloquium was sponsored by IAU Commission 29 (Stellar Spectra), and co-sponsored by IAU Commissions 12 (Radiation and Structure of the Solar Atmosphere), 14 (Fondamental Spectroscopic Data), 36 (Theory of Stellar Atmospheres) and 48 (High Energy Astrophysics). We would like to thank the Presidents of these Commissions for their support.

We are very grateful to the many organizations which also sponsored the Colloquium, the Osservatorio Astronomico di Capodimonte, the Consiglio Nazionale delle Ricerche, the Istituto Astrofisica Spaziale, the Ministero della Pubblica Istruzione, and the Regione Campania.

The scientific organization was undertaken with the enthusiastic help of all the members of the Scientic Organizing Committee, and largely profited for the wide use of the Earnet/Bitnet computer network. From this point of view we believe that this was the first international meeting for which this network has placed such a large role.

We are indebted to the Local Organizing Committee for making perfect arrangements on the island of Capri. We have also greatly appreciated Lidia Barbanera, Teresa Ievolella and Dario Mancini for their helpful cooperation before and during the meeting.

The Editors.

SCIENTIFIC ORGANIZING COMMITTEE

R. Viotti, Frascati (Italy), Chairman
B. Baschek, Heidelberg (Germany Fed. Rep.)
A.M. Boesgaard, Honolulu (USA)
M. Friedjung, Paris (France)
E.A. Gurtovenko, Kiev (USSR)
S. Johansson, Luud (Sweden)
M. Joly, Meudon (France)
C. Jordan, Oxford (United Kingdom)
H. Netzer, Tel Aviv (Israel)
H. Nussbaumer, Zurich (Switzerland)
M. Rigutti, Napoli (Italy)

LOCAL ORGANIZING COMMITTEE

A. Vittone, Chairman
M.T. Gomez
G. Longo
G. Severino

LIST OF PARTICIPANTS

A. Altamore, Roma, Italy
G.B. Baratta, Roma, Italy
B. Baschek, Heidelberg, Germany
T. Brage, Lund, Sweden
G. Busarello, Napoli, Italy
M. Carlsson, Oslo, Norway
K.G. Carpenter, Boulder, USA
A. Cassatella, VILSPA, Spain
F. Castelli, Trieste, Italy
D. de Martino, Napoli, Italy
M.S. Dimitrijevic, Beograd, Yugoslavia
D. Dravins, Lund, Sweden
T. Fernandez Castro, Madrid, Spain
M. Friedjung, Paris, France
R. Gilmozzi, VILSPA, Spain
M.T. Gomez, Napoli, Italy
L. Gratton, Roma, Italy
M. Gros, Paris, France
J.E. Hansen, Amsterdam, Netherlands
S. Johansson, Lund, Sweden
M. Joly, Meudon, France
C. Jordan, Oxford, UK
S. Krawczyk, Torun, Poland
R.L. Kurucz, Cambridge, USA
A. Lesage, Meudon, France
G. Longo, Napoli, Italy
C. Marsi, Trieste, Italy
J. Moity, Meudon, France
G. Muratorio, Marseille, France
A. Natta, Firenze, Italy
H. Netzer, Tel Aviv, Israel
A.E. Nilsson, Lund, Sweden
H. Nussbaumer, Zurich, Switzerland
E. Oliva, Firenze, Italy
U. Pauls, Zurich, Switzerland
C. Rossi, Roma, Italy
G. Russo, Garching, Germany
R.J. Rutten, Utrecht, Netherlands
P.L. Selvelli, Trieste, Italy
G. Severino, Napoli, Italy
A. Talavera, VILSPA, Spain
K.L. Vankata Krishna, Norman, USA
R. Viotti, Frascati, Italy
A. Vittone, Napoli, Italy
E.J. Wampler, Garching, Germany
B.J. Wills, Austin, USA
F.J. Zickgraf, Heidelberg, Germany

GENERAL INTRODUCTION TO THE "FE II PROBLEM" *

B. Baschek
Institute of Theoretical Astrophysics
University of Heidelberg
Im Neuenheimer Feld 561
D-6900 Heidelberg
Federal Republic of Germany

ABSTRACT. A largely historical introduction is given to selected problems connected with observations and interpretations of iron lines in astronomical objects.

1.   INTRODUCING THE LINE-RICH AND UBIQUITOUS SPECTRUM OF FE II

When asked to give a general introduction to the topic of this Colloquium, I was reluctant to accept because it seemed to me that everyone but me knew what the "Fe II problem" is. Assured, however, by Dr. Viotti that I would be essentially free to decide what to say and that a historical and even subjective introduction would be welcome, I'm going to try to present to you some old and new problems associated with iron in the universe. In this, I will not say much about to atomic physics aspects of Fe II as this is dealt with by the next speaker.

Why is it that Fe II attracts so much attention that an entire IAU Colloquium is devoted to this ion? First, iron is one of the abundant elements in the universe. With an abundance by number of $3 \cdot 10^{-5}$ of that of hydrogen, it ranks only after H,He,C,N,O, and Ne in the solar composition, and is comparable in abundance to S,Mg, and Si. Then, Fe II has a very line-rich spectrum and the laboratory work on it is by no means completed. Numerous allowed and forbidden lines show up in the optical range and have long been observed in a variety of astronomical sources (cf. the Grotrian diagram by Merrill, 1956). In recent times, also the ultraviolet lines of Fe II became accessible to observation, mostly through satellites.

The richness in lines of Fe II originates in its about half-filled 3d-shell. Let us compare Fe II (ground configuration $1s^2 2s^2 2p^6 3s^2 3p^6 \cdot 3d^6 4s$) with Ca II in the ground state of which ($1s^2 2s^2 2p^6 3s^2 3p^6 \cdot 3d^0 4s$)

---

* This paper is dedicated to the memory of *Johannes Richter* (1925-1986) of the Institut für Experimentalphysik, University of Kiel, whose measurement of oscillator strengths for iron and other elements has helped to put astronomical spectroscopy on a firmer foundation.

R. Viotti et al. (eds.), Physics of Formation of FeII Lines Outside LTE, 1–8.
© *1988 by D. Reidel Publishing Company.*

the 3d-shell is not occupied. While in Ca II between the three lowest terms 3d^{O}4s ^{2}S, 3d ^{2}D, and 3d^{O}4p ^{2}P two multiplets with together five (allowed) line transitions occur, namely the H and K lines and the infrared triplet around 8500 Å, the number of terms in Fe II is increased to 24 for the 3d^{6}4s configuration, to 8 for 3d^6·3d = 3d^7, and to 68 for 3d^{6}4p leading to far more than 10 000 lines if one further accounts for the increased multiplicity (maximum 6).

At last, the excitation and ionization conditions for populating Fe II levels are very favourable in a great number of astronomical objects. Due to the ionization potential of 7.9 eV for Fe I and 16.2 eV for Fe II, moderately hard ultraviolet photons ($\lambda \lesssim$ 1570 Å) are sufficient to ionize neutral iron, and temperatures between about 5000 to 20 000 K are typical for producing strong Fe II lines. Consequently, we observe Fe II absorption lines in stellar photospheres of "intermediate" temperatures, and a few lines absorbed by the interstellar medium. Allowed as well as forbidden Fe II emission lines arise from thin "envelopes" around a variety of objects such as novae and supernovae, Be stars, T Tau stars, symbiotic stars, the solar and stellar chromospheres, H II regions, and the active nuclei of Seyfert 1 galaxies and quasars. Particularly complicated Fe II line profiles comprising emission and absorption components are observed in variable supergiants like S Dor and η Car.

This ubiquity and variety of Fe II lines in the universe makes their analysis such a fascinating topic. Furthermore, the interpretation of the complicated spectrum of Fe II demands "interdisciplinary" collaboration, in particular between atomic physicists and astrophysicists to provide the large amount of atomic data required for the kinetic rate and radiative transfer equations.

2. EARLY SPECTROSCOPY

After I. Newton's discovery of the spectral dispersion of the white sunlight in 1666, it still took about 150 years before spectral *lines* were observed. While in 1802 W. Wollaston could resolve 7 lines (or groups of lines) in the solar spectrum, by 1812/14 thanks to improved instrumentation J. Fraunhofer succeeded in seeing 10 strong lines (named A a B C D E b F G H) and almost 600 fainter absorption lines, among them the "d line" at 4383.6 Å which is - as we know today - the strongest Fe I line.

The first successful photographic recording of the Fraunhofer lines was accomplished by E. Becquerel in 1842. In the same year, Ch. Doppler formulated his law of the change of wavelength according to relative motions which could later be verified by astronomical observations of radial velocities.

The long-standing problem of the origin of the Fraunhofer lines required many steps (e.g. by L. Foucault 1849, A.J. Ångström 1855) before the "discovery of spectral analysis" by G. Kirchhoff and R. Bunsen (1859) in Heidelberg opened the door for quantitative spectroscopy. Kirchhoff and Bunsen themselves succeeded 1860/61 with the first spectroscopic discovery of new elements in terrestrial material, of caesium and rubidium, named after their blue-gray and red lines, respectively.

Already in 1861/63 G. Kirchhoff performed the first analysis of the solar spectrum by comparing the Fraunhofer lines with the laboratory spectra of 33 elements most of which had been recorded by himself. In particular, the coincidence of 60 lines in the spectrum of iron with solar lines established the existence of this element in the photosphere without any doubt. (Fortunately, this procedure was successful because the Sun is a relatively cool star with ionization and excitation conditions not greatly different from those in the laboratory sources).

From the 1860s onward, stellar and nebular spectroscopy made rapid progress due to the work of W. Huggins, W.A. Miller, L.M. Rutherfurd, A.Secchi, G. Airy and others. It was soon recognized that everywhere in the universe the same elements are found as on Earth. The observations of previously unknown spectral lines in astronomical sources were repeatedly attributed to "new" elements. In the case of helium, the "solar element", observed by the 5876 Å line in the chromospheric flash spectrum at the eclipse of 1868, a new element was actually discovered, in other cases, discoveries led to the recognition of unusual physical conditions such as the "nebulium" lines 4957/5007 Å and the "coronium" line 5302 Å which only decades later could be identified as forbidden lines of O III and Fe XIV, respectively.

Shortly before the discovery of the electron in 1897 by J.J. Thomson, N. Lockyer noted that the "white stars" exhibit the socalled enhanced or spark lines and put forward the hypothesis that the atoms are composite and break up at high temperatures to yield proto-elements, e.g. proto-iron or, as we would say today, Fe II. Perhaps we may define 1897 as the birth year of the Fe II problem?

The theory of ionization in stellar atmospheres was not formulated until 1920 by M.N. Saha.

3. EARLY QUANTITATIVE STELLAR ANALYSES

Quantitative spectroscopy became possible when from the 1920s onward quantum mechanics began to provide the necessary physical background, in particular absorption coefficients, oscillator strengths, data on pressure broadening etc.

The extensive pioneering analysis of the solar spectrum by Russell (1929) still had to be based upon eye estimates of the Fraunhofer line intensities which were calibrated by means of relative intensities within a multiplet, furthermore absolute f-values were not yet available. Nevertheless, Russell was able to derive solar abundances which are remarkably close to the modern values. In particular, he found that Fe, Mg and Si have about the same abundance.

Russell's deep insight into the line formation in a stellar analysis may be illustrated by a few quotations from his Halley Lecture (Russell, 1933): "Only a fraction, often a very small fraction indeed, of the atoms of a given element are at work in the production of a given line. They must be not only of the right degree of ionization but in a definite state of excitation, and, even so, some will produce one line and some another, in accordance with definite transition probabilities"..,

"The absence of many familiar elements from the solar spectrum, which was long unexplained, is now easy to understand. It arises, not in the sun at all, but in the earth's atmosphere. A small but mischievous amount of ozone in the higher regions, utterly beyond the reach of air-craft or even of sounding balloons, cuts off all entering radiations shorter than 2,900 Angstroms. This is the great tragedy of terrestrial astrophysics, for the region thus hopelessly barred to us contains more of interest and importance than any other part of the spectrum".

Russell determined an ionization potential of 8.5 eV of "an element whose atoms would be just half ionized" and recognized that e.g. iron is mostly singly ionized in the photosphere : "A great majority of the metals have ionization potentials lower than this, so that they are prepon-derantly ionized in the sun. Yet the solar spectrum is usually described as characterized by the arc lines of the metals. The earth's atmosphere is again to blame. For most of the richer metallic spectra... the stron-gest enhanced lines lie beyond λ 3,000, and are missed out. Could we enter the forbidden region, we would doubtless find the enhanced lines stronger. Only one group of the greater ones is accessible - the H and K lines of Ca^+. These dwarf all other observable lines; but the great magnesium pair at 2795, 2802 - just out of reach - probably much surpass them". A remarkable prediction of ultraviolet line strengths!

Problems with the determination of the abundance of the dominant element H in the sun could be resolved when Wildt (1939) relized the negative hydrogen ion as the main absorber.

The first detailed quantitative analysis of a stellar spectrum, that of the BO V star τ Sco, was performed by Unsöld (1941/44). The interesting question of the iron abundance in such a hot, young star could hot be answered at that time as the spectral resolution was not sufficient to recognize the faint Fe III lines.

4.   REMARKS ON FE II EMISSION LINES

In view of the contributions by Dr. Gratton and Dr. Viotti at this Colloquium, I will restrict myself to only a few remarks on the early quantitative analyses of iron emission lines.

The observations of 1937/39 of the emission lines of H, He I, and Fe II in the Be star γ Cas were analysed by Wellmann (1951). Of the 21 Fe II lines between 4173 and 5317 Å neither b-factors nor absolute f-values were known. By assuming a common b-factor for the levels and applying relative f-values within multiplets and supermultiplets, how-ever, Wellmann could draw the following **conclusion**: the dilution factor W is important for the line strengths; due to the spread in radial velocities there is hardly any self-absorption; the emission originates from a region separated from the star, and iron is dominantly in the form of Fe III. Characteristic parameters are $W \simeq 10^{-2}$, electron density $N_e \simeq 10^9$ cm^{-3}, and $T \simeq 15\ 000$ K.

Regarding the interpretation of forbidden Fe II lines in the sun and stars, e.g. in η Car, I mention the work of Pagel (1968) and Viotti (1969).

5. THE "IRON PROBLEM" OF THE 1960s

Based upon improved observational material Goldberg, Müller and Aller (1960) published a very detailed analysis of the solar photospheric spectrum. In order to derive element abundances they utilized a model atmosphere in combination with weighting functions and collected the measured and calculated oscillator strengths which had by then become available in increasing number. The abundance of iron was determined from Fe I only, using the relative f-values of King and King (1938) and Carter (1949) on the absolute scale of Bell et al. (1958). The result was lg N(Fe)/N(H) = 6.5 with the usual normalization lg N(H) = 12.0 (and the small correction by Zwaan, 1962). In contrast to Russell's (1929) earlier result and to the meteoritic data where Fe and Si are of comparable abundance, this new abundance of iron came out about a factor of 10 lower. The problem became more serious in 1963/67 when – after the pioneering work of Woolley and Allen (1948) – detailed studies of the abundances in the solar corona by Pottasch (1963/64) and Jordan (1966) yielded a high iron abundance of about 7.5. A similar value also resulted from the forbidden photospheric [Fe II] lines (Swings, 1965).

It is interesting to recall that Groth (1961) derived a "high" iron abundance of 7.5 from low-excitation Fe I lines in the supergiant α Cyg (A2Ia). He considered this as a probably real deviation from the solar abundance.

As we have seen, Fe II is the dominant ion in the solar photosphere so that it would be best to use its lines to derive the iron abundance. Oscillator strengths, however, became available only by Roder's (1962) experimental work (besides "astrophysical" f-values). Roder's absolute scale was linked to that of the Fe I socillator strengths. However, since photospheric Fe II lines yielded an iron abundance which differed from that based on Fe I lines, Baschek et al. (1963) suggested a correction to Roder's absolute f-values – unfortunately a wrong conclusion due to their belief in the correctness of King and King's Fe I oscillator strengths.

The solution of the solar iron problem emerged around 1967/69 by improved and new experimental methods to measure f-values, e.g. wall-stabilized arcs (Garz and Kock, 1969), shock-tube measurements (Huber and Tobey, 1968; Grasdalen et al., 1969), and beam-foil techniques (Whaling et al. 1969) and resulted in a revision of the Fe I oscillator strengths of King and King whose high-excitation lines turned out to be systematically too large (corresponding to an error in their furnace temperatures by a few 100 K). With the revised f-values, the photospheric iron abundance now because 7.5, too. New measurements of Fe II oscillator strengths (Baschek et. al.,1970) essentially confirmed Roder's values and resulted in a photospheric iron abundance of 7.6. For more details, see e.g. Garz et al. (1969) and Withbroe (1971).

A different "iron problem" of the 1960s arose from a series of papers by Pecker and collaborators in the Ann. d'Astrophysique (1959/61) where non-LTE effects in the sun of an order of magnitude were claimed for metals such as iron, a result not confirmed by subsequent studies (cf.e.g. Cayrel, 1965). The non-LTE analysis of Athay and Lites (1972) was based on 15 levels of Fe I; modern calculations include of the order

of 100 levels of Fe I/II and show that non-LTE effects are modest for
Fe I and only small for Fe II in A to G stars (e.g. Saxner, 1984; Steenbock, 1985; Gigas, 1986).

6.   OUTLOOK: SOME PRESENT-DAY FE II PROBLEMS

     Line spectroscopy of Fe II during the last years is characterized
on the one hand by observations in the ultraviolet which, for fainter
sources, have become feasible particularly through the IUE satellite
(launch in 1978), on the other hand by the rapid increase of the sensitivity of optical detectors.
     In this concluding section I will first briefly refer to my own
recent work on Fe II, and then sketch some more or less arbitrarily
selected recent Fe II problems in other fields.
     High-resolution ultraviolet spectra of early-type stars obtained
with IUE exhibit numerous absorption lines of which at most half can at
present be identified (cf. e.g. Baschek, 1983). In particular for A stars,
Fe II is very likely to contribute to a larger number of the unidentified
lines. We began a collaboration between the atomicphysics group at Lund
and the theoretical astrophysics group at Heidelberg on laboratory ultraviolet analysis of Fe II based upon new hollow-cathode spectra in the
range 1300 to 3200 Å where many lines including strong ones are as yet
not identified. As a first result we founde some 120 new doubly excited
Fe II lines in the $(3d^54s4d+3d^54p^2)$ configuration system of which many
could be identified in the IUE high-resolution spectrum of the sharp-
lined B 9.5 V star 21 Peg. For further details, we refer to the poster
contributions at this Colloquium by Baschek and Johansson, by Brage
et al., and by Adam et al. and to Adam et al. (1986).
     A very active field of research is the diagnosis based upon Fe II
and [Fe II] lines of low-density plasmas around a variety of different
objects with the aim to obtain a model for the stratification, the
velocity fields, the excitation conditions etc. Of interest are - and
will be discussed at this Colloquium - e.g. stellar chromospheres, Be
stars, ζ Aur systems, S Dor variables, novae and supernovae. Out of
recent discoveries I would like to mention here that [Fe II] lines have
even been observed in the jet of the Herbig-Haro object HH 34 (Bührke
and Mundt, 1986), and that the infrared [Fe II] line at λ 1.644 μm
(a^4F-a^4D) has been detected in the supernova SN 1983 N (Graham et al.,
1986) so that the "formation" of about 0.3 $M_\odot$ of Fe by the decay of
$^{56}Ni \rightarrow ^{56}Co \rightarrow ^{56}Fe$ can be inferred directly.
     Finally, strong Fe II emission in the optical as well as in the
ultraviolet is observed in many quasars and Seyfert 1 galaxies following
the first detection of optical Fe II lines in 3 C 273 by Wampler and
Oke (1967). There seems to be a problem in these "Fe II galaxies" since
the total emission in the Fe II lines exceeds that in Lyman α (cf. e.g.
Wills et al., 1985). Possible excitation mechanisms are radiative excitation or fluorescence either by Lα, by the Mg II lines or by numerous
lines within Fe II itself. The optical thickness of the clouds surrounding
the active galactic nucleus essentially determines the ratio of optical
to ultraviolet Fe emission.

Surely, the detailed discussion and the solution of these problems will be the task of the many contributions by specialists in the various subjects to be presented at this IAU Colloquium.

7. REFERENCES

Adam, J., Baschek, B., Johansson, S., Nilsson, A.E., Brage, T.: 1986,
     Astrophys. J. (Dec. 15), in press
Athay, R., Lites, B.: 1972, Astrophys. J. 176, 809
Baschek, B.: 1983, Highlights Astr. 6, 781
Baschek, B., Garz, T., Holweger, H., Richter, J.: 1970, Astron. Astro-
     phys. 4, 229
Baschek, B., Kegel, W.-H., Traving, G.: 1963, Z. Astrophys. 56, 282
Bell, G.D., Davis, M.H., King, R.B., Routly, P.M.: 1958, Astrophys. J.
     127, 775
Bührke, T., Mundt, R.: 1986, IAU Symp. No. 122 on *Circumstellar Matter*,
     Heidelberg, in press
Carter, W.W.: 1949, Phys. Rev. 76, 962
Cayrel, R.: 1965, Proc. 2nd Harvard-Smithsonian Conference on Stellar
     Atmospheres, Smithsonian Astrophys. Spec. Rep. No. 174, 453
Garz, T., Kock, M.: 1969, Astron. Astrophys. 2, 274
Garz, T., Kock, M., Richter, J., Baschek, B., Holweger, H., Unsöld, A.:
     1969, Nature 223, 1254
Gigas, D.: 1986, Astron. Astrophys. 165, 170
Goldberg, L., Müller, E.A., Aller, L.H.,: 1960, Astrophys. J. Suppl. 5,1
Graham, J.R., Meikle, W.P.S., Allen, D.A., Longmore, A.J., Williams, P.M.:
     1986, Mon.Not.R.astr.Soc. 218, 93
Grasdalen, G.L., Huber, M., Parkinson, W.H.: 1969, Astrophys. J. 156,
     1153
Groth, H.-G.: 1961, Z. Astrophys. 51, 206
Huber, M., Tobey, F.L., Jr.: 1968, Astrophys. J. 152, 609
Jordan, C.: 1966, Mon.Not.R.astr.Soc. 132, 463, 515
King, R.B., King, A.S.: 1935, Astrophys. J. 82, 377; 1938, Astrophys. J.
     87, 24
Merrill, P.W.: 1956, Lines of the Chemical Elements In Astronomical
     Spectra, Carnegie Inst. Washington Publ. 610
Pagel, B.E.J.: 1969, in '*Les transitions interdites dans les spectres
     des astres*', 15. Coll. International d'Astrophys. Liège, p. 189
Pottasch, S.R.: 1963, Mon.Not.R.astr.Soc. 125, 543; Astrophys. J. 137,
     945; 1964, Mon.Not.R.astr.Soc. 128, 73
Roder, O.: 1962, Z. Astrophys. 55, 38
Russell, H.N.: 1929, Astrophys. J. 70, 11
Russell, H.N.: 1933, '*The Composition of the Stars*', Halley Lecture,
     Oxford, Clarendon Press
Saxner, M.: 1984, Doctoral Thesis, Uppsala University
Steenbock, W.: 1985, in '*Cool Stars with Excesses of Heavy Elements*',
     eds. M. Jaschek, P.C. Keenan, p. 231, D. Reidel, Dordrecht
Swings, J.P.: 1965, Ann. d'Astrophys. 28, 703
Unsöld, A.: 1941, Z. Astrophys. 21, 1, 22; 1942, Z. Astrophys. 21, 229;
     1944, Z. Astrophys. 23, 75

Viotti, R.: 1969, Astrophys. Space Sci. 5, 323

Wampler, E.J., Oke, J.B.: 1967, Astrophys. J. 148, 695

Wellmann, P., 1951, Z. Astrophys. 30, 71, 88, 96

Whaling, W., King, R.B., Martinez-Garcia, M.: 1969, Astrophys. J. 158,389

Wildt, R.: 1939, Astrophys. J. 89, 295

Wills, B.J., Netzer, H., Wills, D.: 1985, Astrophys. J. 288, 94

Withbroe, G.L.: 1971, in 'The Menzel Symposium on Solar Physics, Atomic
       Spectra, and Gaseous Nebulae', ed. K.B. Gebbie, NBS Special Publ.
       353, 127

Woolley, R.v.d.R., Allen, C.W.: 1948, Mon.Not.R.astr.Soc. 108, 292

Zwaan, C.: 1962, Bull. astr. Inst. Netherlands 16, 225

FeII REFERENCE CATALOGUE

G.B. Baratta
Osservatorio Astronomico, Roma, Italy

R. Viotti
Istituto Astrofisica Spaziale, Frascati, Italy

As discussed during this conference the FeII problem has a long history of investigation which led to the publication of a very large number of papers. Several of the old papers are of pure historical interest, since for instance they only give some line identification without a quantitative estimate of the line intensity. Yet some old works could provide useful information about the time variability. More recently many new researchs on FeII have been developed by groups working in different fields of Astrophysics, and this led to the publication of a large deal of papers.

In view of this Capri Conference and to give some help to future studies on FeII, we have started to collect all the papers related to the problem in the FeII Reference Catalogue (Viotti and Baratta 1986), and to provide each paper with a number of subject headings. Presently the Catalogue includes about 200 entries listed in alphabetical and chronological order, and published until June 1986. Two more tables give the analytical indices of the articles arranged by subjects and astrophysical categories. The following subjects have been considered:

Atomic data (levels, gf-values, collision strengths, etc.)
Physical processes (excitation, ionization, etc.)
Data analysis
Models (including spectral synthesis)
FeII, [FeII] line intensity, profile, identification
General (statistics, etc.)

The astrophysical categories are:

The Sun
Early type stars
Luminous Blue Variables (Hubble-Sandage variables, P Cyg, S Dor, Eta Car type)
Late type stars (including cool stellar variables)

9

R. Viotti et al. (eds.), Physics of Formation of FeII Lines Outside LTE, 9–10.
© *1988 by D. Reidel Publishing Company.*

Cataclysmic variables and supernovae
Diffuse matter (including SNRs and interstellar matter)
Extragalactic objects

These subjects were widely discussed during this conference, so that the Catalogue may be a useful complement of
these Proceedings.

The Catalogue is available in printed form. It could
also be provided through the Earnet-Bitnet network in form
of two files with about 1000 and 500 records for the main
catalogue and the analytical indices respectively. This
computer version is currently updated and corrected. The
username and node is: UVSPACE@IRMIAS (Att/n Roberto Viotti).

The authors would greatly appreciate suggestions and
additions for the Catalogue.

REFERENCE

Viotti, R., Baratta, G.B.: 1986, The FeII Reference
Catalogue, Internal Report, Istituto Astrofisica Spaziale,
CNR, Frascati

SESSION 1

BASIC ATOMIC DATA

THE LABORATORY SPECTRUM AND ATOMIC STRUCTURE OF FE II

Sveneric Johansson
Department of Physics, University of Lund
Sölvegatan 14, S-223 62 LUND, Sweden

ABSTRACT. The atomic structure of Fe II is presented in terms of
configuration systems, configurations, spectroscopic terms and levels.
Allowed transitions between different pairs of configurations are
deduced from the known term structure and related to the observed
spectrum. The spectral lines are grouped together in transition groups,
transition arrays, supermultiplets and LS-multiplets. Transitions
violating the selection rules are briefly discussed. The representation
of observed and predicted Fe II lines in Multiplet Tables is reviewed.

1. INTRODUCTION

In July 1926, i.e. exactly sixty years ago, Henry N. Russell (1926)
submitted a paper to the Astrophysical Journal, that is regarded to be
the pioneering work on the term analysis of singly ionized iron, Fe II.
He reported on 61 energy levels, belonging to 16 different LS terms,
which were established by means of 214 identified lines in the
wavelength region 1550 - 7712 Å. The starting point of the analysis was
a discovery at NBS by Meggers et al (1924) of two multiplets around 2600
Å, which we now know as UV Multiplets 1 and 64. The suggestion by these
authors that the lower term in UV 1, a^6D, represented the ground state
of Fe II was verified in Russell's work. It was assigned to the $3d^64s$
configuration according to Hund's rule. Russell identified a low ^6S term
as $3d^54s^2$ ^6S and the lowest even ^4F term as belonging to the $3d^7$
configuration and concluded that "the only point of unusual interest is
that three different electronic configurations, s^2d^5, sd^6, d^7, appear to
contribute to the production of the low terms". This is the foundation
for the basic atomic structure of Fe II and will be the starting point
in the present review of the knowledge of Fe II today. Another statement
by Russell, that has been repeated many times since 1926 by other
authors, has been proven to be wrong and it seems to be an unfulfilled
desire for ever - "All the lines of astrophysical importance have been
classified.....".
   The fact that a knowledge of the atomic structure of Fe^+ or term

R. Viotti et al. (eds.), Physics of Formation of FeII Lines Outside LTE, 13–34.
© 1988 by D. Reidel Publishing Company.

system of Fe II is of equal importance for astrophysical purposes as a
list of classified lines, was demonstrated by P.W. Merrill already in
1928, i.e. only two years after Russell's analysis had been published.
Stimulated by Bowen's success with the identification of the nebular
lines Merrill (1928) calculated the wavelengths for parity forbidden
Fe II lines from the known energy levels of the low configurations and
managed to identify about 25 lines in the spectrum of η Carinae as
[Fe II]. As late as nine years ago another 50 emission lines were
identified in the infrared spectrum of η Car (Johansson 1977) as
high-level transitions in Fe II, which introduced a second generation of
Fe II lines in stellar spectroscopy.
The situation today as regards the knowledge of Fe II contra the future
needs of atomic data for astrophysical work might be best illustrated by
the huge number of reports at this meeting on new laboratory
measurements of wavelengths and oscillator strengths, an extended
analysis of the term system, new identifications in stellar spectra and
new calculations on the level structure and transition probabilities or
oscillator strengths. In spite of all experimental and theoretical
efforts on the atomic physics of Fe^+ and other ions, it is still
impossible to explain all radiation from cosmic emission objects and all
opacity in absorption spectra. About 50 % of the individual spectral
features in high-resolution spectra are still unidentified (Baschek
1983). The astrophysical need for Fe II data is still the major reason
to continue the laboratory work on this spectrum. Even if it is
possible, in principle, to calculate the whole spectrum by means of a
conventional semi-empirical theoretical method and a large computer the
result is too rough for identification work and detailed studies in
astrophysical spectra. For unknown levels in known configurations
typical uncertainties of the calculated energies are of the order 10-100
cm^{-1} (0.6-6 Å at 2000Å), while the levels are experimentally determined
with an uncertainty of 0.01-0.1 cm^{-1} (0.0006-0.006Å at 2000Å). For pure
ab initio predictions of unknown configurations the calculated level
values have in general larger uncertainties. The relative uncertainties
for calculated oscillator strengths are in general larger, which may not
be that serious for line identification purposes as for the shape of a
synthetic spectrum.

There are now 675 energy levels known in Fe II, which have been found
in the analysis of laboratory spectra ranging from 900 Å up to 50000 Å.
The "history of Fe II", as regards the development of the term system,
may be found in the reference list in vol II of AEL, Atomic Energy
Levels (Moore, 1952a), for the period between 1926 and 1952 and for the
period up to 1986 in the new, condensed NBS compilation of energy levels
in all iron group elements (Sugar and Corliss 1985). Besides Russell's
work, the significant contributions to the knowledge of Fe II up to 1978
are due to to J.C. Dobbie (1938), L.C. Green (1939) and B. Edlén.
Edlén's unpublished material was included in vol II of AEL. In 1978 the
term system was expanded in the extended analysis by Johansson (1978a).

In this paper we are going to discuss the laboratory spectrum and
atomic structure of Fe^+ and also give some astrophysical applications
connected to recent laboratory work. Besides a general review of the
term system we will give some relations between the atomic structure,

transition arrays and the astronomical multiplet concept. The laboratory
spectrum will be treated in connection with a current project on Fe II
(Johansson and Baschek 1984), where the wavelength region 1700-3250 Å
is under investigation. The studied region coincides more or less with
the wavelength range covered by IUE (the International Ultraviolet
Explorer). A current analysis of the far infrared spectrum, recorded
with the FTS (Fourier Transform Spectrometer) at Kitt Peak, will also be
discussed.

2. THE TERM STRUCTURE

The complexity of the structure of Fe II may be discussed in terms
of the building-up principle for filling the electron shells in atoms of
the iron-group elements. Besides the 3d shell being filled, which may
give as much as five valence electrons or holes, the 4s subshell is also
occupied, indicating that the binding energy for the 4s electron is
somewhat larger than for the 3d electron. This fact can be explained
from a simple model by saying that the penetrating 4s-electron is more
firmly bound to the nucleus than the more shielded 3d-electron. However,
if the atom is ionized, the shielding of the 3d-electron is less
effective and in doubly ionized atoms the $3d^k$ configuration will always
comprise the ground-state. In Fig 1 we have demonstrated this by adding
3d- and 4s-electrons to the $3d^5$ core of Fe IV and put the different
configurations in the $(3d+4s)^k$
complex on a rough energy scale.
By increasing the value of k in
steps by 1 we get the spectra Fe
III- Fe I. We can see that in Fe
III the $3d^k$ configuration (k=6) is
the lowest one while in Fe II
(k=7) it is 0.2 eV above the
ground state, $3d^6$4s. In Fe I the
ground configuration contains two
4s electrons and the $3d^k$
configuration (k=8) appears high
up in the term system.

Another way to illustrate the
competition between 3d- and
4s-electrons is to display the
isoelectronic behaviour of the
$(3d+4s)^k$ complex. In Fig 2 we have
plotted the positions of low
spectroscopic terms in the $3d^k$ and
$3d^{k-2}4s^2$ configurations relative
to the position of $3d^{k-1}$4s for the
first four members of the Mn I
sequence (k=7). As zero level we
have chosen the centre of gravity
for the two 4s terms built on the
lowest parent term, ^{5}D. We can see

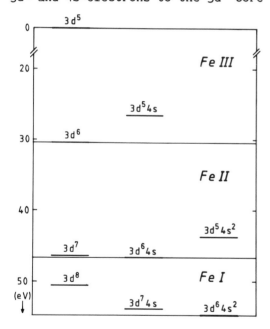

Fig.1. The relative positions of
the low even configurations in
Fe I-III, showing the competition
between the 3d- and 4s electrons.

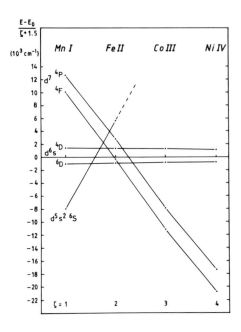

Fig.2. The relative positions of
the low even configurations in
the Mn I isoelectronic sequence.
(The zero line is the centre of
gravity for the lowest pair of
$3d^6 4s$ terms).

how the $3d^7$ configuration comes down
and the $3d^5 4s^2$ configuration moves
up in the system when the effective
nuclear charge ζ increases. (It
could be pointed out that the $3d^k 4s^2$
configuration has not been found in
any third spectrum, except for Ti
III (Edlén and Swensson 1975)).

The most concentrated overlap of
the three configurations in the
$(3d + 4s)^k$ complex occurs in the
second spectra, for which the
relative position of the lowest
terms are given in Fig 3 as a
function of k. In this iso-ionic
diagram we can see that the lowest
state is assigned to the $3d^{k-1} 4s$
configuration in the beginning of
the sequence. As the number of
d-electrons increases the $3d^k$
configuration becomes the ground
configuration up to k=5, i.e. a half
filled 3d shell, where a big
discontinuity occurs. In the second
half of the shell (k= 6-10) the
picture is very similar to that in
the first half (k= 1-5), i.e. the
curves from the first half are more
or less repeated. We can also see in

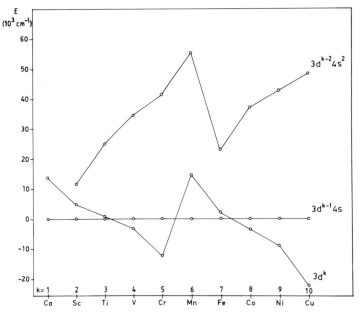

Fig. 3. The relative
positions of the low
even configurations
in the isoionic
sequence of second
spectra (spark
spectra) of the iron
group elements. (Each
point represents the
position of the
lowest term of
respective
configurations).

Fig 3 that the $3d^{k-2}4s^2$ configuration never occupies the ground state in singly ionized atoms.

The three possible combinations of $(3d + 4s)^k$ for each k>1 - $3d^k$, $3d^{k-1}4s$ and $3d^{k-2}4s^2$ - make it suitable to split up the term system of Fe II in three different partial systems. The first one, $3d^6(^ML)nl$, is built on the <u>parent terms</u> ^ML in the ground configuration $3d^6$ of Fe III and will here be designated as the "normal system". The second system, $3d^5(^ML)nln'l'$, has its <u>grand-parent terms</u> ^ML in the ground configuration $3d^5$ of Fe IV (see Fig 1) and will be referred to as the "doubly-excited system". The third system, $3d^44s^2nl$, is based on the parent terms of $3d^44s^2$ in Fe III. It will not be treated here since it is expected to lie high up in Fe II and has so far no experimentally known levels. It is, however, relevant to include that system in the term analysis of first spectra (neutral atoms), at least for elements having the parent configuration, $3d^{k-2}4s^2$, at a relatively low excitation potential (e.g. Sc,Ti,Fe, see Fig 3).

2.1 The normal system, $3d^6(^ML)nl$.

The normal system of configurations, $3d^6(^ML)nl$, is well established for nl= 4s and 4p, fragmentarily known for nl= 4d and 5s and only analysed for the <u>subconfigurations</u> $3d^6(^5D)nl$, i.e. terms built on the ^5D parent term $(^ML=^5D)$, for nl=6s,7s,5p,6p,5d,6d,4f, as illustrated in Fig 4. At the top in the diagram all the terms of the parent configuration $3d^6$ in Fe III are displayed as a function of energy with the ground term ^5D to the left. From each parent term is drawn a vertical line, which terminates in two or more boxes (or horisontal lines) at different excitation energy. The lowest box contains all levels, which are created when a 4s electron is added to the parent term, i.e. a $3d^6(^ML)4s$ subconfiguration, the next lowest a 4p electron, the next one 4d, 5s and so on, as indicated to the left in the figure. The height of a box represents roughly the energy span for the corresponding subconfiguration. Example: If a 4s electron is added to a triplet parent term we get one quartet and one doublet term - maximum 6 levels - all of them enclosed in one box. Those 4s levels, which are based on a singlet parent term, form only a doublet term and are represented by a horisontal line in the figure (the fine-structure splitting is negligible on this energy scale). Dashed lines indicate that all levels are not experimentally known. For the $3d^7$ configuration - i.e. addition of another d electron to the $3d^6$ core - it is not possible to assign the terms to a particular parent term. All the levels of this configuration have therefore been enclosed in one big $3d^7$ box to the left in Fig 4.

From Fig 4 we can realize that the term structure of the $3d^64s$ configuration follows the same pattern as the parent configuration, $3d^6$. Thus, the interaction between the 3d and 4s electrons is not large enough to perturb this general pattern but causes just an increase of the number of levels and of the range of excitation energy, represented by the height of the box. However, there is a small effect on the $3d^6$ core due to the screening of the added 4s electron (Edlén 1972). This effect causes a shrinking of the energy intervals between the parent terms by about 4%, which is too small to be seen in the type of diagram

in Fig 4. The $3d^6$ pattern also dominates the structure of $3d^64p$ as well
as other configurations and we see in Fig 4 that the boxes are smaller
for high configurations, indicating a weaker interaction between the nl
and 3d electrons for high n values. The regular pattern of the $3d^6(^ML)nl$
configurations is manifested in the observed spectrum in the way that
most strong 4s-4p lines appear in the same wavelength region, determined
by the 4s-4p separation. This separation can be derived from Fig 4 to be
about 40000 cm^{-1} corresponding to a transition wavelength of 2500 Å. The
term splittings i.e. the heights of the boxes, stretch the resonance
region by about ±300 Å around 2500 Å.

The energy differences 4p-5s and 4p-4d are nearly the same as the
4s-4p separation (4p-5s ≈ 37000 cm^{-1} and 4p-4d ≈ 43000 cm^{-1}), which
means that all 4p-5s and 4p-4d transitions appear in the same region as
4s-4p with their centres at 2700 Å and 2300 Å, respectively. This
coincidence of energy differences makes the 2000-3000Å region very rich
of strong Fe II lines, corresponding to the 4s-4p, 4p-5s and 4p-4d
transitions. Hence, a reinvestigation of Fe II in this wavelength region
has been undertaken and a progress report is given at this meeting.

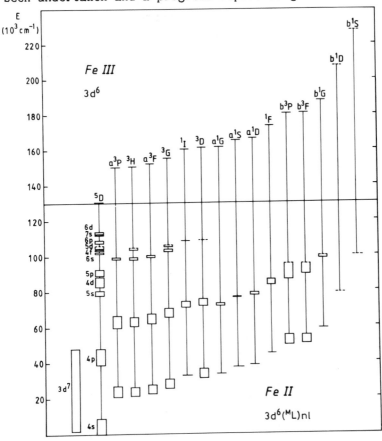

Fig. 4. The normal system of configurations, $3d^6(^ML)nl$,
where ^ML is the parent term (see text).

2.2 The doubly-excited system, $3d^5(^ML)nln'l'$.

The electrostatic interaction between the electrons in the doubly
excited system gives rise to a completely different term structure than
in the normal system. The coupling between the two outer electrons, nl
and n'l', is stronger than the coupling between any of them and the 3d
electron, if n=n'=4. It is therefore appropriate to add a two-electron
configuration, nln'l', to the grand-parent configuration, $3d^5$, in Fe IV.
LS coupling of the nl and n'l' electrons gives different <u>intermediate
terms</u>, e.g. $4s^2$ ^1S; 4s4p ^1P,^3P; 4s4d ^1D,^3D; $4p^2$ ^3P, ^1D, ^1S. In Fig 5 we
have made a diagram of the $3d^5(^ML)nln'l'$ system with a similar design as
we used in Fig 4. From the grand-parent terms of Fe IV vertical lines
are drawn which terminate in horisontal lines representing the $4s^2(^1S)$
terms. Since the filled 4s subshell, $4s^2$, does not contribute any spin
or angular momentum, the grand-daughter term in $3d^54s^2$ will be the same
as the corresponding grand-parent term. For the 4s4p configuration the
terms have been grouped according to the intermediate term, giving one
box for the 4s4p(^3P) terms and one for the 4s4p (^1P) terms. This means
that the $3d^5(^6S)4s4p(^3P)$ box contains three terms, ^8P, ^6P and ^4P,
derived by vector coupling of spin and angular momenta for the
grand-parent ^6S and the intermediate ^3P terms (see Fig 7). Coupling of

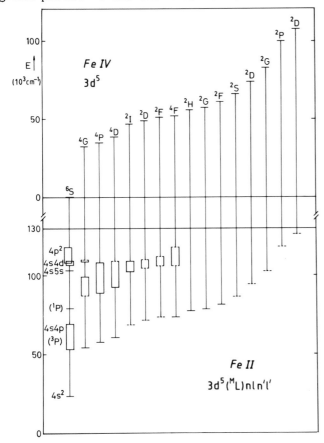

Fig. 5. The system of
doubly-excited configu-
rations, $3d^5(^ML)nln'l'$,
where ^ML is the grand-
parent term (see text).

the grand-parent term ^{4}G to the ^{3}P term creates 9 grand-daughter terms, arranged in three triads: ^{6}H,G,F; ^{4}H,G,F and ^{2}H,G,F. The coupling of an intermediate ^{1}P term to any grand-parent term, except for ^{6}S, gives one triad of grand-daughters with the same multiplicity as the grand-parent.

As can be seen in Fig 5 there is still a number of unknown levels (dashed lines) in the lowest configuration of the doubly-excited system, $3d^5 4s^2$. However, it should then be noted that this configuration extends over a large range of energy i.e. the lowest term is close to the ground-state and the highest one is predicted to lie just below the ionization limit. Even if the 4s4p configuration seems to be fragmentarily known the 4s4p boxes in Fig 5 account for more than 120 levels. (The total number of levels in the $3d^5 4s4p$ configuration is 417!!). For high configurations, e.g. 4s4d, 4p^2, 4s5s, only terms based on the lowest grand-parent, ^{6}S, are known. For other grand-parent terms these configurations will lie well above the lowest ionization limit.

2.3 "One- and two-electron" representation of Fe II

The level structure along a vertical line in Fig 4 and Fig 5 can be illustrated in a partial term diagram with one or two electrons added to a single-limit core. For the normal system the gross structure will be quite similar to the gross structure of a one-electron system, where the 4f orbit is near hydrogenic.

In Fig 6a we have picked out one specific parent term, ^{5}D, in the normal system $3d^6 (^M L)nl$ and arranged the boxes in different l-series for the $3d^6 (^5 D)nl$ subconfigurations. We can see that the boxes shrink with increasing n for all series, converging to the total fine-structure

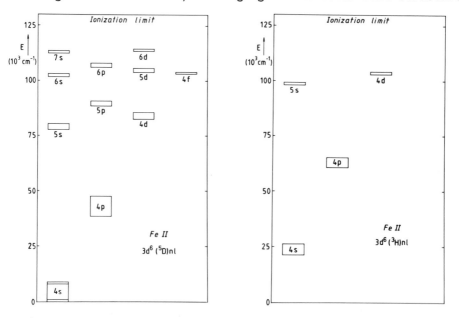

Fig. 6. Partial term diagrams of subconfigurations in the normal system.
   a) $3d^6 (^5 D)nl$ b) $3d^6 (^3 H)nl$

splitting of the parent term. The number of levels enclosed in the boxes is higher for large l values, ns = 9 (⁶¹⁴D), np = 25 (⁶¹⁴PDF), nd = 37 (⁶¹⁴SPDFG) and nf = 45 (⁶¹⁴PDFGH) levels. The total term splitting (the height of the box) is clearly smaller for higher values of l. This term splitting is a measure of the interaction between the 3d- and nl-electrons.

In Fig 6b we illustrate how the gross structure is repeated when the same kind of diagram is constructed for another parent term, ³H. The only difference is that the energy scale is shifted by an amount similar to the energy separation between the parent terms, ³H and ⁵D (≈ 20000 cm⁻¹). A direct comparison of the two diagrams in Fig 6 gives a good estimate of the positions for e.g. the 5p and 4f subconfigurations in Fig 6b.

Simple tools like Ritz diagrams, partial term diagrams and diagrams showing regularities in term- and fine-structure provide a good help in finding missing levels and spotting possible misidentifications. However, this is of course only true when the atom behaves as our simple model predicts and when there are no serious perturbations. More detailed information about these semi-empirical methods are given in a review article in Handbuch der Physik by Edlén (1964).

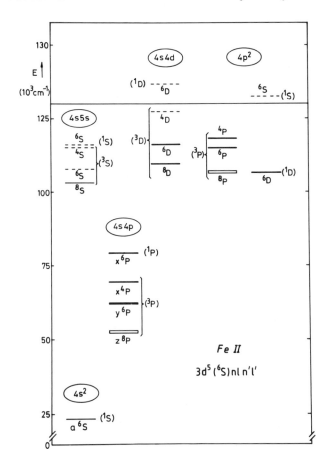

Fig. 7. Partial term diagram of the 3d⁵(⁶S)nln'l' subconfigurations.

In Fig 7 a partial term diagram has been drawn for the doubly-excited subconfigurations, $3d^5(^6S)nln'l'$, illustrating the level structure in a single limit case, where the two nl and n'l' electrons are added to the grand-parent term ^6S. In this diagram the boxes on the vertical line to the left in Fig 5 are spread out horisontally, arranged in different series. The individual terms have been marked in the diagram and the dashed lines indicate predicted positions of unknown terms.

The choice of coupling for low configurations of this system is supported by the ordering of the spectroscopic terms. In the 4s4p configuration the three terms associated to the intermediate ^3P term are grouped at the bottom and the term associated to the intermediate ^1P term on top. This arrangement is a consequence of the relative magnitudes of the different electrostatic interactions with the 4s-4p interaction as the strongest one, and the 3d-4s and 3d-4p as the weaker ones. However, in the 4s5s configuration we see that the high ^6S term is very close to the ^4S term, and the structure is more like a double-limit structure, where a 5s electron is added to the parent terms ^7S and ^5S of $3d^64s$ in Fe III giving the subconfigurations $3d^64s(^7S)5s$ $^6S,^8S$ and $3d^64s(^5S)5s$ $^4S,^6S$. This implies that the 3d-4s interaction dominates over the 3d-5s and 4s-5s interactions. The configuration interaction between 4s4d and 4p² is known to be strong in many spectra and it really affects the term structure of these configurations in Fe II. This is the subject of a separate contribution to this meeting (Brage et al. 1986) and will not be discussed in detail here.

3. THE SPECTRUM

It would have been more consistent to start this paper with a presentation of the experimental data, from which the term diagram has emerged, but we have choosen to discuss the spectral distribution of Fe II lines in view of the term structure. That might also be a better approach for our estimate of those wavelength regions in stellar spectra, where we expect a large contribution of Fe II. In spectra of most cosmic objects the atomic lines appear in absorption, which is not directly convertible for a comparison to a laboratory spectrum in emission. The intensity of an absorption line in a stellar spectrum is mostly dependent on excitation potential of the lower level in the transition and on the oscillator strength. For laboratory intensities, however, there might be other parameters, besides the excitation potential of the upper level and the transition probability, that are important, e.g. such excitation mechanisms, that are selective in the population of levels. This is of course also valid for emission lines in stellar spectra (see below).

3.1 Laboratory recordings

There are mainly three sets of laboratory registrations, that are used in the current analysis of Fe II. All of them have been made with the same type of light source, a hollow-cathode (H-C) lamp, and they cover the wavelength region 900 - 50000 Å. In the VUV region spectrograms have been taken at the 10.7 m normal-incidence spectrograph at the National Bureau of Standards, Washington D.C., from 900 Å up to 3750 Å. The instrument provides a good resolution, and the reciprocal dispersion of 0.78 Åmm^{-1} enables us to measure the lines with an wavelength uncertainty of only a few mÅ. In parts of the optical and photographic infrared regions, 4800 - 11000Å, spectra have been photographed (in higher grating orders) with a 3 m Czerny-Turner spectrograph at Lund. The wavelength uncertainty is of the order of 0.01 -0.02 Å. In the far-infrared region we have used high-resolution spectra recorded by J Brault with the FTS instrument at Kitt Peak, having an extremely high wavenumber accuracy, better than 0.005 cm^{-1}.

The H-C lamp has been run with different carrier gases, argon and neon, both in a pulsed and in an unpulsed mode. In general the second spectrum (spark spectrum) is more readily excited in a pulsed mode, while the first spectrum (arc spectrum) is strongest in an unpulsed discharge. Thus, the sorting of the different ionization stages is partly done by a comparison of the spectra from the two different discharges. This straight-forward method is not quite reliable, since some Fe II lines show up very strongly in the unpulsed H-C lamp, when neon is used as a carier gas. It turns out that these "enhanced" Fe II lines in the Fe-Ne hollow-cathode lamp are assigned either to levels in a particular range of excitation energy or to high levels in the normal system, belonging to the $3d^6(^5D)nl$ subconfigurations (i.e. the levels in Fig 6a). In both cases it is a question of high-level transitions. Two different excitation mechanisms are probably responsible for a selective population of these highly excited levels. Charge-transfer collisions between Fe atoms and Ne$^+$ ions in the light source involve also a transfer of energy from the neon ion to the iron atom, which favours the excitation of levels in a specific energy region (see e.g. Johansson 1978a, or Johansson and Litzén 1978). Radiative recombination should favour the population of levels in the $3d^6(^5D)nl$ configurations, since these are created when an outer electron is added to the ground term of the parent ion.

3.2 Classification of transitions in groups.

Before we look in detail on the distribution of spectral lines in different wavelength regions we will make a classification of the analysed transition arrays into different groups, I-VIII, in Table 1. The grouping has not been made according to any stringent physical criteria but has emerged from the practical work on the term analysis. The ordering reflects some kind of hierarchy in prominence of the transitions, which may be expressed in terms of intensity, completeness, excitation potential and so on. In addition, the wavelength ranges for the various transition groups are quite narrow. The following concepts (Cowan 1981, Condon and Shortley 1963) are used in connection with transitions:

Transition array: The set of all possible transitions between all levels of one configuration and all levels of some other configuration.

Supermultiplet: The set of all lines arising in transitions between all levels of one subconfiguration and all levels of some other

subconfiguration. (We will always think of an LS-allowed supermultiplet, i.e. the subconfigurations are based on the <u>same parent</u> term).
  <u>Multiplet</u>: Lines arising from transitions between levels of two specific terms.
    In Table 1 we have for each group given the associated transition arrays, the wavelength regions for the corresponding supermultiplets, the present state of the knowledge and the presence of the <u>known</u> individual LS multiplets in the astronomical Multiplet Tables (Moore 1945,1952b) graded from A (practically complete) to D (fragmentary). The last column gives the range of the excitation potential for the lower level in the transitions. Data are given both for the normal system, $3d^6(^NL)nl$, and the doubly-excited system, $3d^5(^NL)nln'l'$, abbreviated to N and DE in the table.

Table 1. Observed transition arrays of Fe II arranged in different
          groups. For each group are given wavelength region, state of
          analysis, presence in the Multiplet Tables and range of the
          excitation potential for the lower level. Information is
          given both for the normal system (N) and the doubly-excited
          system (DE). Key to the grades is given below.

Group	Transition arrays	Wavelength region(Å) Normal system N	Doubly-exc system DE	State of analysis N	DE	Present in M.T. N	DE	Exc.P. (eV)
Ia	4s-4p	2300-2800	≈ 1800	A	B	A	C	0- 7
Ib	3d-4p	1600-7000	900-2300	A	B	A	D	0- 6
IIa	4p-4d	2200-2700	≈ 1750	B	D	B	-	5- 8
IIb	4p-5s	2500-3100	≈ 2000	B	D	B	-	5- 9
III	5s-5p	7500-9500		C	-	-	-	≈10
	5p-6s,5p-5d	6000-9500		C	-	-	-	≈11
IV	4d-4f	4800-6700		C	-	-	-	10-11
V	4d-5p	far		C	-	-	-	10-11
	6s-6p,5d-6p	infrared		C	-	-	-	≈13
	6p-7s,6d	(see fig 8)		C	-	-	-	≈13
VI	5s-6p,4d-6p	3300-4500		D	-	-	-	10-11
	5p-7s,5p-6d	4000-4500		D	-	-	-	≈11
VII	4p-6s,7s	< 1800		C	-	-	-	5- 6
	4p-5d,6d	< 1700		C	-	-	-	5- 6
VIII	4s-5p,6p	< 1150		C	-	D	-	0- 1

Key: A = Transition array complete or nearly complete
     B = Data known for more than the lowest subconfigurations
     C = Supermultiplets between lowest subconfigurations complete or
         nearly complete
     D = Only a few lines known
     - = No transitions known

In Fig 8 we have made a Grotrian diagram of the $3d^6(^5D)nl$ system and indicated the wavelength for the strongest line in each supermultiplet. In view of this digram and Table 1 we will give more detailed information about the different transition groups below. In the discussion of the doubly excited system we refer to Figs 5 and 7. As we see in Table 1 data are only available for the first four groups in this system.

3.3 Line distribution within the normal system

The strongest lines in Fe II, as well as in spectra of all iron group elements of low and medium ionization stages, are attributed to 4s-4p and 3d-4p transitions. The lines can in general be arranged in pure LS-multiplets (see Sec 4) even if mixing allow for intercombination lines.

3.3.1. <u>Group Ia, 4s-4p transitions</u>. As discussed at the end of Sec 2.1 the 4s-4p supermultiplets appear around 2500 Å. From Fig 4 we can realize that inter-parent 4s-4p transitions (transitions between LS terms built on different parent terms) extend the wavelength region far

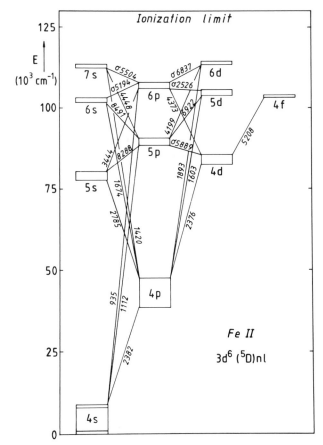

Fig. 8. Partial Grotrian diagram of the $3d^6(^5D)nl$ subconfigurations. Wavelengths in Ångström. Infrared lines are marked by a "σ" and the wavenumber in cm^{-1}.

down in the vacuum ultraviolet (VUV) and up in the optical region,
($\lambda > 3000$ Å), verified e.g. by the optical multiplets 27, 37, 49 and 73.
These multiplets consist of transitions between the triad of quartet
terms in the lowest 4p subconfiguration (the z^4F, z^4D and z^4P terms,
built on the ^5D parent) and quartets in those 4s subconfigurations,
which are based on the parents a^3P, ^3H, a^3F and ^3G (see Fig 4).

The still unknown levels of the 4s and 4p configurations are
predicted to have high excitation energies and will probably not give
any significant contribution to any cosmic spectrum.

3.3.2 Group Ib, 3d-4p transitions.

A configuration of only equivalent
electrons, like $3d^7$ in Fe II, is not analogous to the $3d^6nl$
configurations as concerns the term structure, due to a cancellation of
terms predicted by the Pauli principle. Hence, there is no similar way
to trace the parent of a term and thus no supermultiplet structure that
characterizes the spectrum. The 3d-4p transition array covers a larger
wavelength region and is dominated by combinations with the two low
quartets, a^4F and a^4P at 0.2 and 1.7 eV, respectively. The strongest
multiplets appear in the UV region with the optical multiplets 6,7 and 8
(a^4P - z^4P,z^4D and z^4F) at the highest wavelength, 3000 - 3200 Å.
Transitions from a^4F to the same 4p terms, z^4D and z^4F, are the strong
multiplets UV 35 and 36 just below 2400 Å. The jump in energy (≈ 2.5 eV)
to the next triad of 4p quartet terms (see Fig 4) brings down the
corresponding 3d-4p transitions (from a^4F) to about 1700 Å. The lower
limit for the prominent Fe II spectrum is set at 1550 Å by this type of
transitions.

All levels in the $3d^7$ configuration are known, which means that
the transition array 3d-4p is nearly complete.

3.3.3. Group II, 4p-4d and 4p-5s transitions.

Since the 4s-4p spacing is
very similar to the 4p-4d and 4p-5s intervals, the latter transitions
will have nearly the same wavenumbers as the 4s-4p lines. In fact, the
4p-4d lines appear at somewhat shorter and the 4p-5s lines at somewhat
longer wavelengths than the 4s-4p lines (see Table 1). The 5s and 4d
configurations are only fragmentarily known to other limits (parent
terms) than the lowest one, for which all levels are known. In the
current work on Fe II in the 1700-3250 Å region (Johansson and Baschek
1984) a great number of unidentified lines show up in the 4p-4d and
4p-5s regions and some of them are certainly combinations between 4p and
missing 4d and 5s levels.

3.3.4. Group III, 5s-5p, 5p-5d and 5p-6s transitions.

With one possible
exception these transitions are only known within the $3d^6(^5D)nl$ system.
The lines appear in the photographic infrared region and have been found
in emission in some cosmic sources. Thus, the strongest unidentified
lines in Thackeray's list (Thackeray 1962) of η Carinae were identifed
as 5s-5p transitions of Fe II (Johansson 1977). The highly excited
5p-5d and 5p-6s lines (E.P. of lower level ≈ 11eV) were found in the
absorption spectrum of ν Sgr (Johansson 1978b).

3.3.5. Group IV, 4d-4f transitions.

The 4d- and 4f subconfigurations are

known for the lowest parent term ^{5}D in the normal system. The 4d configuration is best described in LS coupling, but the level structure of 4f is different from all other known configurations in Fe II, having a pattern typical for JK coupling (Johansson and Litzén 1974). This means that it is not meaningful to split the 4d-4f-transitions into LS Multiplets. Since it is difficult to find a more proper way to arrange spectral lines for astrophysical applications than in LS multiplets we could treat all the known 481 4d-4f lines as a super-multiplet and use the notation "4d-4f" or "(^{5}D)4d-(^{5}D)4f" in stellar line lists.

In astrophysical spectra 4d-4f transitions were first identified in ν Sgr (Johansson 1978b) but have in a recent publication (Johansson and Cowley 1984) been thoroughly studied in a number of Ap stars.

3.3.6. Group V, 4d-5p, 6s-6p, 5d-6p, 6p-7s and 6p-6d transitions. These transitions fall in the far infrared region and are surprisingly strong on FTS recordings. As discussed in Sec 3 this probably depends on particular excitation conditions in the direct-current H-C lamp. The combination of a selective excitation in the light source and the extremely high wavenumber accuracy achieved with the FTS instrument has enabled us to analyse these high configurations.

The astrophysical importance of the transitions in this group is difficult to estimate but so far they have not been identified in any high-resolution far-infrared spectrum of emission line objects. However, the lines provide accurate level values for highly excited subconfigurations as well as possibilities to predict wavelengths for the transitions in the next three groups, which may be of more astrophysical interest.

The transitions involved in this group are restricted to levels based on the lowest parent term.

3.3.7. Group VI, 5s-6p, 4d-6p, 5p-7s and 5p-6d transitions. The wavelengths for lines in this group can be calculated from known level values in the corresponding 3d^6(^{5}D)nl subconfigurations, established by means of the infrared lines in Group V.

The supermultiplets in this group appear in the blue region, where no extended analysis of the laboratory spectrum has been done. The lines are probably present on the FTS recordings, but these have not been investigated for analysis purposes below 10 000 Å. However, the lines that are expected to be the strongest ones are clearly present in the line lists of Ap stars and identification work is in progress (Johansson and Cowley 1986).

3.3.8. Group VII, 4p-6s,7s and 4p-5d,6d transitions. Lines to higher members of the ns- and nd-series occur in the VUV region in Fe II and have been observed in the laboratory spectrum. Even if the oscillator strength is expected to be smaller for transitions to high members in a series the lines in this group might appear in stellar absorption spectra because of the low excitation potential for the lower level.

3.3.9. Group VIII, 4s-5p,6p transitions. The comments given for Group

VII lines are also valid for this group. Moreover, these transitions
would be more likely to appear in absorption spectra, even in cool
stars, since the lower level belongs to the ground configuration. As a
matter of fact some of the lines have been found among the interstellar
lines in stellar spectra (Nussbaumer et al. 1981). The 4s-6p transitions
appear around 930 Å and have not yet been published. Presumptive
interstellar lines were reported at the 4th IUE conference in Rome
(Johansson 1984).

3.4 Line distribution within the doubly-excited system.

According to the coupling conditions in the doubly-excited system
$3d^5(^{M}L)nln'l'(^{M'}L')$ (see Sec 2.2) verified by the intensities of
observed lines, the intermediate term $^{M'}L'$ acts as a good quantum
number, for which the LS selection rules are valid. It is therefore
convenient to define a subconfiguration in this system as all terms,
that are based on the same grand-parent term and have the same
intermediate term $^{M'}L'$. Since the two outer electrons form either a
singlet or a triplet intertermediate term, we will distinguish between
singlet subconfigurations and triplet subconfigurations.

A supermultiplet will consequently be defined as all possible
transitions between two subconfigurations.

3.4.1. Group Ia, 4s-4p transitions. There are two types of 4s-4p
transitions known in the doubly excited system, $3d^5(^{M}L)4s^2 - 3d^5(^{M}L)4s4p$
and $3d^5(^6S)4s4p - 3d^5(^6S)4p^2$.

The strongest combinations between the $3d^54s^2$ and $3d^54s4p$
configurations are attributed to singlet subconfigurations, i.e. the
$3d^5(^{M}L)4s^2(^1S) - 3d^5(^{M}L)4s4p(^1P)$ supermultiplets, since $4s^2$ has only a
singlet term. Only one multiplet of this kind is given in the Multiplet
Tables, viz UV 191, which turns out to be prominent in many stellar
spectra and of current interest as regards its excitation (see the paper
by Johansson and Hansen in this book). Multiplet 191 consists of three
lines around 1780 Å and is ascribed to the $3d^5(^6S)4s^2(^1S)^6S -
3d^5(^6S)4s4p(^1P)^6P$ supermultiplet, abbreviated $a^6S - x^6P$ in Fig 7. The
energy difference $3d^5(^{M}L)4s^2(^1S) - 3d^5(^{M}L)4s4p(^1P)$ is expected to be
nearly constant for the various grand-parent terms, ^{M}L, and the
corresponding lines are predicted to appear around 1800 Å. This type of
singlet supermultiplets is only known for one more grand-parent term,
^4G, simply because no other $4s4p(^1P)$ terms are known in Fe II (see Fig
5).

There is also a 4s-4p jump in the transition between 4s4p and $4p^2$,
for which we can distinguish between a singlet and a triplet
subconfiguration for each grand-parent term. The $4p^2$ configuration is
known only for the grand-parent term ^6S, which is displayed in Fig 7.
The triplet subconfigurations of 4s4p and $4p^2$ have the same set of LS
terms, ^8P, ^6P and ^4P, which means that the supermultiplet consists of
three individual LS multiplets, $^8P-^8P$, $^6P-^6P$ and $^4P-^4P$. The
corresponding lines have wavelengths around 1800-1900 Å. These
transitions as well as transitions within the singlet supermultiplets
are presented at this meeting in two separate contributions by Brage et

al. and Adam et al.

3.4.2. Group Ib, 3d-4p transitions.

The selection rules in the doubly-excited system do not allow levels of
$3d^5 4s4p$, based on an intermediate triplet term, to make electric dipole
transitions to levels of the $3d^5 4s^2$ configuration. (The rules are of
course not strictly obeyed and in reailty it means that the lines are
weak). However, the $3d^5 4s4p$ levels can decay in 3d-4p transitions to the
$3d^6 4s$ configuration in the normal system. If this configuration is
written in the "two-electron representation" (see Sec 2.3.) as
$3d^5 (^"L)4s3d(^{1'3}D)$ we realize that 3d-4p transitions can be treated as
4s3d-4s4p transitions in the two-electron representation. Since the 3d
electron is equivalent to the core electrons the Pauli principle limits
the number of channels in the $3d^5 (^"L)4s3d - 3d^5 (^"L)4s4p$ transition
array. This is manifested in a number of metastable states in the
$3d^5 4s4p$ configuration, some of them with high excitation potential
and still unknown. We expect e.g. a $3d^5 (^3H)4s4p(^3P)^"K_{17, 2}$ level (odd
parity) around 103000 cm^{-1} ($\approx$ 13 eV) but there is no even-parity level
with lower energy having J>13/2!!

The 3d-4p transitions discussed above account for a number of
strong lines in the Fe II spectrum below the "limit" at 1550 Å discussed
above. Because of their relatively low excitation potential they should
have a good chance to show up in stellar absorption spectra.

A rewriting of $3d^6 4p$ as $3d^5 3d4p$ may illustrate another 3d-4p jump
in the doubly-excited system, the $3d^5 (^"L)3d4p - 3d^5 (^"L)4p^2$ transition
array. The corresponding lines have been found in the spectrum of 21 Peg
(Adam et al. 1986) and are reported at this meeting.

3.4.3 Group II, 4p-5s and 4p-4d transitions.

The 4s5s- and 4s4d-
configurations in the doubly-excited system are partly known for those
triplet subconfigurations, which are built on the lowest grand-parent
term ^6S (see Fig 7). The most prominent lines are assigned to the
octet transitions, $4s4p\ ^8P - 4s5s\ ^8S$ at 1980 Å and $4s4p\ ^8P - 4s4d\ ^8D$
between 1750 and 1800 Å. The structure of the $3d^5 4s5s$ and $3d^5 4s4d$
configurations is illustrated in Fig 7 and discussed in Sec 2.3.

3.5 Lines ascribed to selection-rule violations.

In the previos sections we have only considered the distribution of
spectral lines, appearing as the result of a completely allowed
LS transition. Without going into details we will here briefly discuss
the occurrence of lines associated to different selection-rule
violations.

3.5.1 "Two-electron jumps".

We have seen in Sec 3.4.2. that 3d-4p
transitions form a link between the normal system and the doubly-excited
system without violating any selection-rules. However, other kinds of
transitions between the two systems should correspond to two-electron
jumps and should be forbidden. The occurrence of seeming two-electron
transitions implies that there is an interaction between two

configurations, one from each system. The strongest case is the optical multiplet 42 around 5000 Å, which corresponds to the $3d^5 4s^2$ - $3d^6 4p$ transition a^6S-z^6P. The multiplet is a consequence of the $3d4p$-$4s4p$ interaction between the configurations $3d^5 4s4p$ and $3d^6 4p$. The quartet terms of $3d^5 4s^2$ - b^4G, d^4P and c^4D around 7 eV, cannot decay in any one-electron transition but combine strongly with the lowest odd-parity levels of $3d^6 4p$, z^4F, z^4D and z^4P at 5 eV. The corresponding lines fall in the photographic infrared and were found in the spectrum of η Car (Johansson 1977). The z^4F - b^4G multiplet turns out to be one of the strongest infrared multiplets in emission line stars and is used for diagnostic purposes. The multiplets are not present in the Multiplet Tables.

3.5.2. Inter-parent transitions (Forbidden supermultiplets).
Inter-parent transitions i.e. combinations between subconfigurations that are not built on the same parent or grand-parent term, appear both in the normal and doubly-excited systems. They are given in the Multiplet Tables as ordinary LS-multiplets and cannot be distinguished there from allowed supermultiplets, since no information is given about the parent or grand-parent. The Fe II spectrum contains many lines, some with relatively high intenity, that are due to inter-parent transitions, which implies a substantial mixing of levels either in the parent- or daughter configuration. In complex spectra this fact reduces the significance of the associated selection rule, saying that a change in parentage is forbidden. The inter-parent transitions stretch the wavelength regions of the transition arrays (see Table 1), discussed in connection with the 4s-4p transitions (see Sec. 3.3.1)

3.5.3. Intercombination lines. Intercombination lines, i.e.
combinations between levels of different multiplicity, are very frequent in Fe II as a consequence of deviations from pure LS coupling. They will be further commented in Sec. 4.

3.5.4. Forbidden lines (parity-forbidden). In order to make the list
complete we have also included the parity-forbidden lines, i.e. electric-quadrupole (E2) or magnetic dipole transitions (M1). These lines are of course of astrophysical importance as mentioned in the introduction but are not observed in the laboratory spectrum. A special paper in this volume by Hansen et al is dealing with the forbidden lines, [Fe II].

4. THE MULTIPLET NUMBERING SYSTEM

In the very first analyses of complex spectra the spectroscopic terms were sorted according to their parity and excitation energy. From their transitions it was often possible to give a LS notation for these terms. When terms of the same type and same parity appeared they were distinguished by a lower case letter as a prefix in order to make the assignment unique. An alphabetic order of the prefixes was established both for even parity and odd parity levels, starting with a,b,c,,,.

Later on the odd parity levels were given prefixes starting with
z,y,x,w,,,, and that is the notation, currently used in compilations of
complex iron group spectra.

The allowed transitions between levels of one even-parity term and
one odd-parity term form a multiplet(LS multiplet). In general the
levels are not pure LS states, particularly at higher excitation
potentials, but may be expressed as a linear combination of several LS
states, sometimes with different multiplicity. Theoretical calculations
of the level structure provide the composition of each level in the LS
basis. For low levels the major LS component generally accounts for more
than 90%. In some cases, at medium excitation potentials, a level may be
described as a 60-40% mixing of two different LS states, which is
revealed by its radiative transitions. If one of these states is a
doublet state and the other one a quartet state we will observe strong
intercombination lines from this particular level i.e. doublet-quartet
transitions. For highly excited levels the major component may be less
than 30% and in such cases the LS notation is only used as a label of
the level and has no real physical meaning.

The Multiplet Tables have been arranged according to the excitation
potential of the lower level of the multiplet. There is no demand that
the transitions strictly obey the LS selection rules. There is one Table
for the optical region ($\lambda>3000$ Å) and one for the region below 3000 Å.
In the latter one the multiplet numbers are accompanied by the prefix
"UV". Multiplets that obey the parity rule and have been observed
completely or partly in any laboratory source are included. In some
cases even predicted multiplets are given. The multiplets are numbered
starting with all combinations to the ground term, then to the next
lowest term and so on. This means that the numbering system breaks down
when new high terms are found, which combine with low levels. Extra
numbers have to be inserted for these new multiplets, if the old
numbering system should be retained. This has been done in updated
multiplet tables for other elements in form of a decimal system (see
e.g. Moore 1965) and is proposed to be done for Fe II as well.

For highly excited configurations, in particular hydrogen-like, the
LS approximation gives a poor description of the structure. This means
that the associated transitions can not be grouped in multiplets, unless
the levels are relabelled in LS notation. This problem is discussed in
Sec 3.2. in the case of $3d^6 4d - 3d^6 4f$ transitions (Group IV in Table
1).

In order to give some relation between the $3d^6(^x L)nl$ system and the
multiplet numbering system we have in Fig 9 inserted the individual UV
multiplet numbers of the supermultiplets in the 4s-4p and 4p-5s
transition arrays. The supermultiplets, based on the quintet parent,
consist of 6 individual LS multiplets and those based on a triplet
parent of 3 LS multiplets. The numbers given in italics in Fig 9 refer
to the highest multiplicity, i.e. sextet multiplets for the quintet
parent term and quartet multiplets for the triplet parent terms. We see
that there are complete supermultiplets in the 4s-4p transition array up
to the $a^1 S$ parent term, indicating that the high level system was only
fragmentarily known at the time when the Multiplet Tables were
constructed.

The question of an updated version of the Fe II multiplet table is discussed by Baschek and Johansson in another article in this book.

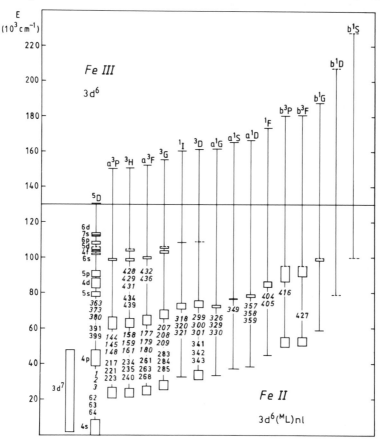

Fig. 9. The normal system of configurations with individual LS multiplet numbers inserted for the 4s-4p and 4p-5s supermultiplets. (Highest multiplicity in italics).

5. CONCLUSIONS

In this article the gross structure of Fe II has been reviewed and the term system has been split up gradually in smaller units in order to get a simplified picture of the level structure and the related spectrum. The major contributions to the line spectrum have been attributed to so called supermultiplets, described in view of an LS coupled ion. Deviations from this model are clearly verified by the fact that a lot of inter-parent multiplets are observed particularly in the optical region. As a matter of fact the majority of the optical multiplets consist of inter-parent transitions and in many cases intercombination lines. Only 36% of the optical multiplets are allowed according to the ΔL and ΔS selection rules.

Before space missions allowed for astronomical observations below the atmospheric cut-off around 3000 Å the most studied wavelength region in stellar spectroscopy with ground-based instruments seems to have been the 3200 -4800 Å range. This range happens to coincide with the only "window" of the total experimentally studied region (900 - 50000 Å), where the Fe II spectrum is "transparent", i.e. where the line density is low. This "window" is located between the resonance region (λ<3200Å) and the dense region of high-level transitions (λ>4800Å) and does not contain any prominent Fe II lines!

This concluding remark implies that the newly opened wavelength regions in astronomy, the far-UV and far-infrared, probably will be much more dominated by Fe II than the optical region ever has been. There is a future for Fe II even in the next period of sixty years.

6. REFERENCES

Adam, J., Baschek, B., Johansson, S., Nilsson, A.E., and Brage, T. 1986, *Astrophys. J.*, in press.

Baschek,B. 1983, in *Highlights of Astronomy* (Ed.R.M.West) **6**, 781.

Brage, T., Nilsson, A.E., Johansson, S., Baschek, B., and Adam, J. 1986, *J. Phys. B.*, in press.

Condon, E.U., and Shortley, G.H. 1963, *The Theory of Atomic Spectra,* University Press, Cambridge.

Cowan, R.D. 1981, *The theory of Atomic Structure and Spectra.* University of California Press, Berkeley.

Dobbie, J.C. 1938, *Ann.Solar.Phys.Obs.Cambridge*, **5**, Part I.

Edlén, B., 1964, in S.Flúgge,ed., *Handbuch der Physik* (Springer-Verlag, Berlin), Vol XXVII.

------- 1972, *Optica Pura y Applicada*, **5**, 101.

Edlén, B., and Swensson, J.W. 1975, *Physica Scripta* **12**, 21.

Green, L.C. 1939, *Phys. Rev.* **55**, 1211.

Johansson, S. 1977, *Mon. Not. R. astr. Soc.* **178**, 17P.

------- 1978a, *Physica Scripta* **18**, 217.

------- 1978b, *Mon.Not.R.astr.Soc.*, **184**, 593.

------- 1984, *Proc. 4th IUE Conference*, Rome, ESA SP-218.

Johansson, S., and Baschek, B. 1984, Atomic Spectroscopy Lund University Ann. Rept. p. 32.

Johansson, S., and Cowley, C.R. 1984, *Astron. Astrophys.*, **139**, 243.

------- 1985, Atomic Spectroscopy Lund University Ann. Rept. p. 43.

Johansson, S., and Litzén, U. 1978, *J. Phys. B.* **11**, L703.

Meggers, W.F., Kiess, C.C., and Walters, F.M.Jr. 1924, *J.O.S.A.*, **9**, 355.

Merrill, P.W. 1928, *Astrophys. J.*, **67**, 391.

Moore, C.E. 1945, *Princeton Obs.Contr.* No.20.

------- 1952a, *Atomic Energy Levels,* U.S.Natl.Bur.Stand.Circ. No. 467, Vol II.

------- 1952b, *An Ultraviolet Multiplet Table,* NBS Circular No. 488, Section 2.

------- 1965, *Selected Tables of Atomic Spectra*, NSRDS-NBS 3, Sec. 1 (SiI-IV).

Nussbaumer, H., Pettini, M., and Storey, P.J. 1981, *Astron. Astrophys.*
                102, 351.
Russell, H.N. 1926, *Astrophys. J.*, 64, 194.
Sugar, J., and Corliss, C.C. 1985, *J. Phys. Chem. Ref. Data*, 14, Suppl.
                No.2.
Thackeray, A.D. 1962, *Mon.Not.R.astr.Soc.* 124, 251.

NEW LABORATORY RECORDINGS OF FE II IN THE IUE REGION

Bodo Baschek
Institute of Theoretical Astrophysics
University of Heidelberg
HEIDELBERG, West-Germany

Sveneric Johansson
Department of Physics
University of Lund
LUND, Sweden

ABSTRACT. An investigation of the laboratory spectrum of Fe II in the wavelength region 1300 - 3250 Å is in progress. We present here a preliminary report on the 1700 - 3250 Å range, where high-resolution spectra from a hollow-cathode lamp contain some 11 500 lines of Fe I and Fe II. We plan to extend the term analysis of Fe II and present all identified lines in an updated multiplet table.

1. INTRODUCTION

The present knowledge of Fe II in the IUE region, $1200 < \lambda < 3200$ Å, is probably sufficient for making identifications of most strong Fe II lines in stellar spectra of moderate or low resolution. However, current work on line identification in high-resolution IUE spectra of the sharp-lined star 21 Peg (see the paper by Adam et al in this volume) indicates that there is definitely a need for an extended analysis of Fe II, since all presently known Fe II lines have been found in that particular stellar spectrum. Moreover, it is generally noticed that only about 50 % of the individual lines in stellar spectra can be identified by means of present compilations of laboratory spectra (Baschek 1983). The wavelength data for Fe II in the Ultraviolet Multiplet Tables are from work by Dobbie, Green and Edlén (see the paper by Johansson in this volume), based on experimental material in the VUV region, 896-2495 Å (Green 1939), and in the air region, 2150 - 6627 Å (Dobbie 1938).

The extended analysis by Johansson (1978) was basically focussed on the infrared region but spectrograms in the VUV region were used in order to confirm new levels. It turned out that all new levels in the doubly-excited system accounted for strong lines below 2000 Å. However, many lines in the wavelength region 1050 - 2300 Å of the laboratory spectrum could not be attributed to any Fe II transition and are therefore still unidentified. Some of them have very recently been assigned to high-level transitions in the doubly-excited system of configurations (Adam et al 1986, Brage et al 1986).

In this paper we report on a current analysis of spectra from an iron hollow-cathode lamp in the region 1300 - 3250 Å, recorded at the

35

R. Viotti et al. (eds.), Physics of Formation of FeII Lines Outside LTE, 35–39.

10.7 m normal incidence spectrograph at the National Bureau of
Standards, Washington. So far we have worked on a line list in the 1700
-3250 Å range, where a total of 11500 lines have been measured, most
of them being Fe II or Fe I. Before starting a systematic search for new
levels in Fe II the line lists have to be cleared from impurity lines,
Fe I lines and already known Fe II lines. That work is now more or less
finished. As a final step we intend to arrange the identified Fe II
lines in an updated multiplet table.

2. EXPERIMENTAL DATA

The "ideal" line list of a laboratory spectrum contains only lines from
that particular element one wants to study. All lines should be
attributed to one particular ionization stage and measured with a
feasible accuracy. The list can be divided into two parts - one with
identified lines, i.e. assigned to an electronic transition, and one
with the remaining unidentified lines. The latter list is used for a
search of new levels by looking for their combinations with already
known levels. The analysis is in general facilitated if the spectral
lines are characterized by some distinguishing feature that permits a
classification of them into different categories. Such "finger prints"
may concern differences in the profile or the intensity of a line that
appears in different light sources.

In the current study of Fe II in the VUV region we have improved
the wavelength accuracy compared to earlier recordings by using the 10.7
m normal incidence spectrograph at the National Bureau of Standards,
Washington D.C. The reciprocal dispersion is 0.78 Åmm^{-1} which limits the
wavelength uncertainty to less than 0.006 Å. We have used a pulsed
hollow cathode lamp with pure iron, run in a low pressure atmosphere of
argon or neon. Spectra have also been recorded from a continuously
burning light source. In general we get lines from Fe I, Fe II and Fe
III and from the carrier gas spectra, especially Ar I-II and Ne I-II.
Some "impurity" lines from carbon, oxygen and hydrogen are also present
on the plates. In principle all impurity lines and gas lines can be
immediately sorted out, since they are known and tabulated. From Fe III
we only see the strongest known lines. The remaining lines on the
spectrograms are believed to belong to Fe I or Fe II. The Fe II lines
are generally stronger on spectra from the pulsed source irrespective of
the choice of carrier gas. The Fe I lines dominate the spectrum from the
continuous hollow cathode and could consequently be distinguished from
Fe II by comparing spectra from the pulsed and the unpulsed source.
However, collisions between gas ions and iron atoms involve transfer of
charge and energy and leave the iron ions in excited states followed by
subsequent radiative deexcitation. This charge-transfer process is very
dominant in the <u>unpulsed</u> light source. Hence, the continuous Ar-Fe lamp
gives a spectrum where some Fe II lines are enhanced due to the
selective excitation of Fe II levels around 60 000 cm^{-1}. The Ne-Fe lamp
yields a large population of Fe II levels around 110 000 cm^{-1} (Johansson
1978) reflected in a number of strong lines from these levels. This
means that the method of sorting Fe I from Fe II by comparing spectra

from a pulsed and an unpulsed discharge is not quite reliable or
sufficient. However, the sorting may be done by comparing the laboratory
emission spectrum with a laboratory absorption spectrum, in which the Fe
I spectrum is expected to dominate. We have got access to absorption
data of iron recorded at Naval Research Laboratory, Washington D.C., and
communicated by Dr C Brown (1985). A comparison with that list gives an
unambiguous classification of all Fe I lines that can be attributed to
low Fe I levels. Absorption lines from Fe I levels with a somewhat
larger excitation potential (>2 eV) do not appear in the laboratory
absorption source but are definitely present in the solar spectrum. Thus
a comparison with the new solar table in the wavelength region 2095 -
3069 (Moore et al 1982) enables us to distinguish Fe I from Fe II in
that region. Moreover, it has been found that some stars act as
effective Fe I filters in the way that Fe II is very dominating but Fe I
is absent in the spectrum (Adam et al 1986). All these considerations
together provide a nearly complete sorting of the lines.

By means of a simple computer program we have matched our final
line list with a list of Fe II lines, predicted from the known energy
levels under consideration of the selection rules. The coincidences are
checked as regards intensity relations and the presence of similar lines
before the identifications are accepted. The final step in the
processing will be to arrange the identified lines in LS multiplets
according to the scheme in the Ultraviolet Multiplet Table (Moore 1952)
and once more check the consistency in the identifications. New
multiplets will be numbered according to a decimal system, which
preserves the old numbering system.

3. THE ANALYSIS

A thorough discussion of the laboratory spectrum of Fe II is given in
the paper by Johansson in this volume, where the large density of Fe II
lines in the wavelength region 2300 - 3100 Å is ascribed to the 4s-4p,
4p-4d and 4p-5s transition arrays. As can be concluded from that paper
(Sec. 3.3.3) there are still a large number of unknown 4d and 5s levels,
built on excited parent terms.

In Table 1 we give a detailed picture of the spectral distribution
of the 12000 lines in the 1700 - 3250 Å region, observed from the
hollow-cathode lamp. Below 1970 Å we have so far only regarded lines
from the continuous laboratory source, while we have a merged list of
lines from the pulsed and unpulsed hollow cathode above 1970 Å. In the
table we divide the lines into three different groups: identified Fe II
lines, identified Fe I lines and unidentified lines. The identified Fe
II lines are predictable from known energy levels and the major part is
included in the Multiplet Tables (Moore 1945, Moore 1952). For Fe I a
great number of new levels have been found, which means that the number
of identified Fe I lines exceeds that in the Multiplet Tables. In the
last column of Table 1 we give the number of unidentified lines, where
very faint and questionable features are excluded. In spite of extended
analyses of both Fe II and Fe I the number of unidentified lines is
surprisingly large. Most of the lines appear in the wavelength range
2200 - 2700 Å, where we expect to find the unknown 4p-4d transitions of

Fe II.

In the previous section we discussed the possibilities to distinguish Fe I from Fe II and make separate line lists even for the unidentified lines. As yet we have not done any systematic comparison with the solar tables but just marked lines which appear in the laboratory absorption source. At this stage we can not give any numbers of the ratio between unidentifed Fe I and Fe II lines but we believe that a large fraction of the unidentified lines belong to Fe I. An extended analysis of the Fe I spectrum is also in progress.

Table 1.   Counts of measured lines in different wavelength regions of the hollow-cathode spectrum of iron and the present state of the analysis.

Wavelength region (Å)	Observed lines	Identified Fe II lines	Identified Fe I lines	Unidentified lines**
1700 - 1800	360*	105*	35*	160*
1800 - 1900	360*	135*	70*	100*
1900 - 2000	350*	100*	60*	65*
2000 - 2100	455	165	55	90
2100 - 2200	615	200	150	125
2200 - 2300	705	285	110	175
2300 - 2400	1070	345	60	465
2400 - 2500	970	365	105	280
2500 - 2600	1115	295	125	385
2600 - 2700	1010	220	110	340
2700 - 2800	970	190	185	325
2800 - 2900	880	190	130	240
2900 - 3000	835	140	150	280
3000 - 3100	805	95	165	260
3100 - 3250	1040	110	325	265
Total	11540	2940	1835	3555

*   Below 1970 Å we have so far only data from the unpulsed hollow-cathode lamp, meaning that Fe II lines might be mssing
**   Very faint and questinable lines are excluded

4. REFERENCES

Adam, J., Baschek, B., Johansson, S., Nilsson, A.E., and Brage, T. 1986, *Astrophys. J.*, in press.
Baschek,B. 1983, in *Highlights of Astronomy* (Ed.R.M.West) **6**, 781.
Brage, T., Nilsson, A.E., Johansson, S., Baschek, B., and Adam, J. 1986, *J. Phys. B.*, in press.
Brown, C.M., Ginter, M.L., and Tilford, S.G., 1985, Absorption Spectra of Fe I, 1550 - 2800 Å, unpublished material.
Dobbie, J.C. 1938, *Ann.Solar.Phys.Obs.Cambridge*, **5**, Part I.

Green, L.C. 1939, *Phys. Rev.* **55**, 1211.
Johansson, S. 1978, *Physica Scripta* **18**, 217.
Johansson, S., and Baschek, B. 1984, Atomic Spectroscopy Lund
          University Ann. Rept. p. 32.
Moore, C.E. 1945, *Princeton Obs.Contr.* No.20.
------- 1952, *An Ultraviolet Multiplet Table,* NBS Circular No. 488,
          Section 2.
Moore, C.E., Tousey, R., and Brown, C.M. 1982, *The Solar Spectrum
          3069-2095 Å*, NRL Report 8653.

SEMIEMPIRICAL DETERMINATION OF FE II OSCILLATOR STRENGTHS

Robert L. Kurucz

Harvard-Smithsonian Center for Astrophysics
Cambridge, Massachusetts, U.S.A.

SUMMARY. The "missing ultraviolet opacity" is caused by lines of
iron group elements that go to excited configurations that have not
yet been seen in the laboratory. For the first 10 ions I am compu-
ting allowed and forbidden transition arrays for all known even and
odd configurations plus as many predicted configurations as I can
fit in the computer. A-sums, Stark and van der Waals broadening
constants, and the Lande g value are computed for each energy level.
These data will be published as SAO reports and distributed on
magnetic tape. For Fe II there are over one million lines.

In 1983, working with Lucio Rossi from Frascati and with John
Dragon and Rod Whitaker at Los Alamos, I finally completed line lists
for all diatomic molecules that produce important opacity in G and K
stars. Once the line data were ready, I computed new opacity tables
for solar abundances. The calculations involved 17,000,000 atomic
and molecular lines, 3,500,000 wavelength points, 50 temperatures,
and 20 pressures, and took a large amount of computer time.

As a test the opacities were used to compute a theoretical solar
model, to predict solar fluxes and intensities from empirical models,
and (with fudging) to produce improved empirical models that are able
to match the Ca II H and K line profiles and both the UV and IR intensities formed near the temperature minimum. The work on empirical
models is in collaboration with Avrett and Loeser.

There are several regions between 200 and 350 nm where the predicted solar intensities are several times higher than observed, say
85% blocking instead of the 95% observed. The integrated flux error
of these regions is several per cent of the total. In a flux constant
model this error is balanced by a flux error in the red. The model
predicts the wrong colors. After many experiments with convection and
opacities, and after synthesizing the spectrum in detail, I have determined that this discrepancy is caused by missing iron group atomic
lines that go to excited configurations that have not been observed in
the laboratory. Most laboratory work has been done with emission
sources that cannot strongly populate these configurations. Stars,
however, show lines in absorption without difficulty.

R. Viotti et al. (eds.), Physics of Formation of FeII Lines Outside LTE, 41–43.
© *1988 by D. Reidel Publishing Company.*

I have used Bob Cowan's Hartree-Fock programs at Los Alamos to compute Slater single- and configuration interaction integrals for the lowest 50 configurations of the first 10 stages of ionization for elements up through Zn and for the first 5 for heavier elements. These calculations allow me to determine eigenvectors by combining least squares fits for levels that have been observed with computed integrals (scaled) for higher configurations. Each least squares iteration takes a significant amount of time on a Cray and many iterations are required. Thus far I have completed new line lists only for Fe I and II, but they produce the strongest effect on the spectrum. Radiative, Stark, and van der Waals damping constants and Lande g values are automatically produced for each line. The complexity of these calculations is illustrated by this table,

	Fe I		Fe II	
	even	odd	even	odd
number of configurations	26	20	22	16
number of levels	5401	5464	5723	5198
largest Hamiltonian matrix	1069	1094	1102	1049
number of least squares parameters	963	746	729	541
(many fixed at scaled HF)				
total number of lines saved	583,814		1,112,322	

The figure on the next page shows the blocking in the solar ultraviolet spectrum produced by Fe lines. The calculation was done once with the old line data mentioned above and again with the newly calculated data. The increase in opacity is dramatic. I expect there to be similar effects in hotter stars.

The National Science Foundation has given me a large grant of time on the Cray XMP/48 at the San Diego Supercomputer Center. For the first 10 ions of the iron group elements I expect to compute all the known even and odd configurations plus as many predicted configurations as I can fit in the computer. Then I will compute allowed and forbidden transition arrays. The sum of the A values and Stark and van der Waals broadening constants will be computed for each energy level. As these sums will be complete up to rather high energy, the lowest level of the lowest omitted configuration, the A-sums will easily yield lifetimes and branching ratios for each line. SAO reports will be prepared that present the gf and lifetime data and compare with a compilation of all measurements. These reports will be maintained on the computer and printed on demand. All the calculations will be available on magnetic tape. As new laboratory analyses become available the calculations will be revised.

Once the line data are complete, I will recompute the opacity tables for a range of abundances, then compute new grids of models and synthesize spectra for comparison to observed spectra.

This work is partially supported by NASA grants NSG-7054 and NAG5-824, and by NSF grant AST-8518900.

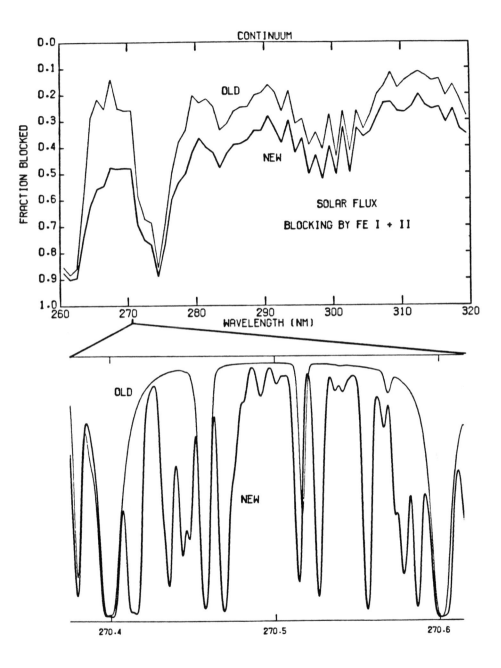

Theoretical determinations of Fe II atomic parameters

H. Nussbaumer
Institute of Astronomy
ETH Zentrum
8092 Zürich (Switzerland)

Summary: The present status of theoretical calculations of radiative transition probabilities, electron collision cross sections and ionization parameters for Fe II is reviewed. A progress report is given on the calculation of radiative probabilities for the forbidden infrared transitions among the four lowest terms a^6D, a^4F, a^4D, a^4P. Examples of density sensitive flux ratios among these infrared lines will be shown.

1. Introduction

In a large number of astronomical objects Fe II lines are present as prominent and numerous features. They show up as emission and absorption lines. A lack of accurate atomic data is one of the serious obstacles to exhaustive diagnostic exploitation of Fe II lines. Although the multitude of excitation and decay paths, including excitation by fluorescence, is a serious source of uncertainty, the lack of atomic data is certainly the most annoying barrier against progress in the interpretation of Fe II spectra. In this brief review I shall present the status of theoretical calculations of atomic data.

2. Past calculations

The first proper calculation of Fe^+ collision cross sections was published only a few years ago (Nussbaumer and Storey, 1980). At that time a new generation of fast computers with virtual Megabyte memories — we were used to kilobytes — had become available . For us it was the IBM 3033. Still, in addition to the many hours of computer time one had to be prepared to invest a few months of ones own time in a venture that might easily have failed. Radiative probabilities, in particular of forbidden transitions, had been calculated much earlier by Garstang (1962) and later by Nussbaumer and Swings (1970). Electric dipole probabilities were calculated by Kurucz and Peytremann (1975). These probability calculations were semi-empirical in that the Slater integrals and the spin-orbit integrals were obtained by fitting to observed energy separations. Nussbaumer and Storey(1980) did an

45

R. Viotti et al. (eds.), Physics of Formation of FeII Lines Outside LTE, 45–50.

ab initio calculation to obtain collision cross sections and radiative transition probabilities among the four terms $3d^6 4s$ ^6D, $3d^7$ ^4F, $3d^6 4s$ ^4D, $3d^7$ ^4P (a^6D, a^4F, a^4D, a^4P). They obtained their collision strengths with a target basis of two configurations ($3d^6 4s$, $3d^7$) and the transition probabilities in a three configuration basis ($3d^6 4s$, $3d^7$, $3d^5 4s^2$). One year later Nussbaumer, Pettini and Storey (1981) published further calculations, with the main emphasis on the sextet transitions between the ground term, $3d^6 4s$ ^6D, and a few of the lowest odd sextet terms. The transition probabilities were mainly calculated in view of their belonging to interstellar absorption lines. — Although the results of both calculations could be termed a success, the authors pointed out serious shortcomings in the quality of their atomic eigenfunctions.

3. Present work

To my knowledge there are at present only two theoretical projects dealing with Fe II:

(1) The atomic physics group at Belfast (Baluja et al. 1986) has done a calculation of collision strengths between the ground term, a^6D, and the first three excited terms a^4F, a^4D, a^4P. These are the atomic data which were calculated for the first time by Nussbaumer and Storey(1980). Baluja et al.(1986) work with a more extended configuration basis and a correspondingly improved set of eigenfunctions. They also allow for the existence of resonance states. Whereas the collision strength of the ^6D – ^4F transition changes by only 10% , the total strengths of a^6D – a^4D and a^6D – a^4P change by a factor two in opposite directions. Although the configuration basis of Baluja et al.(1986) is still unsatisfactory, we can now assume with more confidence that the uncertainty in the collision cross sections should not exceed a factor two.

(2) P.J. Storey and H. Nussbaumer are at present engaged on a calculation of radiative probabilities for the infrared transitions among the four terms a^6D, a^4F, a^4D, a^4P. The corresponding infrared lines span the wavelengths from $\approx 7500\,\text{Å}$ to $\approx 90\mu$.

4. Focus on infrared lines

Development of infrared observing techniques have led to a growing interest in infrared lines. They are observed in objects as different as Seyfert galaxies and Herbig-Haro objects. Some of the lines from the a^4F – a^4D multiplet, which lie in the range $1.5 - 2.0\mu$, have been observed in the emission of supernova remnants. The exponential decay in type I supernova light curves has led to speculation that the light curve is governed by decay of short lived ($\tau_{1/2} \approx 6$ days) radioactive ^{56}Ni synthesised during the supernova event; ^{56}Ni decays eventually to ^{56}Fe. According to this line of thought a young supernova shell might contain something of the order of $1 M_\odot$ of Fe. It is now a matter of proving the presence of that much iron and, if present, of calculating its effect on the energy balance during the shell evolution. Kirshner and Oke(1975) drew attention to the very high iron abundance of the young type I supernova 1972 e in NGC 5253. They wondered about the source of the excitation energy. Meyerott (1978,1980), after preliminary work of Colgate and Mc Kee

(1969), suggessted this to be radioactive decay producing primary electrons which mainly ionize He. The secondary electrons produced in the ionization–recombination process, and line radiation from He^0 and He^+ ionize the heavier elements. The radioactive energy input is balanced by cooling through forbidden transitions. The transition $a^4P_{5/2} \rightarrow a^4F_{9/2}$ at 8617 Å has the largest probability of the transitions within the 4 energetically lowest terms (A=0.0299 s^{-1}). At densities of 10^7 cm^{-3} all the levels of these 4 terms are in Boltzmann distribution. At temperatures of 3000K or lower, energy is mainly lost through the fine structure transition $a^6D_{7/2} \rightarrow a^6D_{9/2}$ at 26μ and also through $a^4F_{9/2} \rightarrow a^6D_{9/2}$ at 5.34μ. At 5000K or hotter and $N_e \gtrsim 10^6$ cm^{-3} the biggest loss occurs through $a^4P_{5/2} \rightarrow a^4F_{9/2}$ at 8617 Å.

There are several pairs of infrared [Fe II] lines which can serve for determining electron densities. As example I show in Figure 1 flux ratios of the prominent lines at 1.602μ and 1.644μ as well as the 9267.6 Å/8617.0 Å. The figure is based on the calculations of Nussbaumer and Storey (1986). The upper levels of these lines have their Boltzmann population at $N_e \gtrsim 10^5$ cm^{-3}.

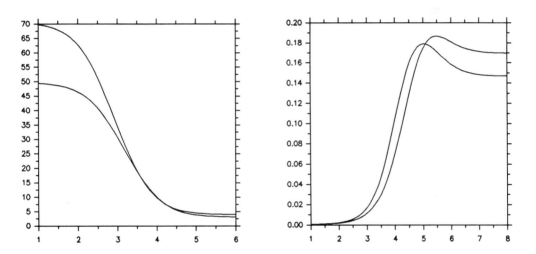

Fig. 1. Ratios of line emissivities. The y-axis shows $\epsilon(\lambda_1)/\epsilon(\lambda_2)$ as a function of the electron density, log $N_e[cm^{-3}]$ is given on the x-axis. ϵ is the energy emitted per unit volume and unit time. (left) The behaviour of $\epsilon(1.602\mu)/\epsilon(1.644\mu)$. The upper curve corresponds to $T_e=$ 3000K, the lower curve represents $T_e=$ 10'000K. (right) Ratios of line emissivities 9267.6 Å/8617.0 Å. The upper curve corresponds to $T_e=$ 10000K, the lower curve represents $T_e=$ 3000K.

If the emission occurs in a gas where the upper levels are statistically populated, as seems to be the case at some stage in the supernova evolution, one can obtain iron abundances from these lines. Let us assume that from an excited level SLJ a line λ is emitted. Neglecting interstellar or intergalactic absorption we observe a flux from an emission region with volume V at a distance d

$$f_\lambda = \frac{1}{4\pi d^2} \int_V N(SLJ) \cdot A_\lambda \cdot h \cdot \frac{c}{\lambda} \cdot dV.$$

In the well known expansion

$$N(SLJ) = \frac{N(SLJ)}{N(\mathrm{Fe}^+)} \cdot \frac{N(\mathrm{Fe}^+)}{N(\mathrm{Fe})} \cdot N(\mathrm{Fe}) = g \cdot k \cdot N(\mathrm{Fe})$$

the relative population g is easily calculated if a Boltzmann distribution is assumed and if the total Fe population can be expressed as the sum of the levels from a^6D, a^4F, a^4D and a^4P. The total mass of Fe, M_{Fe}, is given as

$$M_{\mathrm{Fe}} = \int_V N(\mathrm{Fe}) \cdot m_{\mathrm{Fe}} \cdot dV$$

where m_{Fe} is the mass of the iron atom. The measured flux is then

$$f_\lambda = \frac{M_{\mathrm{Fe}}}{4\pi d^2 m_{\mathrm{Fe}}} \cdot g \cdot k \cdot h \cdot \frac{c}{\lambda} \cdot A_\lambda.$$

The calculated iron mass depends thus inversely on A_λ. Graham et al. (1986) find $\approx 0.3 M_\odot$ of iron in SN 1983 n which occured in the galaxy M 83. They derive that mass from the interpretation of the 1.644 μ line. This is the transition $a^4D_{7/2} \rightarrow a^4F_{9/2}$. Its probability is 0.00438 s^{-1} in the eight configuration calculation of Nussbaumer and Storey (1986). It had half that value in the early three configuration calculation (Nussbaumer and Storey, 1980).

5. Main problems and outlook

An awkward obstacle that occasionally creeps up in theoretical spectroscopic work is the lack of reliable information on the atomic energy structure. Thanks to the efforts of Johansson (1978) this is not the case for those bound states of Fe^+ which are now at the center of astrophysical interest. However, the problem is there for the high lying and autoionizing states of Fe^0 and Fe^+. (Johansson and others might have something to add on doubly excited states.) No reliable ionization and recombination calculations can be done without a knowledge of autoionization states. Up to now no such calculations have been performed for Fe II. (Arnaud and Rothenflug (1985) rely on the measurements of Montague et al.(1983)).

Charge transfer can strongly influence ionization and recombination. Meyerott (1978) suggests that charge transfer reactions

$\mathrm{He}^+ + \mathrm{Fe} \rightarrow \mathrm{He}^0 + \mathrm{Fe}_\alpha^+$

or

$\mathrm{He}^+ + \mathrm{Fe} \rightarrow \mathrm{He}^0 + \mathrm{Fe}_\alpha^{+2} + \mathrm{e}^-$

may be of importance in young SN, α signifies an autoionizing state. Needless to say that these processes have not yet been calculated.

Whether the atomic eigenfunctions are constructed according to the method of Eissner et al.(1974) or Crees et al.(1978), they always rely on a fast convergence of a configuration expansion. But whereas in a $n = 2$ case, such as C^{+2}, high quality eigenfunctions can be obtained by brute force (basis of $\approx$ 100 configurations for a few terms), demands on computer time and storage (even with the Megabytes of virtual memory) become prohibitive if such methods are tried on the lowly ionized $3d^k$, $3d^{k-1}4l$ systems. In the $n = 2$ and the $3s^u3p^v$ systems, a large number of terms and energy levels can be treated in one pass. For Fe II the forementioned limitations, and also the numerical methods employed force the atomic physicist to concentrate his attention to a limited number of transitions among a few terms.

In our earlier work (Nussbaumer and Storey 1980) the transition probabilities were calculated in a three configuration approximation: $3d^64s$, $3d^7$, $3d^54s^2$. In our work mentioned in Sections 3 and 4, we did extensive preliminary investigations which resulted in an eight configuration basis: $3d^64s$, $3d^7$, $3d^64d$, $3d^65d$, $3d^65s$, $3d^54s^2$, $3d^54p^2$ and $3d^54s4d$. Again I do not want to bore you with details. For the important transitions differences against earlier three configuration work can be as large as a factor two for some cases, or just a few percent in others.

Zeippen(1980), and Eissner and Zeippen(1981) have drawn attention to the importance of relativistic corrections to the magnetic dipole operator. These corrections can become important when spin-orbit coupling between terms vanishes, leaving only very small two-electron spin-orbit and orbit-orbit effects. Although direct spin-orbit coupling does vanish between $3d^64s$ and $3d^7$, the combination of configuration and spin-orbit interaction leads to amplitudes for the ordinary magnetic dipole operator which are large enough that relativistic corrections are unlikely to be important.

Ab initio calculations of Fe II atomic data are still in their infancy. To my knowledge only radiative transition probabilities and electron–ion collision strengths are being calculated at the moment, but no charge exchange, ionization or recombination data. No fast progress is foreseen unless a group should decide to dedicate a major fraction of their activity to Fe II.

Acknowledgment: I thank Dr. P.J. Storey for his comments.

References:

Arnaud, M., Rothenflug, R.: 1985, Astron.Astrophys.Suppl.Ser. **60**, 425

Baluja, K.L., Hibbert, A., Mohan, M.: 1986, J.Phys. B (submitted)

Colgate, S.A., Mc Kee, C.: 1969, Astrophys. J. **157**, 623

Crees, M.A., Seaton, M.J., Wilson, P.M.H.: 1978, Computer Phys.Commun. **15**, 23

Eissner, W., Jones, M., Nussbaumer, H.: 1974, Computer Phys.Commun. **8**, 270

Eissner, W., Zeippen, C.J.: 1981, J.Phys. B.**14**, 2125

Garstang, R.H.: 1962, Monthly Notices Roy.Astron.Soc. **124**, 321

Graham, J.R., Meikle, W.P.S., Allen, D.A., Longmore, A.J., Williams, P.M.: 1986, preprint

Johansson, S.: 1978, Physica Scripta **18**, 217

Kirshner, R.P., Oke, J.B.: 1975, Astrophys. J. **200**, 574

Kurucz, R.L., Peytremann, E.: 1975, Smithsonian Astron.Obs.Spec. Rept. No. 362

Meyerott, R.E.: 1978, Astrophys. J. **221**, 975

Meyerott, R.E.: 1980, Astrophys. J. **239**, 257

Montague, R.G., Diserens, M.J., Harrison, M.F.A.: 1983, J.Phys. B **17**, 2085

Nussbaumer, H., Pettini, M., Storey, P.J.: 1981, Astron.Astrophys. **102**, 351

Nussbaumer, H., Storey, P.J.: 1980, Astron.Astrophys. **89**, 308

Nussbaumer, H., Storey, P.J.: 1986, Astron.Astrophys. (to be submitted)

Nussbaumer, H., Swings, J.P.: 1970, Astron.Astrophys. **7**, 455

Zeippen, C.J.: 1983, J.Phys. B **13**, L485

CALCULATIONS OF TRANSITION PROBABILITIES FOR FORBIDDEN LINES IN THE $3d^7$ CONFIGURATION IN Fe II.

J.E. Hansen and G.M.S. Lister
Zeeman Laboratory, University of Amsterdam
Plantage Muidergracht 4
1018 TV Amsterdam
The Netherlands

ABSTRACT. We have used a new method to include higher-order configuration-interaction effects in calculations of transition probabilities for magnetic dipole and electric quadrupole transition probabilities for transitions within the $3d^7$ configuration in Fe II. This is the first step towards a re-evaluation of forbidden transition probabilities for transitions within $3d^N$ and $3d^N4s$ configurations in iron-group elements.

This paper reports a reevaluation of the forbidden transition probabilities within the $3d^7$ configuration of Fe II.

We have used the recently introduced set of orthogonal operators for d^N configurations [1,2] to construct wavefunctions which include electrostatic configuration interaction effects to infinite order in perturbation theory and magnetic effects to second order in connection with an effective quadrupole operator which takes electrostatic configuration interaction effects into account. Only a brief description of the method is given here and details will be published elsewhere.

For the magnetic dipole transition probabilities, which depend exclusively on the angular SLJ composition of the states, no effective operator is necessary and the transition probabilities can be calculated in the usual way [3] using the accurate $|SLJ>$ wavefunctions obtained as result of parametric fitting based on the orthogonal operators. The quadrupole transition probabilities depend in addition on the radial part of the wavefunction and an effective operator which takes the mixing with perturbing configurations into account is necessary. We have included the second order effects due to the perturbing configurations $3s3d^8$, $3d^64s$, $5s$, $6s$, $4d$, $5d$, $6d$ and $5g$ as well as the third-order effects due to the overlapping $3d^64s$ configuration. The second-order effects due to the $5s$, $6s$, $4d$, $5d$, $6d$ and $5g$ configurations are small. However, it is somewhat doubtful whether the effects of the overlapping $3d^64s$ configuration can be taken into account by perturbation theory and we are considering to include these

51

R. Viotti et al. (eds.), Physics of Formation of FeII Lines Outside LTE, 51–61.
© *1988 by D. Reidel Publishing Company.*

effects to infinite order in a future publication by using the
diagonalised $3d^7 + 3d^64s$ interaction matrix as lowest order
approximation.

Table 1 shows the results obtained for the M1 and E2
transition probabilities. The results of the previous calculation
by Garstang are also included [4]. Table 1 gives the results for all
multiplets in the $3d^7$ configuration. The multiplets are identified
by multiplet number [5] (if available), term designation [6], J value
and energy of the initial and final states. Wavelengths smaller
than 10μ are tabulated, below 2000 Å in vacuum and above in air.
The wavelengths have been supplied by S. Johansson and are
obtained from the most accurate energy level values presently
available [7]. They should be accurate to about 0.001 Å in many
cases (see ref. 7).

Two previous calculations of these transition probabilities
exist. The one due to Garstang [4] was mentioned already and is
included in Table 1. A later calculation by Nussbaumer and Swings [8]
is in good agreement with Garstang's work for the strong lines
except in some cases which are tabulated in ref. 8. The
compilation [9] by Smith and Wiese is based on these two calculations
and Smith and Wiese have chosen not to report values which deviate
strongly between the two calculations. We have found that our
transitions [8] probabilities for the lines tabulated by Nussbaumer
and Swings [8] in all cases are in much better agreement with the
results due to Garstang [4] than with those due to Nussbaumer and
Swings.

Our differences with Garstang's results are in general rather
small although they can amount to a factor of 4 even for strong
(electric quadrupole) transitions. For the magnetic dipole
transitions the differences are smaller. This insensitivity of
particularly the magnetic dipole transition probabilities to the
exact eigenvector composition is one reason that the fairly simple
(by today's standards) calculations due to Garstang have been so
successful in the past (see ref. 9). We believe, however, that our
values are more accurate and this should make it possible to
reduce the uncertainties in abundance determinations based on these
lines. As already mentioned we hope to calculate also lines within
the $3d^64s$ configuration and between $3d^7$ and $3d^64s$ in the future as
well as extending the calculations to other ions in the iron group.

ACKNOWLEDGMENT.

We thank Dr. S. Johansson for providing us with the latest values
of the wavelengths for the forbidden $3d^7$ lines.

REFERENCES.

1. B.R. Judd, J.E. Hansen and A.J.J. Raassen, J. Phys. B $\underline{15}$,
    1457 (1982).
2. J.E. Hansen and B.R. Judd, J. Phys. B $\underline{18}$, 2327 (1985).
3. J.E. Hansen, A.J.J. Raassen and P.H.M. Uylings, Astroph. J.
    $\underline{277}$, 435 (1984).
4. R.H. Garstang, Mon. Not. R. Astr. Soc. $\underline{124}$, 321 (1962).
5. C.E. Moore, Multiplet Table of Astrophysical Interest, Contr.
    Princeton Univ. Obs., No. 20 (1945).
6. C. Corliss and J. Sugar, J. Phys. Chem. Ref. Data $\underline{11}$, 135
    (1982).
7. S. Johansson, Phys. Scripta $\underline{15}$, 183 (1977), and private
    communication.
8. H. Nussbaumer and J.P. Swings, Astron. & Astrophys. $\underline{7}$, 455
    (1970).
9. M.W. Smith and W.L. Wiese, J. Phys. Chem. Ref. Data $\underline{2}$, 85
    (1973).

Table 1

No.	Designation	J_f	J_i	λ (Å)	E_f cm^{-1}	E_i cm^{-1}	A(M1)(sec^{-1}) This work	Garstang	A(E2)(sec^{-1}) This work	Garstang
	$a^4F - a^4F$	9/2	7/2		1873	2430	5.838×10^{-3}	5.8×10^{-3}	2.251×10^{-10}	1.9×10^{-10}
		9/2	5/2		1873	2838	–		4.010×10^{-10}	3.3×10^{-10}
		7/2	5/2		2430	2838	3.917×10^{-3}	3.9×10^{-3}	6.933×10^{-11}	5.9×10^{-11}
		7/2	3/2		2430	3117	–		1.369×10^{-10}	1.1×10^{-10}
		5/2	3/2		2838	3117	1.411×10^{-3}	1.4×10^{-3}	1.213×10^{-11}	1.0×10^{-11}
13F	$a^4F - a^4P$	9/2	5/2	8616.952	1873	13474	–		2.745×10^{-2}	1.7×10^{-2}
		7/2	5/2	9051.948	2430	13474	3.485×10^{-4}	3.0×10^{-4}	6.871×10^{-3}	3.9×10^{-3}
		7/2	3/2	8891.912	2430	13673	–		1.659×10^{-2}	1.0×10^{-2}
		5/2	5/2	9399.045	2838	13474	1.475×10^{-4}	1.2×10^{-4}	1.275×10^{-3}	7.0×10^{-4}
		5/2	3/2	9226.617	2838	13673	1.068×10^{-5}	1.7×10^{-5}	9.885×10^{-3}	6.0×10^{-3}
		5/2	1/2	9033.496	2838	13905	–		1.200×10^{-2}	7.5×10^{-3}
		3/2	5/2	9652.705	3117	13474	4.613×10^{-5}	3.4×10^{-5}	1.238×10^{-4}	6.5×10^{-5}
		3/2	3/2	9470.935	3117	13673	1.178×10^{-5}	1.1×10^{-5}	2.849×10^{-3}	1.7×10^{-3}
		3/2	1/2	9267.563	3117	13905	4.964×10^{-6}	4.2×10^{-6}	1.596×10^{-2}	9.9×10^{-3}
14F	$a^4F - a^2G$	9/2	9/2	7155.157	1873	15845	1.372×10^{-1}	1.5×10^{-1}	4.376×10^{-5}	3.3×10^{-5}
		7/2	9/2	7452.538	2430	15845	4.426×10^{-2}	4.8×10^{-2}	2.325×10^{-6}	3.6×10^{-6}
		5/2	9/2	7686.230	2838	15845	–		4.438×10^{-6}	5.3×10^{-6}
		9/2	7/2	6896.175	1873	16369	4.654×10^{-3}	5.2×10^{-3}	5.934×10^{-6}	5.4×10^{-6}

Multiplet	J	J	λ	E	E				
	7/2	7/2	7172.004	2430	16369	5.298×10^{-2}	5.6×10^{-2}	9.566×10^{-6}	5.2×10^{-6}
	5/2	7/2	7388.178	2838	16369	4.040×10^{-2}	4.3×10^{-2}	1.427×10^{-6}	2.8×10^{-6}
	3/2	7/2	7544.011	3117	16369	—		8.500×10^{-6}	1.2×10^{-5}
15F $a^4F - a^2P$	7/2	3/2	6275.512	2430	18361	—		1.744×10^{-3}	1.3×10^{-3}
	5/2	3/2	6440.400	2838	18361	1.864×10^{-2}	2.3×10^{-2}	7.577×10^{-4}	3.2×10^{-4}
	3/2	3/2	6558.497	3117	18361	1.289×10^{-2}	1.6×10^{-2}	2.150×10^{-4}	5.4×10^{-5}
	5/2	1/2	6229.261	2838	18887	—		7.592×10^{-4}	7.1×10^{-4}
	3/2	1/2	6339.675	3117	18887	2.697×10^{-4}	3.1×10^{-4}	4.262×10^{-4}	1.0×10^{-4}
16F $a^4F - a^2H$	9/2	11/2	5413.345	1873	20340	9.980×10^{-5}	1.1×10^{-4}	5.177×10^{-5}	9.6×10^{-5}
	7/2	11/2	5581.860	2430	20340	—		4.501×10^{-6}	8.8×10^{-6}
	9/2	9/2	5280.257	1873	20806	1.075×10^{-3}	6.6×10^{-4}	8.077×10^{-6}	1.7×10^{-5}
	7/2	9/2	5440.465	2430	20806	5.279×10^{-4}	4.7×10^{-4}	1.508×10^{-5}	1.9×10^{-6}
	5/2	9/2	5563.960	2838	20806	—		4.050×10^{-6}	2.0×10^{-5}
17F $a^4F - a^2D$	9/2	5/2	5362.052	1873	20517	—		2.482×10^{-4}	1.0×10^{-4}
	7/2	5/2	5527.338	2430	20517	2.636×10^{-1}	2.7×10^{-1}	1.166×10^{-4}	4.8×10^{-6}
	5/2	5/2	5654.856	2838	20517	2.905×10^{-2}	3.0×10^{-2}	1.970×10^{-6}	1.0×10^{-5}
	3/2	5/2	5745.698	3117	20517	1.253×10^{-2}	1.3×10^{-2}	2.595×10^{-5}	5.8×10^{-6}
	7/2	3/2	5295.714	2430	21308	—		2.043×10^{-4}	3.8×10^{-5}
	5/2	3/2	5412.654	2838	21308	2.645×10^{-1}	2.1×10^{-1}	2.239×10^{-4}	3.3×10^{-5}
	3/2	3/2	5495.824	3117	21308	1.415×10^{-1}	1.4×10^{-1}	4.481×10^{-7}	2.8×10^{-6}

Table 1 (cont. 1)

No.	Multiplet Designation	J_f	J_i	λ (Å)	E_f cm^{-1}	E_i cm^{-1}	A(M1)(sec^{-1}) This work	A(M1)(sec^{-1}) Garstang	A(E2)(sec^{-1}) This work	A(E2)(sec^{-1}) Garstang
27F	$a\,^4F - b\,^2F$	9/2	5/2	3339.136	1873	31812	—		1.276×10^{-4}	1.1×10^{-4}
		7/2	5/2	3402.500	2430	31812	8.335×10^{-3}	1.2×10^{-2}	1.268×10^{-4}	4.1×10^{-4}
		5/2	5/2	3450.397	2838	31812	6.834×10^{-3}	8.7×10^{-3}	3.776×10^{-4}	1.0×10^{-4}
		3/2	5/2	3484.008	3117	31812	3.592×10^{-2}	4.6×10^{-2}	1.619×10^{-3}	2.3×10^{-3}
		9/2	7/2	3318.384	1873	31999	3.582×10^{-2}	4.2×10^{-2}	8.326×10^{-4}	3.7×10^{-4}
		7/2	7/2	3380.955	2430	31999	6.312×10^{-3}	6.2×10^{-3}	1.203×10^{-4}	1.1×10^{-4}
		5/2	7/2	3428.243	2838	31999	1.472×10^{-2}	1.7×10^{-2}	7.481×10^{-4}	8.4×10^{-4}
		3/2	7/2	3461.442	3117	31999	—		1.691×10^{-4}	6.8×10^{-5}
	$a\,^4F - d\,^2D$	7/2	3/2	2209.518	2430	47675	—		3.855×10^{-3}	
		5/2	3/2	2229.619	2838	47675	3.976×10^{-2}		8.454×10^{-5}	
		3/2	3/2	2243.607	3117	47675	2.540×10^{-2}		1.460×10^{-4}	
		9/2	5/2	2165.392	1873	48039	—		1.664×10^{-2}	
		7/2	5/2	2191.865	2430	48039	3.498×10^{-2}		5.738×10^{-4}	
		5/2	5/2	2211.644	2838	48039	5.004×10^{-3}		5.389×10^{-6}	
		3/2	5/2	2225.407	3117	48039	2.138×10^{-3}		1.373×10^{-6}	
	$a\,^4P - a\,^4P$	3/2	1/2		13673	13905	5.518×10^{-4}	5.5×10^{-4}	1.340×10^{-11}	8.7×10^{-12}
		5/2	1/2		13474	13905	—		2.642×10^{-9}	1.6×10^{-9}
		5/2	3/2		13474	13673	1.864×10^{-4}	1.9×10^{-4}	3.838×10^{-11}	2.3×10^{-11}

Transition	J	J'	λ						
$a^4P - a^2G$	5/2	9/2	42178.336	13474	15845	—	—	4.257×10^{-10}	1.9×10^{-6}
	5/2	7/2	34533.504	13474	16369	1.044×10^{-9}	—	9.788×10^{-11}	1.7×10^{-7}
	3/2	7/2	37079.468	13673	16369	—	—	1.300×10^{-9}	7.1×10^{-10}
$a^4P - a^2P$	5/2	3/2	20460.070	13474	18360	6.136×10^{-2}	6.3×10^{-2}	3.524×10^{-6}	2.0×10^{-6}
	3/2	3/2	21327.689	13673	18360	3.101×10^{-2}	3.2×10^{-2}	5.175×10^{-7}	1.0×10^{-6}
	1/2	3/2	22436.425	13905	18360	1.409×10^{-2}	1.4×10^{-2}	2.545×10^{-9}	—
	5/2	1/2	18471.154	13474	18887	—	—	3.704×10^{-6}	—
	3/2	1/2	19175.388	13673	18887	4.516×10^{-5}	3.7×10^{-5}	9.665×10^{-7}	—
	1/2	1/2	20066.959	13905	18887	7.827×10^{-2}	8.1×10^{-2}	—	—
$a^4P - a^2H$	5/2	9/2	13636.307	13474	20806	—	—	8.771×10^{-10}	—
$a^4P - a^2D$	5/2	5/2	14195.524	13474	20517	2.371×10^{-2}	1.9×10^{-2}	4.529×10^{-6}	2.6×10^{-6}
	3/2	5/2	14607.826	13673	20517	8.542×10^{-3}	7.3×10^{-3}	2.199×10^{-7}	4.6×10^{-8}
	1/2	5/2	15119.574	13905	20517	—	—	1.189×10^{-8}	2.4×10^{-7}
	5/2	3/2	12761.984	13474	21308	3.529×10^{-4}	1.7×10^{-3}	6.191×10^{-6}	4.2×10^{-6}
	3/2	3/2	13094.244	13673	21308	1.177×10^{-2}	1.1×10^{-2}	1.417×10^{-6}	6.8×10^{-7}
	1/2	3/2	13503.950	13905	21308	5.355×10^{-3}	5.4×10^{-3}	4.432×10^{-8}	2.8×10^{-7}
$a^4P - b^2F$	5/2	5/2	5451.817	13474	31812	4.391×10^{-4}	3.4×10^{-4}	3.312×10^{-5}	2.9×10^{-4}
	3/2	5/2	5511.562	13673	31812	2.492×10^{-5}	1.3×10^{-5}	4.340×10^{-4}	3.5×10^{-4}
	1/2	5/2	5582.859	13904	31812	—	—	4.083×10^{-4}	6.8×10^{-7}

Table 1 (cont. 2)

No.	Multiplet Designation	J_f	J_i	λ (Å)	E_f cm^{-1}	E_i cm^{-1}	A(M1)(sec^{-1}) This work	A(M1)(sec^{-1}) Garstang	A(E2)(sec^{-1}) This work	A(E2)(sec^{-1}) Garstang
		5/2	7/2	5396.716	13474	31999	5.328×10^{-5}	4.2×10^{-5}	1.420×10^{-4}	8.6×10^{-8}
		3/2	7/2	5455.253	13673	31999	–		1.552×10^{-3}	8.2×10^{-5}
	$a^4P - d^2D$	5/2	3/2	2923.095	13474	47675	3.966×10^{-2}		5.128×10^{-8}	
		3/2	3/2	2940.184	13673	47675	1.239×10^{-1}		3.808×10^{-2}	
		1/2	3/2	2960.352	13905	47675	3.111×10^{-2}		1.742×10^{-2}	
		5/2	5/2	2892.279	13474	48039	3.127×10^{-1}		1.290×10^{-5}	
		3/2	5/2	2909.009	13673	48039	$5.496 \cdot 10^{-2}$		6.331×10^{-2}	
		1/2	5/2	2928.751	13904	48039	–		5.439×10^{-3}	
	$a^2G - a^2G$	9/2	7/2		15845	16369	2.138×10^{-3}	2.1×10^{-3}	1.274×10^{-10}	1.0×10^{-10}
	$a^2G - a^2P$	7/2	3/2	50205.114	16369	18361	–		4.910×10^{-8}	
	$a^2G - a^2H$	9/2	11/2	22237.655	15845	20340	1.380×10^{-2}	1.3×10^{-2}	3.115×10^{-5}	4.1×10^{-5}
		7/2	11/2	25176.085	16369	20340	–		9.685×10^{-7}	1.3×10^{-6}
		9/2	9/2	20151.238	15845	20806	4.079×10^{-2}	3.8×10^{-2}	4.406×10^{-7}	2.4×10^{-6}
		7/2	9/2	22534.598	16369	20806	1.296×10^{-2}	1.2×10^{-2}	2.275×10^{-5}	3.2×10^{-5}

	Transition	J	J'	λ	E_l	E_u	(1)	(2)	(3)	(4)
	$a^2G - a^2D$	9/2	5/2	21396.849	15845	20517			4.292×10^{-5}	
		7/2	5/2	24103.752	16369	20517	5.389×10^{-7}		2.319×10^{-6}	
		7/2	3/2	20242.799	16369	21308	—		4.694×10^{-5}	
44F	$a^2G - b^2F$	9/2	5/2	6261.118	15845	31812	—		1.050×10^{-2}	1.4×10^{-2}
		7/2	5/2	6473.862	16369	31812	2.745×10^{-2}	2.8×10^{-2}	2.307×10^{-2}	3.5×10^{-2}
		9/2	7/2	6188.552	15845	31999	3.151×10^{-2}	2.9×10^{-2}	2.511×10^{-2}	1.0×10^{-1}
		7/2	7/2	6396.312	16369	31999	5.078×10^{-2}	4.5×10^{-2}	2.260×10^{-3}	1.1×10^{-4}
	$a^2G - d^2D$	7/2	3/2	3193.418	16369	47675	—		4.409	
		9/2	5/2	3105.225	15845	48039	—		4.407	
		7/2	5/2	3156.675	16369	48039	3.544×10^{-4}		3.189×10^{-1}	
	$a^2P - a^2P$	3/2	1/2		18361	18887	2.471×10^{-3}	2.4×10^{-3}	2.468×10^{-9}	2.4×10^{-9}
	$a^2P - a^2D$	3/2	5/2	46362.793	18361	20517	6.886×10^{-3}	8.6×10^{-3}	1.442×10^{-6}	6.6×10^{-7}
		1/2	5/2	61326.198	18887	20517	—		7.136×10^{-8}	2.9×10^{-8}
		3/2	3/2	33919.026	18361	21308	4.791×10^{-2}	5.8×10^{-2}	1.939×10^{-6}	7.9×10^{-7}
		1/2	3/2	41289.551	18887	21308	8.696×10^{-3}	1.1×10^{-2}	4.646×10^{-7}	4.9×10^{-8}
47F	$a^2P - b^2F$	3/2	5/2	7432.247	18361	31812	4.622×10^{-4}	4.2×10^{-4}	1.219×10^{-2}	5.9×10^{-3}
		1/2	5/2	7734.710	18887	31812	—		1.107×10^{-2}	4.4×10^{-3}
		3/2	7/2	7330.218	18361	31999	—		1.374×10^{-2}	1.4×10^{-3}

Table 1 (cont. 3)

No.	Multiplet Designation	J_f	J_i	λ (Å)	E_f (cm⁻¹)	E_i (cm⁻¹)	A(M1)(sec⁻¹) This work	A(M1)(sec⁻¹) Garstang	A(E2)(sec⁻¹) This work	A(E2)(sec⁻¹) Garstang
	a^2P – d^2D	3/2	3/2	3410.352	18361	47675	2.343×10^{-4}		9.135×10^{-1}	
		1/2	3/2	3472.682	18887	47675	2.917×10^{-6}		1.026	
		3/2	5/2	3368.481	18361	48039	1.791×10^{-2}		1.690	
		1/2	5/2	3429.277	18887	48039	–		5.149×10^{-1}	
	a^2H – a^2H	11/2	9/2		20340	20806	1.465×10^{-3}	1.5×10^{-3}	7.310×10^{-11}	3.2×10^{-11}
49F	a^2H – b^2F	9/2	5/2	9083.417	20806	31812	–		2.463×10^{-2}	3.0×10^{-2}
		9/2	7/2	8931.481	20806	31999	1.303×10^{-4}	3.4×10^{-5}	1.962×10^{-3}	3.4×10^{-3}
		11/2	7/2	8574.895	20340	31999	–		3.271×10^{-2}	3.1×10^{-2}
	a^2H – d^2D	9/2	5/2	3670.927	20806	48039	–		4.327×10^{-2}	
	a^2D – a^2D	5/2	3/2		20517	21308	7.500×10^{-3}	7.4×10^{-3}	1.617×10^{-9}	1.9×10^{-9}
	a^2D – a^2H	5/2	9/2		20517	20806	–		1.055×10^{-13}	
52F	a^2D – b^2F	5/2	5/2	8851.153	20517	31812	9.482×10^{-3}	6.5×10^{-3}	3.122×10^{-3}	1.7×10^{-3}
		3/2	5/2	9517.769	21308	31812	2.904×10^{-3}	2.0×10^{-3}	6.436×10^{-3}	3.1×10^{-3}

Transition	J	J'	λ	E	E'				
a²D – b²F	5/2	7/2	8706.826	20517	31999	4.329×10^{-3}	2.8×10^{-3}	1.437×10^{-2}	3.7×10^{-3}
	3/2	7/2	9351.089	21308	31999	–		2.872×10^{-3}	7.6×10^{-4}
a²D – d²D	5/2	3/2	3681.141	20517	47675	1.502×10^{-1}		8.139×10^{-3}	
	3/2	3/2	3791.589	21308	47675	6.946×10^{-7}		1.474×10^{-1}	
	5/2	5/2	3632.404	20517	48039	8.255×10^{-4}		3.884×10^{-2}	
	3/2	5/2	3739.905	21308	48039	7.988×10^{-2}		4.294×10^{-2}	
b²F – b²F	5/2	7/2		31812	31999	7.573×10^{-5}		6.450×10^{-13}	
b²F – d²D	5/2	3/2	6302.275	31812	47675	7.433×10^{-2}		1.334×10^{-1}	
	7/2	3/2	6377.548	31999	47675	–		1.797×10^{-2}	
	5/2	5/2	6160.762	31812	48039	1.387×10^{-1}		2.216×10^{-2}	
	7/2	5/2	6232.673	31999	48039	7.435×10^{-2}		1.412×10^{-1}	
d²D – d²D	3/2	5/2		47675	48039	5.211×10^{-4}		1.513×10^{-10}	

ACCIDENTAL DEGENERACY BETWEEN DOUBLY EXCITED STATES IN FE II

T. Brage, A. E. Nilsson and S. Johansson
Department of Physics
University of Lund
LUND, Sweden

B. Baschek and J. Adams
Institute of Theoretical Astrophysics
University of Heidelberg
HEIDELBERG, West-Germany

The structure of singly ionized iron, Fe II, is quite well described by an independent particle model, with LS-coupled basis functions. Deviations from this approach appear in a theoretical calculation as mixings between different levels with the same parity and J-value. These interacting levels share each others properties, e.g. allowed decay channels.

In the following we will distinguish between two different types of mixings:

1) Long-range effects. States with similar symmetry, e.g belonging to the same type of LS-term, often interact even if they are well separated in energy. An interaction of this type is always present in a many-electron system, since it represents correlation effects.

2) Short-range effects. States with quite different properties might interact if they, by accident, are close in energy. As an example, spin-orbit interaction might induce mixing between states belonging to overlapping terms of different types, even though relativistic effects are relatively weak in Fe II. This type of interactions are hard to predict in a theoretical calculation, since they are unexpected and very sensitive to the accuracy of the calculated relative energies of the two levels involved.

In this report we give some examples of this latter type of effect. We start by discussing one specific case, a $^6D-^8P$ mixing, in more detail and by presenting some results from a theoretical investigation of this system (Brage et al 1987).

Two of the three fine-structure levels of the $3d^5(^6S)4p^2$ ^8P-term were known from earlier analysis (Johansson 1978). The fine-structure splitting within this term is expected to be the same as for $3d^54s4p$ ^8P, and the interval between the two known levels, $^8P_{7/2}$ and $^8P_{9/2}$, agrees

63

R. Viotti et al. (eds.), Physics of Formation of FeII Lines Outside LTE, 63–65.

very well with the corresponding interval in the 4s4p octet. However, while searching for the missing J=5/2 level of ^{8}P, five new levels, which seemed to form a ^{6}D-term with irregular fine-structure, were found. The missing ^{8}P-level was also found, but some 150 cm$^-$1 from the expected position.

A closer investigation of this system gave us examples of both types of interactions mentioned above and also showed how delicate the balance could be between the two effects.

Firstly, there are long-range interactions between the three ^{6}D:s of the 3d^5(^{6}S)(4s4d + 4p^2)-complex. Even though these terms are separated by about 10000 cm$^-$1, they are almost completely mixed, with a weight of the second component in the eigenvector expansion of about 30-45%.

Secondly, the overlap between the lowest of these ^{6}D:s and the ^{8}P induces a short-range mixing, caused by spin-orbit interaction.

A limited theoretical model, only including those terms of the 3d^5(4s4d + 4p^2)-complex belonging to the lowest grand-parent term 3d^5 ^{6}S of Fe IV, was used to investigate this system further. By multiplying the off-diagonal energy matrix elements of the ^{6}D:s with a scaling factor α, we were able to study the dynamics of the ^{6}D-^{8}P interaction. It was found that a change of only 5-10% in α, and thereby in the strength of the interaction between the ^{6}D:s, alters the picture drastically. For a scaling factor around 0.97, where 1.0 corresponds to the "physical" situation, the ^{8}P and ^{6}D are completely mixed, while for value of α around 1.05 the two term do not coincide and the mixing has almost disappeared.

This type of mixing has a large effect on the fine-structure of the two terms (Brage et al 1987), but it also gives rise to a number of unexpected lines. In figure 1 we show the calculated spectrum for α=1.0 and 1.05. The number of lines and their relative intensities are seen to be highly affected by the interaction, which is not present in the second case (lower picture).

Some other examples of short-range effects in Fe II have earlier been reported, e. g.:

1. The two terms 3d^{5}4s(^{5}S)4p ^{6}P and 3d^6(^{1}D)4p ^{2}P, designated x^{6}P and w^{8}P in AEL, coincide at an excitation energy of about 79000 cm^{-1}, causing strong interaction between the two J=3/2 states (Johansson 1978). Since the two important multiplets UV 191 (a^{6}S - x^{6}P) and UV 110 (a^{2}P - w^{2}P) originate from these terms, their mixing will give rise to extra components (a^{6}S - w^{2}P$_{3/2}$ and a^{2}P - x^{6}P$_{3/2}$). These spin-forbidden lines are not predicted by large-scale calculations (Kurucz 1981) but they have almost the same observed intensities as the allowed components. The reason for this is that the ^{2}P-^{6}P mixing is very sensitive to the accuracy in their calculated relative position.

2. Another interesting type of unexpected interaction is the mixing between different members of the same Rydberg series. This becomes possible in Fe II, since the total energy span of the different parent terms of 3d^6 in Fe III is of the same order of magnitude as the 4p-5p splitting. Therefore some high 3d^{6}4p terms overlap with low 3d^{5}5p. The first reported example is the 3d^6(^{5}D)5p ^{4}P - 3d^6(b^{3}P)4p ^{4}P interaction (Johansson et al 1980).

In this report we have given a number of examples of important short-range effects, that could be very hard to represent in a large-scale calculation. To get a good understanding of these effects it is therefore essential to make a careful study of the mixed states and their decay channels.

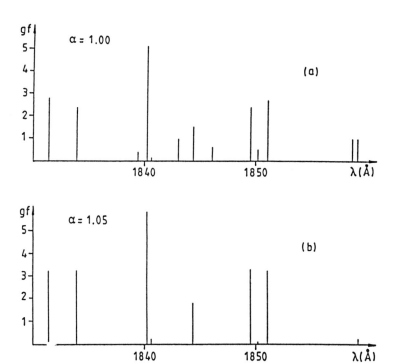

Figure 1. Synthetic spectra for two different α:s, showing transitions from $3d^5(^6S)4p^2$ ^8P and ^6D to $3d^5(^6S)4s4p$ ^8P. For α=1.00 the mixing between the ^8P and the ^6D is strong, and a number of "forbidden" lines appear. For α=1.05 the two terms are well separated and only the $^8P-^8P$ transitions are present.

REFERENCES

Brage, T., Nilsson, A.E., Johansson, S., Baschek, B., and Adam, J. 1987, *J. Phys. B.*, in press.
Johansson, S. 1978, *Physica Scripta* 18, 217.
Johansson S., Litzén U., Sinzelle J. and Wyart J.-F. 1980, *Physica Scripta* 21, 324.
Kurucz R. L. 1981, *Smithsonian Astrophys. Obs. Special Report* 390

EXPERIMENTAL DETERMINATION OF FeII gf-VALUES :
REVIEW OF AVAILABLE DATA AND WORKS IN PROGRESS

J. MOITY
Observatoire de Paris
Laboratoire de Spectroscopie du DEPEG
92195 Meudon, CEDEX, France

ABSTRACT. A review of the experimentaly determined FeII gf-values is given. After recalling the fundamentals of the methods used to measure these data, a critical discussion of the available measurements is done. Finally, the work in progress at Meudon is presented.

1. INTRODUCTION : BIBLIOGRAPHY OF AVAILABLE EXPERIMENTAL FeII gf-VALUES

In 1981, a critical compilation of atomic transition probabilities for allowed lines of iron, cobalt and nickel, in all stages of ionization, was published by the N.B.S. /1/.

In the chapter devoted to FeII, only 127 gf-values were given, derived from seven papers among which five are experimental ones /2-6/ and the two others /7-8/ are solar data analyses.

When compared with the amount and the quality of the data published for FeI (about 1700 lines, a number of them with a precision of about 5%), the lack of precise and reliable experimental FeII gf-values at that time is obvious (the mean precision lay between 25-50%). Even nowadays this problem persists (especially in quantity), not only for FeII but more generally for all once ionized 3d-elements. As Richter /9/ pointed out in 1984, the reason for this problem is simple : "the excitation of ions requires very hot light-sources which should be, if possible, thermal sources. These conditions can be fulfilled by electrical arcs, especially cascade arcs. But it is difficult to prove the existence of LTE for the ionic species, and it is sometimes very hard to introduce the desired element with a sufficient concentration into the arc".

The above mentioned papers were not the only to deal with experimental determination of FeII gf-values before 1981. One should mention the pioneer work of Roder in 1962 /10/ so as two other spectroscopic works /11-12/. Always before 1981, a number of authors had measured some FeII level lifetimes,using collisional excitation technics : either electron-beam excitation /13-14/ or beam-foil spectroscopy

67

/5,15-16/. The results from /13-14/ agreed for the low-lying resonance
levels and could be used to calibrate the relative f-values measurements
/4/, so that the absolute scale was fairly well established at that time but for the strongest resonance lines in the near-UV (2500 A -
3000 A).

In 1983 we published transition probabilities for about 500 FeII
lines in the wavelength range 2550 A - 5300 A /17/. The emission spectroscopy and the photographic technic used limited the precision to
about 25%, but for the first time all the above mentioned works could
be tied together.

Since that time, only four experimental papers appeared, typical
of the recent technical developments in that field. The lifetimes of
the low-lying resonance levels have been remeasured by laser fluorescence method /18-19/, and the new results confirm the previous ones /13-
14/. Relative f-values of lines originating from these levels have been
measured (or remeasured) either by the branching-ratio technic /20/ or
by a combination of independent emission and absorption measurements
/21/. The use of the FTS spectrometer at Kitt Peak /22/ for emission
measurements /20-21/ has led for the first time to some FeII f-values
with a precision of about 5%.

2. CLASSICAL METHODS FOR MEASURING TRANSITION PROBABILITIES

The classical methods of emission and absorption spectroscopy have
yielded the great majority of measured gf-values for neutral and once
ionized species. Because it is a well-known matter, only the fundamentals and the main requirements of each methods are recalled in this
chapter. More details can be found in the general review by Huber and
Sandeman /23/.

It must be emphasized that, generally, the spectroscopic methods
are used to determine a set of relative f-values. Then, the absolute
calibration of this set must be done in an independent way, as will be
seen afterwards.

Rather than the usual distinction between emission and absorption
spectroscopy, it seems more convenient nowadays to distinguish between
methods needing a thermal light-source and those needing no knowledge
of the thermal state of the plasma source.

2.1. Methods needing a thermal plasma

Each time the classical spectroscopic methods are used in their original form to measure f-values of lines originating from various levels,
a Boltzmann distribution of the excited levels has to be fulfilled
(and checked !) and the excitation temperature of the species must be
determined.

2.1.1. Emission measurements. In emission spectroscopy, one has to
determine the total radiance in a spectral line. For a transition from
an upper level u to a lower ℓ and in the case of an homogeneous optically thin plasma, this total radiance is related to the transition probability $A_{u\ell}$ via the well-known relation :

$$\int_{line} I_\lambda \, d\lambda = I = N_u \; A_{u\ell} \; h\nu \, L \tag{1}$$

in which N_u is the upper level population, $h\nu$ is the photon energy
for the given transition and L is the length of the emitting layer.
      Therefore, in order to directly derive the A-value, one has to
perform an absolute radiance measurement, to determine the upper level
population by an independent method and to measure the length of the
emission zone. It is obvious that this method, with such a number of
requirements, can hardly lead to precise and reliable results.
      That is why relative measurements are generally prefered, accor-
ding to the following relation between the intensities of two transitions u → ℓ and m → n :

$$\frac{I_{u\ell}}{I_{mn}} = \frac{A_{u\ell}}{A_{mn}} \; \frac{g_u}{g_m} \; \frac{\lambda_{mn}}{\lambda_{u\ell}} \; exp\left(\frac{E_m - E_u}{k \, T}\right) \tag{2}$$

      The requirements for using relation (2) are less stringent than
previously : only a relative spectral sensitivity calibration of the
experimental device is required for comparing the intensities at various wavelengths. Furthermore, particle densities have not to be known.
Only the existence of a partial LTE for the studied species must be
checked and its excitation temperature has to be determined to convert
relative intensities into relative A-values. Of course, as all along
this chapter, we suppose that the plasma source is homogeneous and
optically thin.

2.1.2. Absorption measurements. Two main methods are commonly used to
derive f-values from absorption spectra : the classical spectroscopy
and the interferometric measurement of the anomalous dispersion in the
vicinity of spectral lines.
      - Classical spectroscopy. In absorption spectroscopy, one measures
the well-known equivalent width of spectral lines. Always in the case
of an homogeneous optically thin source, this quantity is tied to the
oscillator strength $f_{\ell u}$ of the transition ℓ → u, at wavelength λ, by
the relation :

$$\int_{line} \frac{I_\lambda^c - I_\lambda}{I_\lambda^c} \, d\lambda = W_\lambda = \frac{\pi e^2}{m c^2} \; \lambda^2 \, N_\ell \, f_{\ell u} \, L \tag{3}$$

I_λ^c is the intensity of the irradiating continuum, N_ℓ is the lower level
population and L is the length of the absorbing layer. The physical
constants e and m are respectively the electron charge and mass, and c
is the light velocity. Just like in emission method, relative measurements are prefered since they prevent from determining particle densities and plasma length. Then, the requirements to obtain a set of relative f-values are the same as is emission except that no spectral sen-

sitivity calibration is needed. This is due to the fact that an equivalent width is obtained by comparing the light weakening in a line profile with the local intensity of the irradiating continuum.
  - Interferometric measurements (hook-method). The refractive index
n of a plasma, at a wavelength λ close to an absorption line λ_o, is related to the oscillator strength $f_{\ell u}$ of the line by the formula :

$$n-1 = \frac{e^2}{4\pi m c^2}\ N_\ell\ f_{\ell u}\ \frac{\lambda_o^3}{\lambda - \lambda_o}\ \left(1 - \frac{N_u}{N_\ell}\frac{g_\ell}{g_u}\right) \qquad (4)$$

Symbols and indexes have the same meaning as in previous formulae.
  Anomalous dispersion measurements are usually done by the hook-
method : details and references about this method can be found in /23-
26/. Briefly, a two-beam interferometer is illuminated by a continous
background light-source and is combined with a spectrograph so that a
fringe pattern is observed in its focal plane. When a cell of an absorbing vapour (or plasma) is put in the test arm of the interferometer,
the fringe pattern maps the refractivity n-l of this sample. In particular, the fringes exhibit hooks on each side of the absorption lines,
due to the large refractivity gradient at wavelengths close to the
spectral lines. An example of a hook-spectrum of TiI is shown in Fig.1:
the measured quantity is the distance Δ (in wavelength units) between
hooks on both sides of a line. It is tied to the oscillator strength
by the relation :

$$\Delta^2 = \frac{r_o}{\pi}\ N_\ell\ f_{\ell u}\ L\ \frac{\lambda_o^3}{K} \qquad (5)$$

in which r_o is the classical electron radius, K is the "hook- constant"
(see e.g. /23/) and the other symbols have the same meaning as in equation (4)

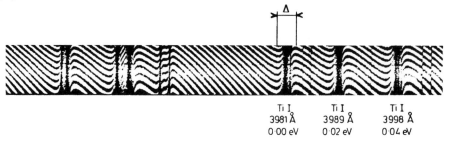

	Ti I	Ti I	Ti I
	3981 Å	3989 Å	3998 Å
	0·00 eV	0·02 eV	0·04 eV

Figure 1. Hook-spectrum of TiI from /26/.

  With this method, no photometry is needed since the measured parameter is a distance. Furthermore, the hook-method is insensitive to saturation and line broadening phenomena, so that a dynamic range of
about 10^5 in intensity can be reached between weak and strong lines
simultaneously measurable (to be compared with the dynamic range of
about 50 of classical spectroscopic methods in which particle densities

must be changed to fulfill the requirement of optically thinness in the measured lines). Of course, the constraints about partial LTE in the studied medium are the same as in the other methods in so far as lines from various levels are investigated.

2.2. Methods which do not need a thermal plasma

It is evident that methods which do not rely on the knowledge of the thermal state of the plasma are, in principle, more reliable.

2.2.1. Branching ratio measurements. Equation (2) of emission measurements shows that relative A-values determination for lines originating from a common upper level do not need any excitation temperature determination. Moreover, when the intensities are expressed in terms of photon fluxes :

$$\Phi_{u\ell} = N_u A_{u\ell} \qquad \text{(in photon.s}^{-1}) \qquad (6)$$

the ratio of a given line intensity $(u \to \ell')$ to the sum of the intensities for all possible decay channels from the common upper level (see Fig.2) can be expressed as :

$$R_{u\ell'} = \frac{\Phi_{u\ell'}}{\sum_\ell \Phi_{u\ell}} = \frac{A_{u\ell'}}{\sum_\ell A_{u\ell}} \qquad (7)$$

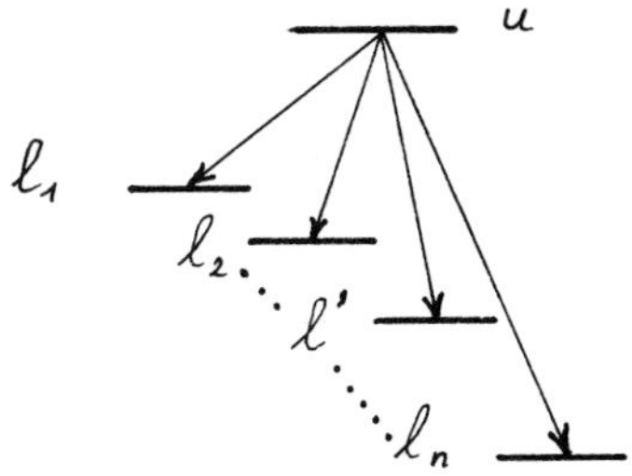

Fig. 2. Schematic diagram of the decay channels from a common upper level.

$R_{u\ell'}$ is called the branching fraction for the transition $u \to \ell'$. Strictly speaking, the term branching ratio is preferably used when the sum in the denominator do not contain all the possible lines because only some of the decays could be measured /27/.

Branching fraction determination can provide the absolute calibration of a set of relative A-values when the lifetime τ_u of the common upper level is known, since :

$$\tau_u = \frac{1}{\sum_\ell A_{u\ell}} \qquad \text{(in s)} \qquad (8)$$

Then : $A_{u\ell'} = \dfrac{R_{u\ell'}}{\tau_u}$ (9)

2.2.2. The "bowtie" method. This method was introduced by Cardon et al. /28/. In its simplest form, it refers to two upper levels (x,y) and two lower levels (a, b) (see Fig.3).

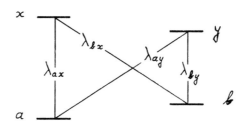

Figure 3. Elementary bowtie with the four interconnecting transitions.

Each of the four transitions indicated on Fig.3 must be measured in emission and in absorption. In both cases, the derived ratio of f-values : $(f_{ax}f_{by})/(f_{bx}f_{ay})$ is independent of level populations. The value of this ratio obtained from the emission measurements should be equal to the value derived from the absorption measurements within li-mits set by the experimental uncertainties : this is a check of the photometry (or hook measurements).
        Furthermore, the eigth measurements provide an overdetermined set of equations from which optimized relative f-values can be derived.

2.2.3. The "leap frogging" method. Refering to Fig.4, when absorption measurements of lines from a common lower level are combined with emis-sion measurements of lines from a common upper level, and if they have one line in common, both sets may be linked together and give a single set of relative f-values without any assumption about the plasma state.

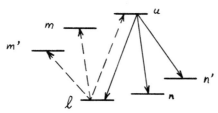

Figure 4. Schematic representation of the "leap-frogging" method in an elementary case.

        Such an idea was pointed out by Ladenburg /29/ and extended to excited levels for TiII measurements in /30/; and for FeII measurements in /21/.

3. THE MEASUREMENTS OF FeII LEVEL LIFETIMES

3.1 Determining the absolute scale.

In the previous chapter were presented various methods used to measure relative transition probabilities. Some independent methods allow to convert these quantities into absolute values.

Vapor pressure data, or partial pressure measurements (/3/,/12/), when they are reliable, can be used to determine particle densities in equations (1),(3) or (5) when the thermal state of the plasma is known.

In the case of FeII, as for other ionic species, two methods have been mainly used. The first one consists in relating the relative ionic scale to the known neutral absolute scale, by use of the Saha equation in the case of a thermal plasma (/2/,/6/,/10/,/17/). It is possible since both spectra appear simultaneously. The second method was explained in paragraph 2.2.1 and consists in combining branching ratio and lifetime measurements. This method is more and more commonly used (/4-5/, /20-21/) and has given the much precise FeII gf-values /20-21/. This is due to the development of lifetime measurements after selective excitation of the levels by tunable lasers, so as to the use of a Fourier Transform Spectrometer for branching-ratio measurements. Furthermore, this method do not require a thermal plasma.

3.2. Presentation of all measured FeII lifetimes

Table I : Comparison of all measured FeII lifetimes : results are in ns and the uncertainty on the last digit is given in parenthesis.

LIFETIMES IN NS

TERM	MEAN ENERGY (kK)	LASER		ELECTRON BEAM		BEAM-FOIL			
		HANN. 1983	SALIH 1983	ASS. 1972	BRZOZ. 1976	SMITH 1973	JOHAN. 1981	ANDER. 1970	DOLBY 1979
z ^{6}D°	38.8	4.0(2)		3.9(4)	4.0(4)				
z ^{6}F°	42.2			3.2(3)	3.5(4)				
z ^{6}P°	43.1			3.8(4)	4.0(4)				
z ^{4}F°	44.7	4.0(3)	3.9(2)						
z ^{4}D°	44.8	3.5(4)							
z ^{4}P°	47.2	3.8(4)							
z ^{4}I°	61.4							4.2(3)	
z ^{2}G°	62.2						4.8(3)	4.8(4)	
z ^{2}I°	62.4					4.7(7)	4.6(1)	4.8(3)	
y ^{2}G°	65.0					4.0*(2)	3.8(1)	4.3(3)	
z ^{2}H°	65.4					3.6(1)	3.6(2)	3.8(6)	
y ^{2}H°	67.7						4.4(1)		
z ^{2}K°	71.2					4.6(3)	4.3*(2)	4.5(3)	
w ^{2}G°	73.1								1.4(3)
x ^{6}P°	79.3								0.88(4)
y ^{6}F°	87.6								1.9(2)

All measured FeII lifetimes are presented in Table I. Generally the values for the individual J-levels are very close together (except those marked by an *) and we have reported the mean values. The references in the Table are as follows : Hann. 1983 /18/; Salih 1983 /19/; Ass. 1972 /13/; Brzoz. 1976 /14/; Smith 1973 /5/; Johan. 1981 /16/; Dolby 1979 /15/; and Ander. 1970 is a private communication reported in/5/.

One should note the good agreement for the low-lying levels between the old measurements by collisional excitation and the recent ones by selective excitation (laser). As for high-lying levels, till now we only dispose of beam-foil measurements which are in good agreement together. However this technic is very sensitive to cascading and blending problems which may lengthen the observed lifetimes.

4. DISCUSSION ABOUT THE AVAILABLE EXPERIMENTAL FeII gf-VALUES

A complete critical discussion about all available experimental FeII gf-values prior to 1983 was presented in our paper /17/, so that a summary of this discussion is just presented in Table II. In this Table are given the mean differences between our log gf in /17/ and those from previous works, so as the mutual scatters defined as the rms deviations of these previous data around their mean deviation from our results.

Table II. Comparison of our FeII log gf (1983) /17/ with previous experimental data.

SOURCE	REFERENCES	Δ LOG GF = LOG GF(OTHER) - LOG GF(MOITY)		
		MEAN DIFFERENCE Δ LOG GF	SCATTER (DEX)	
ARC	/10/ RODER (1962) .... :	+0.12	±0.07	(19%)
SHOCK-TUBE (Abs)	/12/ GRASDALEN ET AL. (1969) .... :	-0.04	±0.17	(47%)
ARC	/2/ BASCHEK ET AL. (1970) ...... :	-0.09	±0.05	(12%)
SHOCK-TUBE (Em.)	/3/ WOLNIK ET AL. (1971) .... :	+0.12	±0.06	(14%)
ARC	/4/ BRIDGES (1973) .... :	0.00	±0.04	(11%)
SHOCK-TUBE ("HOOKS")	/6/ HUBER (1974) .... :	+0.19	±0.05	(11%)
BEAM-FOIL	/5/ SMITH ET AL. (1973) .... :	-0.28	±0.07	(17%)
SUN	/8/ BLACKWELL ET AL. (1980) .... :	-0.06	±0.05	(13%)
QUANTUM CALCULATIONS	/31/ NUSSBAUMER ET AL. (1981) ... :			
	MULT. UV-1	+0.05	±0.04	
	" 42	-0.08	±0.04	
	" UV-190	-1.16	±0.28	

One should note in this Table that the scatters are generally rather low and that the absolute scale is defined to within ± 0.1 dex in the mean. The disagreement with the results by Smith et al. (1973) /5/ still remains unexplained : several causes may add their effects. First,

possible contributors to this discrepancy may be either cascading problems in the beam-foil measurements /5/ which would lengthen the observed lifetimes, and then reduce the gf-values, or a deviation from LTE
in our arc source (or a wrong temperature determination) which would
result in an overestimate of our gf-values for lines from high-lying
levels. Second, a probable cause may be a wrong photographic calibration of low intensities in our work /17/ which trend to an overestimate
of the faint line intensities. This is suggested by a detailed analysis
of the various results for multiplet UV-1, so as by Figure 5 in which
our results /17/ are graphically compared with the recent results by
Whaling (1984) /20/ : a trend appears which shows that our results for
small gf-values are probably overestimated.

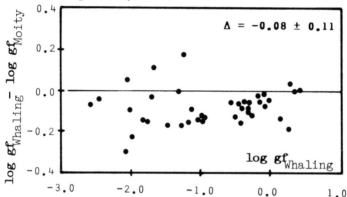

Figure 5. Plot of (log gf /20/ - log gf /17/) versus log gf /20/.

What is troublesome is that an opposite trend appears in Figure 6
in which our results are compared with the recent ones by Kroll et al.
(1986) /21/. A similar slight trend exists on Figure 7 in which the
two recent works /20-21/ are compared. These problems must remind us
that great cares have to be taken to ensure a linear response of the
detectors, a reliable spectral sensitivity calibration and the optical
thinness of the studied plasmas.

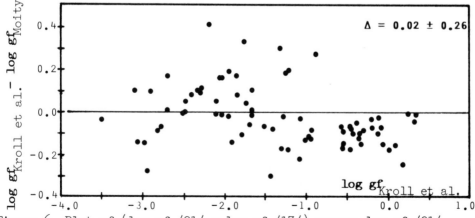

Figure 6. Plot of (log gf /21/ - log gf /17/) versus log gf /21/.

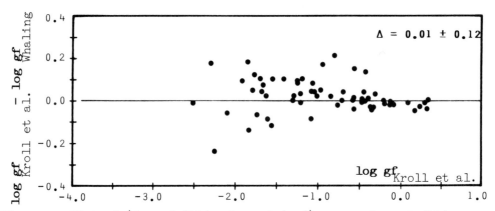

Figure 7. Plot of (log gf /21/ - log gf /20/) versus log gf /21/.

    In Table II is also given a comparison between our results /17/
and quantum calculations /31/ which shows rather good agreement for
low-lying multiplets. Comparison with the semi-empirical calculations
by Kurucz (1981) /32/ also shows a fairly good agreement for numerous
L-S coupling permitted lines from low levels (mean deviation : - 0.08
dex; scatter : ± 0.23 dex, see /17/).

5. NEW MEASUREMENTS IN PROGRESS AT MEUDON

The aim of our new experiments, now in progress, is to measure FeII
gf-values of visible lines in the wavelength range 5000 A - 8000 A.
Such lines are very promising for the study of solar photospheric abundance of iron since most of them appear in the solar spectrum as faint
isolated lines, generally lying on the linear part of the curve of
growth, and standing out against a well-defined continuous background.
It is well-known that more than 90% of the iron in the solar atmosphere exists as FeII, so that an abundance determination based on this
species should be quite independent of the model atmosphere used. Furthermore, comparison with abundances derived from FeI would be a test
for the various models so as for possible departures from LTE in the
photosphere.
    gf-values for 42 FeII lines in the wavelength range 4124 A -
7712 A had been derived by Blackwell et al. (1980) /8/ from a solar
spectrum analysis. But, till now, there has never been any publication
of laboratory FeII gf-values above the line at 5316.61 A.
    A work similar to ours is near completion at ETH-Zürich, but only
for three very faint lines in the wavelength range 7200 A - 7500 A
(see the contribution of U. Pauls in this volume).
    Our measurements are done by analyzing the light emitted by a
stable cascade arc in the center part of which iron is introduced as
the volatil compound $Fe(CO)_5$. Diaphragms limit the observed aperture
to a very low value so that only regions near the axis of the arc are
analyzed, for homogeneity and LTE to be ensured. The spectrometer has
a typical inverse dispersion of 1.5 A/mm and its photomultiplier is
cooled at - 30° C for a low-noise detection : with a wide opened exit-

slit, the intensities of the sharp FeII lines are directly integrated. The spectrometer is under the control of a micro-computer which receives and stores the measurements and also controls the setting of the grating on the line positions that have been previously stored in its memory. In that way, the experiments are fully automatized and digitalized under the control of the computer program. Furthermore, all along the measurements, an additional monochromator measures a given FeII line intensity in order to control the stability of the light-source. Its measurements are also stored by the computer and taken into account when treating the data.

The possibility of programming the experimental sequences allows us to easily treat the lines in pairs, so that a great number of independently measured relative intensities of line pairs are obtained. This overdetermined system is then optimized by a least square fitting method, so that optimized relative transition probabilities can be derived. This work is now in progress and, at the present time, preliminary relative intensities of the lines in multiplets 27 and 38 have been measured with a precision of 4%.

6. CONCLUSION

The obtainment in a near future of precise FeII gf-values in the red part of the spectrum, based on lifetimes and reliable relative measurements, should provide a new tool for solar and stellar abundance studies. In conjunction with the precise FeI gf-values available, this should allow the study of fine phenomena like departures from LTE in stellar atmospheres.

The discrepancy between /5/ and /17/ about the lines from high-lying levels will be solved only when new measurements of these level lifetimes, obtained by laser excitation, are available. Such measurements are urgently needed.

The extension of the FeII gf-value measurements towards vacuum-UV is also urgently needed : this region is very important for opacity calculations and we only have theoretical values at our disposal. The confrontation with experiments is fundamental for the development of new theoretical calculations which can only provide the enormous number of data required. Moderatly precise works ($\sim$ 15%) would be sufficient, and such experiments would also help to solve a number of identification problems in this region.

REFERENCES

/1/ Fuhr, J.R., Martin, G.A., Wiese, W.L., Younger, S.M. : 1981, J. Phys. Chem. Ref. Data **10**, 305.

/2/ Baschek, B., Garz, T., Holweger, H., Richter, J. : 1970, Astron. Astrophys. **4**, 229.

/3/ Wolnik, S.J., Berthel, R.O., Wares, G.W. : 1971, Astrophys. J. **166**, L31.

/4/ Bridges, J.M. : 1973, Proceedings 11th Int. Conf. Phen. Ion. Gases, Ed. Stoll, I., Czech. Acad. Sci., Inst. Phys., Prague, p. 418.

/5/ Smith, P.L., Whaling, W. : 1973, Astrophys. J. **183**, 313.

/6/   Huber, M.C.E. : 1974, Astrophys. J. **190**, 237.
/7/   Phillips, M.M. : 1979, Astrophys. J. Suppl. Ser. **39**, 377.
/8/   Blackwell, D.E., Shallis, M.J., Simmons, G.J. : 1980, Astron.
      Astrophys. **81**, 340.
/9/   Richter, J. : 1984, Phys. Scr. **T8**, 70.
/10/ Roder, O. : 1962, Z. Astrophys. **55**, 38.
/11/ Warner, B. : 1967, Mem. R. Astron. Soc. **70**, 165.
/12/ Grasdalen, G.L., Huber, M.C.E., Parkinson, W.H. : 1969, Astrophys.
      J. **156**, 1153.
/13/ Assousa, G.E., Smith, W.H. : 1972, Astrophys. J. **176**, 259.
/14/ Brzozowski, J., Erman, P., Lyyra, M.,Smith, W.H. : 1976, Phys.
      Scr. **14**, 48.
/15/ Dolby, J.S., Mc Whirter, R.W.P., Sofield, C.J. : 1979, J. Phys.
      B : At. Molec. Phys. **12**, 187.
/16/ Johansson, S., Litzén, U., Lundin, L., Mannervik, S. : 1981,
      Phys. Scr. **24**, 30.
/17/ Moity, J. : 1983, Astron. Astrophys. Suppl. Ser. **52**, 37.
/18/ Hannaford, P.,Lowe, R.M. : 1983, J. Phys. B : At. Mol. Phys.
      16, L43.
/19/ Salih, S., Lawler, J.E. : 1983, Phys. Rev. **A28**, 3653.
/20/ Whaling, W. : 1984, Technical Report 84A, California Institute
      of Technology, Pasadena.
/21/ Kroll, S., Kock, M. : 1986, Submitted to Astron. Astrophys. Suppl.
      Ser.
/22/ Brault, J.W. : 1976, J. Opt. Soc. Amer. **66**, 1081.
/23/ Huber, M.C.E., Sandeman, R.J. : 1986, Reports in Progress
      Physics, in press.
/24/ Marlow, W.C. : 1967, Appl. Opt. **6**, 1715.
/25/ Huber, M.C.E. : 1971, Modern Optical Methods in Gas Dynamic
      Research, ed. Dosanjh, D.S., New York, Plenum, 85.
/26/ Kock, M.,Kühne, M. : 1977, J. Phys. B : At. Molec. Phys. **10**,
      3421.
/27/ Tozzi, G.P., Brunner, A.J., Huber, M.C.E. : 1985, Mon. Not. R.
      Astr. Soc. **217**, 423.
/28/ Cardon, B.L., Smith, P.L., Whaling, W. : 1979, Phys. Rev. **A20**,
      2411.
/29/ Ladenburg, R. : 1933, Rev. Mod. Phys. **5**, 243.
/30/ Danzmann,K., Kock, M. : 1980, J. Phys. B : At. Molec. Phys. **13**,
      2051.
/31/ Nussbaumer, H., Pettini, M., Storey, P.J. : 1981, Astron.
      Astrophys. **102**, 351.
/32/ Kurucz, R.L. : 1981, Special Report 390, Smithsonian Astrophysi-
      cal Observatory, Cambridge, Mass.
note: the ref.: Ander. 1970, in Table I, is a private communication by
      Andersen T., 1970; reported in /5/.

A PHOTOSPHERIC SOLAR IRON ABUNDANCE FROM WEAK FE II LINES

U. Pauls[*1], N. Grevesse[◦] and M. C. E. Huber[*1]
* Institut für Astronomie
Eidgenössische Technische Hochschule
ETH–Zentrum
CH – 8092 Zürich, Switzerland
◦ Institut d'Astrophysique
Université de Liège
B – 4200 Ougrée–Liège, Belgium

ABSTRACT. The main difficulties encountered in determining a photospheric solar iron abundance are considered and a way is shown to circumvent most of them. Branching fractions of weak Fe II lines are measured from a hollow–cathode discharge to derive accurate transition probabilities from a previously determined lifetime. These transition probabilities — together with accurate equivalent widths from the solar spectrum — are subsequently used to calculate a photospheric iron abundance. The choice of suitable lines results in an abundance value nearly independent of any assumption concerning the temperature model and line–broadening parameters.

1. INTRODUCTION

The solar iron abundance has repeatedly been determined over several decades. Still, uncertainties and discrepancies remain and we do not consider the problem to be solved.

The history of the solar iron abundance has been described by numerous authors (see e.g. Garz et al. 1969; Withbroe 1971; Blackwell 1974; Huber 1977; Grevesse 1984a,b). The topic has also been covered by Baschek (1987) in this conference. Thus we will only summarize the lessons we have learned from history.

An enumeration of the main problems of abundance determinations must emphasize three areas. First, the line–formation theory: the derivation of a curve of growth needs input parameters like damping constants and microturbulence; the former parameters will vary from line to line (cf. Foy 1972; O'Neill & Smith 1980) and the latter must still be considered as being of uncertain physical origin (cf. Gray 1978). The next point to consider are the deviations from LTE: ionization as well as excitation cannot necessarily be represented by the Saha and Boltzmann–equations, respectively. And last but not least there is the problem of the transition probabilities. Their accuracy has improved considerably in the last years (cf. Huber and Sandeman 1986), but they are frequently not available for the best atom species nor for the lines that are most suitable for an abundance determination.

---

[1] U.P. and M.C.E.H. were guest observers at the National Solar Observatory, Kitt Peak, National Optical Astronomy Observatories, operated by the Association of Universities for Research in Astronomy, Inc., under contract with the U.S. National Science Foundation.

R. Viotti et al. (eds.), Physics of Formation of FeII Lines Outside LTE, 79–82.

2. ABUNDANCE DETERMINATION

2.1. General Procedure

In the following we describe how most of the problems just mentioned can be avoided.

The most promising way leads over weak Fe II lines and an accurate determination of transition probabilities in an emission measurement.

We have decided to use the Fe II instead of the Fe I spectrum because singly–ionized iron is 20 to 25 times more abundant in the solar photosphere than neutral iron. The use of Fe II lines thus has the advantage to render our abundance result insensitive to the particular atmospheric model adopted as well as to deviations from the ionization equilibrium (to overionization, for example). We also note that Blackwell et al. (1984) reported an abundance anomaly for a certain group of Fe I lines and that Steenbock (1985) showed that non–LTE conditions slightly affect Fe I but not Fe II lines. We note further that the iron abundance derived by Blackwell et al. (1984) with the aid of accurate ($\pm 1\%$) Fe I transition probabilities is about 40 % higher than the meteoritic value (cf. Grevesse 1984a).

It is important to restrict the choice of solar Fe II lines to well–measurable but weak ones. And with weak we mean that their equivalent widths lie on the linear part of the curve of growth. The relation between equivalent width and abundance therefore remains independent of line–broadening parameters like microturbulence and damping constants.

Two of the three major sources of error in abundance determination — namely line–formation theory and model atmosphere — can thus be circumvented and we now must concentrate on the accuracy of the transition probabilities.

The most direct method to determine transition probabilities is to measure branching fractions (BF) and to convert them by means of the lifetime of the upper level (τ_i) into absolute transition probabilities, i.e. Einstein A–values. The branching fraction of a transition between the upper level i and the lower level k is defined as

$$BF_{ik} = A_{ik} / \sum_m A_{im} \; , \quad \text{where} \quad \sum_m A_{im} = 1/\tau_i \; ,$$

if the sum is taken over all lower levels m that can be reached by radiative transitions from level i. One thus obtains

$$A_{ik} = BF_{ik}/\tau_i \; .$$

The branching fractions can be measured from any emission light–source, as long as the transitions of interest are optically thin: in that case the branching fraction $BF_{ik} = \Phi_{ik}/\sum_m \Phi_{im}$ is obtained by measuring the relative photon fluxes of the relevant transitions.

Fe II lines suitable for our abundance determination had to fulfil three criteria: they had to be unblended in the solar spectrum, had to lie in a region with a well–defined continuum and be weak enough to belong to the linear part of the solar curve of growth. For a valid laboratory measurement two further requirements had to be added: the upper level had to have a known lifetime and all important decay channels (including the weak line in question) had to be unblended in the spectrum of the emission light–source employed.

Out of a dozen suitable Fe II lines we selected three for this initial investigation; their equivalent widths are 1.5 and 2 pm [as measured from the Jungfraujoch–atlas (Delbouille et al. 1973)] and they lie near 720 and 750 nm, i.e. in a region of the solar spectrum that has a well–defined continuum and is not crowded. The lifetime of the upper level has been determined by Hannaford and Lowe (1983); they measured the radiative decay after selective laser excitation.

2.2. Laboratory Measurement

A solar spectral line of the abundant species Fe^+ that has an equivalent width of only 1.5 or 2 pm does imply a very small transition probability. Such lines are extremely difficult to detect and measure in the laboratory. This demands a suitable source, a good resolution and, particularly, a recording with a high signal–to–noise ratio.

The light source used was a hollow–cathode discharge with an iron electrode. The running conditions were optimized for a large ratio of ionized over neutral iron and especially for the population of the upper levels in question.

The spectrometer ideally suited for our purposes was the solar Fourier–transform spectrometer (FTS) on Kitt Peak (cf. Brault 1985). A FTS does not scan the single spectral elements in sequence but observes the entire spectrum at once; in our case this means that variations in the population of the upper levels and hence in the intensity of the emission lines do not affect our measurements of branching fractions . A second point in favour of the FTS is its resolution; by use of path differences ranging from one to several decimeters we did always resolve the line profiles and consequently had to deal with inherent line blends only.

To increase the signal–to–noise ratio we used narrowband filters centered on the weak Fe II lines; this is helpful because the noise of an FTS is proportional to the square root of the total number of incoming photons. The effect of such a filter is illustrated in figure 1.

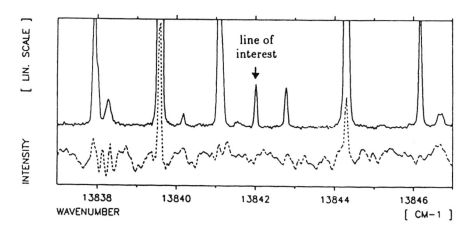

Fig. 1: Comparison of two recordings of the same spectrum by a Fourier–transform spectrometer (FTS). In one case (lower trace) the spectrum was recorded over a wide wavelength range (ca 200 nm to 1 μm), while in the other case (upper trace) the bandwidth of the light admitted to the FTS was restricted by a narrowband filter (full width at half maximum: $FWHM = 1.5$ nm). The scale of the upper trace is expanded by a factor of 10; accordingly, the actual increase in the signal–to–noise ratio is of the order of 100.

To calibrate the spectral response function of the FTS we used known branching ratios of transitions that have common upper levels and belonged to the spectra of the carrier gases, in particular Ar and Kr. Spectra with argon were calibrated with lines published by Adams and Whaling (1981); suitable krypton ratios have been measured specifically for this purpose on a 3–m grating spectrometer with the aid of an argon mini–arc (Bridges and Ott 1977) as radiometric standard. The filter functions of the narrowband spectra were calibrated with a tungsten ribbon lamp. Fe I lines present on broad as well as on narrow–

band scans (and taken with the same source conditions) served as intensity links between them.

3. CONCLUSIONS

By judicious choice of weak Fe II lines it is possible to obtain a solar photospheric iron abundance whose accuracy is essentially dependent on the transition probabilities only. These, however, are very small and correspondingly difficult to measure. A Fourier–transform spectrometer and narrowband filters are essential to reach the required signal–to–noise ratio to detect and measure such lines in the laboratory.

A final value for the solar iron abundance will be released after control measurements have confirmed our preliminary results.

ACKNOWLEDGEMENT.
We thank J.W.Brault, R.Hubbard and J.Wagner for their kind and expert support during the FTS runs. Part of this research was supported by the Schweizerischer Nationalfonds.

REFERENCES.
Adams DL & Whaling W 1981 *J.Opt.Soc.Am.* **71** 1036-1038
Baschek B 1987 in this volume
Blackwell DE 1974 *Q.Jl R.astr.Soc.* **15** 224-245
Blackwell DE, Booth AJ & Petford AD 1984 *Astron.Astrophys.* **132** 236-239
Brault JW 1985 in *High Resolution in Astronomy* 1-61 15th Advanced Course of the Swiss
    Society of Astronomy and Astrophysics, Saas–Fee (Sauverny: Obs. de Genève)
Bridges JM & Ott WR 1977 *Appl.Opt.* **16** 367-376
Delbouille L, Neven L & Roland G 1973 *Photometric Atlas of the Solar Spectrum from* λ
    3000 to 10000 Å (Liège: Institut d'Astrophysique de l'Université)
Foy R 1972 *Astron.Astrophys.* **18** 26-38
Garz T, Holweger H, Kock M & Richter J 1969 *Astron.Astrophys.* **2** 446-450
Gray DF 1978 *Solar Phys.* **59** 193-236
Grevesse N 1984a *Phys. Scripta* **T8** 49-58
Grevesse N 1984b in *Frontiers of Astronomy and Astrophysics* 72-81 proceedings of the 7th
    European Regional Astronomy Meeting (Firenze: Italian Astron.Soc.)
Hannaford P & Lowe R 1983 *J.Phys.B: At.Mol.Phys.* **16** L43-L46
Huber MCE 1977 *Phys.Scripta* **16** 16-30
Huber MCE & Sandeman RJ 1986 *Rep.Prog.Phys.* **49** 397-490
O'Neill JA & Smith G 1980 *Astron.Astrophys.* **81** 108-112
Steenbock W 1985 in *Cool Stars with Excesses of Heavy Elements* 231-235 Jaschek M &
    Keenan PC (eds) (Dordrecht: Reidel)
Withbroe GL 1971 *Nat.Bur.Stand.(U.S.)Spec.Publ.* **353** 127-148

EMPIRICAL GF-DETERMINATION FROM THE SOLAR SPECTRUM

Robert J. Rutten
Sterrewacht "Sonnenborgh", Utrecht, The Netherlands

Roman I. Kostik
Main Astronomical Observatory, Kiev, USSR

ABSTRACT. We test the reliability of Fe I and Fe II oscillator strengths determined empirically from optical solar lines by comparing fits of the observed line widths and line depths for various formation parameters.

1. INTRODUCTION

Ideally one uses precise oscillator strengths in any analysis requiring them. In practice, this is usually not possible. Fig. 1 illustrates this for iron abundance determination. The upper term diagram consists of Fe I lines for which precise oscillator strengths have been measured at Oxford (Blackwell et al. 1982 and references therein). The lower term diagram consists of actual solar Fe I lines suitable for abundance determination. The Oxford lines are mostly strong ultraviolet lines while in abundance determination one selects weak optical lines; the latter tend to lie higher in the term diagram. It would be better to use Fe II lines wherever Fe II is the dominant ionization stage, but Fe II lines have not been measured at Oxford.

Therefore a familiar recipe has been to derive empirical oscillator strengths form the solar spectrum. The classic example is Holweger's (1967) fitting of the equivalent widths of the stronger Fe I lines in the blue by adjusting their oscillator strengths, in order to derive his well-known LTE model photosphere from their observed depths. More recently, Gurtovenko and Kostik (1981a) used the updated version of this model (Holweger and Müller 1974, HOLMUL) to find the gf-values of the 865 solar Fe I lines shown in Fig. 1 by fitting their observed depths. For a smaller subset, they also fitted the equivalent widths (Gurtovenko and Kostik 1981b). The differences between the two fits were analysed by Rutten and Kostik (1982), and led to a NLTE formulation of the Fe I curve of growth by Rutten and Zwaan (1983). The latter was used by Rutten and Van der Zalm (1984) to compile 256 Fe I oscillator strengths from equivalent widths in the Jungfraujoch Atlas (Delbouille et al. 1973).

In this paper, we use line widths and line depths taken from

R. Viotti et al. (eds.), Physics of Formation of FeII Lines Outside LTE, 83–92.
© *1988 by D. Reidel Publishing Company.*

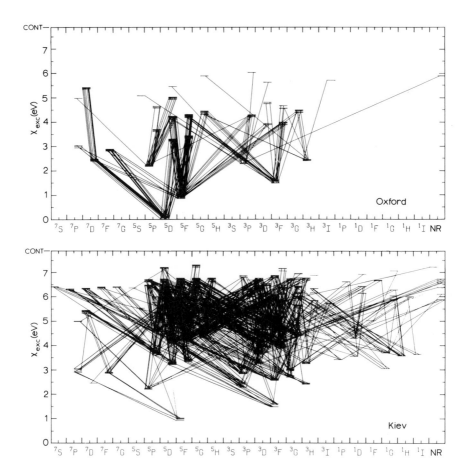

Fig. 1.   Term diagrams for Fe I lines, respectively measured at Oxford
(top) and observable in the optical solar spectrum (bottom). Each term
label designates both the even-parity term (lefthand interval) and the
odd-parity term (righthand interval). Level designations are from the
RMT (Moore 1959); NR and NR° at right combine all number designations
in the RMT.

Rutten and Van der Zalm (1984) for 354 Fe I lines and 22 Fe II lines, which are all the clean iron lines present in the solar spectrum between 4000 and 8000 Å. Since all parameters and assumptions of solar line formation enter into the reliability of solar-fitted oscillator strengths, we compare "gf_W" fits to the equivalent widths with "gf_D" fits to the depths of these lines for various combinations of input parameters. The resulting scatter diagrams provide a measure of the precision obtainable.

2. ATMOSPHERIC MODEL

Fig. 2 compares gf_D and gf_W LTE fits for two model atmospheres. The HOLMUL model (top) is the favourite choice of LTE abundance deter- miners because the spread it produces is smaller than when the well- known HSRA model (Gingerich et al. 1971, bottom) is used. This is demonstrated by the difference between the top and bottom panels. Fits with the HSRA assuming LTE show increasing discrepancies between gf_D and gf_W for larger height of line formation, implying that the HSRA- computed lines are too deep for their observed areas. This depth excess is due to the NLTE departures that go together with cool models as the HSRA (see Rutten, elsewhere in these proceedings). For Fe I, assuming LTE results in too large computed line opacities, thus in line-core formation that is located too high, thus in too low line- core brightness temperatures. For Fe II, assuming LTE produces too low line source functions in the upper photosphere, thus too low line-core brightness temperatures also.

The top panel of Fig. 3 shows that the discrepancies largely vanish when NLTE departures computed from the HSRA model are taken into account. The bottom panel is for the recent MACKKL model (Maltby et al. 1986), which is close to the HOLMUL model in the upper photo- sphere. Its concomitant NLTE departures are small; when taken into account (thanks to specification in a private communication by E. H. Avrett), a scatter diagram results that is similar to the top panel. This similarity demonstrates that the choice between either a hot model with small NLTE effects or a cool model with large NLTE departures cannot be settled from the observed Fe I and Fe II lines.

3. CONTINUUM OPACITY

Figs. 4 and 5 display typical errors due to erroneous continuum intensities. They show results for an increase by 10% in the continu- ous opacity, respectively for the line widths and for the line depths. Experience learns that the continua produced by different spectrum synthesis codes can differ by as much as 10%, even in the optical where H$^-$ dominates. Also, most lines do not reach the "true" continuum but only a "local" continuum affected by unresolved weak lines and far wings of nearby strong lines.

For Fe I lines, the differences in the fitted gf_W values are <u>smaller</u> than the opacity increment (10% = 0.04 dex) for weak lines on

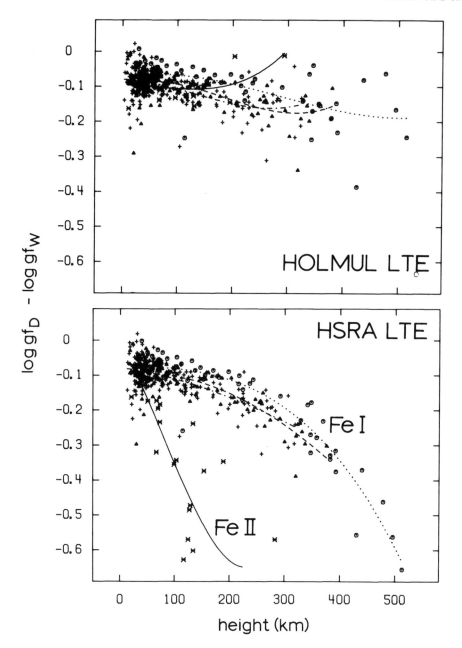

Fig. 2. Logarithmic differences between gf_W and gf_D for LTE modeling against the line-center height of formation for the gf_W fit.
Fe II lines: asterisks, fitted with the solid curve.
Fe I lines: split into low excitation lines (circles, dotted fit), middle excitation lines (triangles, dashed fit) and high excitation lines (plusses, dot-dashed fit) according to Rutten and Kostik (1982). Microturbulence: MACKKL model. Macroturbulence: 1.3 km/s. Damping: radiative broadening and Van der Waals broadening computed from Warner's (1969) square radii and increased by 30%.

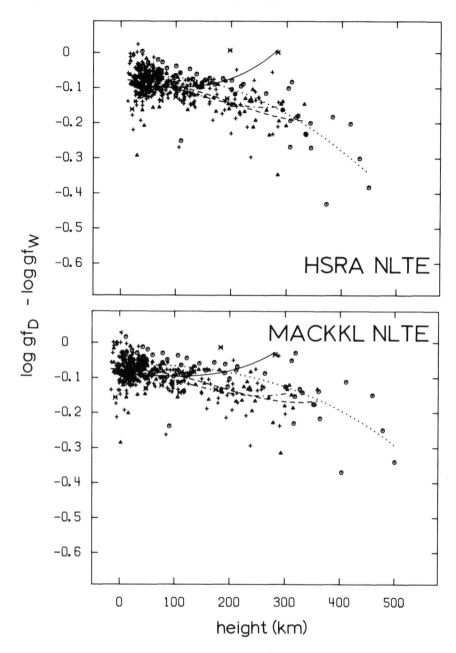

Fig. 3.   Logarithmic differences between gf_D and gf_W for NLTE
modeling. Top: HSRA model, Fe I and Fe II NLTE departures as in Rutten
and Kostik (1982), taken from Lites and White (1973) and Cram et al.
(1980), respectively.
Bottom: MACKKL model; Fe I departures from Avrett, Fe II in LTE.
Symbol coding and other parameters as in Fig. 2

the linear part of the curve of growth ($\log W/\lambda < -5.5$, bottom panel
Fig. 5). This is due to the rapid increase of the Fe I line extinction
relative to the continuous extinction with height (Rutten and Van der
Zalm 1984 Fig. 3). The Fe I increments computed here in LTE for the
HOLMUL model are slightly larger than the values tabulated by Rutten
and Van der Zalm for NLTE HSRA modeling[1].

The differences increase with excitation energy at given
formation height because for higher-excitation lines the outward
increase in line-to-continuum extinction ratio is reduced by their
larger sensitivity to the decreasing temperature. A reversed split is
seen in the bottom panel of Fig. 4 against oscillator strength; it
simply follows the Boltzmann population factor. This figure illustrates important selection effects. At high excitation, most lines
selected for clean observability are already on the flat part of the
curve of growth (upward curves here) and are therefore very sensitive
to the turbulence parameters. The strongest high-excitation lines are
the only optical Fe I lines with large transition probabilities. Their
photon losses result in upper-level under- excitation and have led to
overestimation of their gf-values by Gurtovenko and Kostik (1981a).

Fig. 5 shows the effects of increased continuous opacity on the
line depths. The gf_D increments for weak lines are about the same as
for the fits of the widths, but they are twice as large for lines from
the shoulder of the curve of growth. They would be very large for
saturated lines ($\log W/\lambda > -4$), but such lines are not at all present
when one selects clean solar lines.

4. TURBULENCE

Finally we show in Fig. 6 two examples in which the assumed turbulence
differs. In the top panel we compare gf_D fits for the HOLMUL model and
LTE, respectively with constant turbulence and with the height-
dependent model specified in the MACKKL model (identical to the microturbulence in Vernazza, Avrett and Loeser 1976, 1981). In the bottom
panel we compare gf_D fits using the height-dependent vertical microturbulence specified by Lites (1973) with fits using the MACKKL microturbulence, for the HSRA and NLTE. The differences in the top panel
are large for the stronger lines which feel the outward decrease of
the MACKKL microturbulence. The bottom panel shows large differences
already for weak lines because Lites' microturbulence is quite large
in the deep photosphere. Thus, the turbulence parameters, often used
as ad-hoc fudge parameters to enforce fits to observations, provide
large uncertainty.

---

1) The specification given in Rutten and Van der Zalm (1984) contains
printing errors. The formula in their Table II should read:
$$x_1 = c_7 - 0.5 \, c_5 \, [\tanh(c_1(x_1 - c_2)) + 1] - c_6 \, \exp[-((x_1 - c_3)/c_4)^2].$$

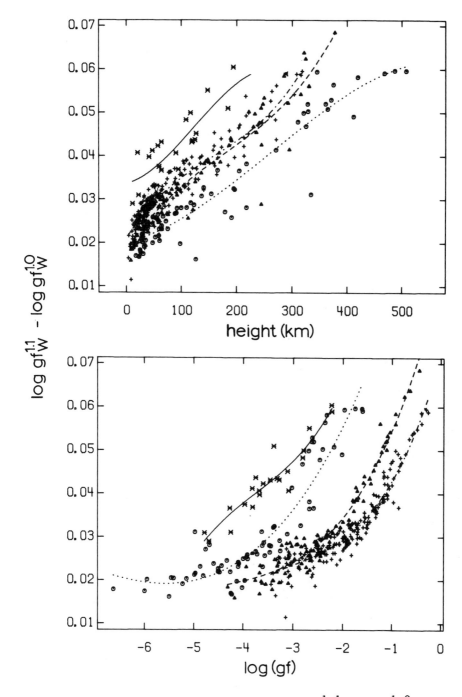

Fig. 4. Logarithmic differences between $gf_W^{1.1}$ and $gf_W^{1.0}$, i.e. the effect on the equivalent widths of a 10% (0.04 dex) increase on the computed continuous opacity, for the HOLMUL model and LTE, against height (top) and log(gf) (bottom). Other parameters and symbol coding as in Fig. 2.

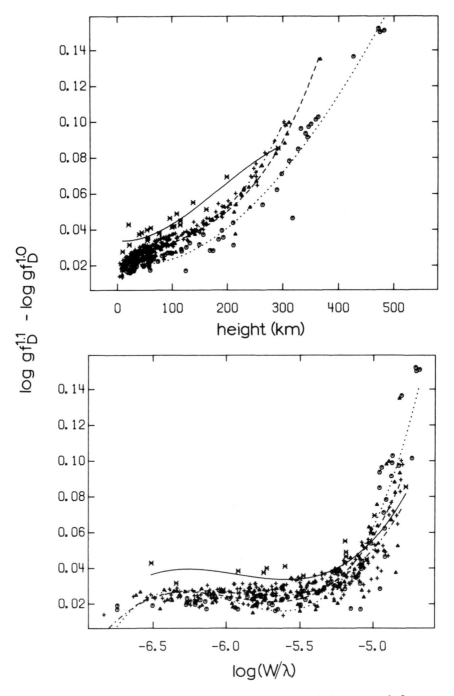

Fig. 5. Logarithmic differences between $gf_D^{1.1}$ and $gf_D^{1.0}$, i.e. the effect on the line depths of a 10% (0.04 dex) increase in the computed continuous opacity, for the HOLMUL model and LTE, against height (top) and the logarithm of the observed line strength (reduced equivalent width, bottom). Other parameters and symbol coding as in Fig. 2.

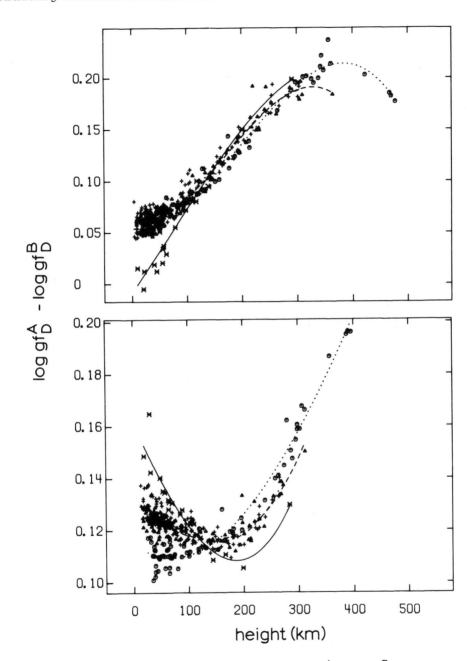

Fig. 6.   Logarithmic   differences   between gf_D^A and gf_D^B for   different
microturbulence.
Top: HOLMUL model and LTE, gf_D^A for constant microturbulence (0.9 km/s)
and macroturbulence = 1.75 km/s, gf_D^B for the height-dependent MACKKL
microturbulence and macroturbulence = 1.3 km/s.
Bottom: HSRA model and NLTE, gf_D^A for Lites' (1973) vertical microturbulence   and   macroturbulence   =   1.75   km/s, gf_D^B for   the   MACKKL
microturbulence and macroturbulence = 1.3 km/s. Symbol coding as in
Fig. 2

5. CONCLUSION

The various scatter diagrams show that empirical gf determination from the solar spectrum works – to some extent, and only for truly weak lines. The precision obtainable is about 0.1 dex or 25% at best, and one has to be careful about selection effects. The results depend sensitively on the ad-hoc choice of the turbulence parameters. Obviously, the advantage of solar gf-determination is that one obtains oscillator strengths for the lines needed. In addition, if other stars are like the Sun and the same computer program and assumptions are used, there may be even some fortuitous cancellation of errors in using solar values for stellar applications.

REFERENCES

Blackwell, D.E., Petford, A.D., Shallis, M.J., Simmons, G.J.: 1982, Mon. Not. R. Astr. Soc. 199, 43
Cram, L.E., Rutten, R.J., Lites, B.W.: 1980, Astrophys. J. 241, 374
Delbouille, L., Roland, G., Neven, L.: 1973, "Photometric Atlas of the Solar Spectrum from λ3000 to λ10000", Institut d'Astrophysique, Liège
Gingerich, O., Noyes, R.W., Kalkofen, W., Cuny, Y.: 1971, Solar Phys. 18, 347
Gurtovenko, E.A., Kostik, R.I.: 1981a, Astron. Astrophys. Suppl. 46, 239
Gurtovenko, E.A., Kostik, R.I.: 1981b, Astron. Astrophys. Suppl. 47, 193
Holweger, H.: 1967, Zeitschr. f. Astrophysik 65, 365
Holweger, H., Müller, E.A.: 1974, Solar Phys. 39, 19
Lites, B.W.: 1973, Solar Phys. 32, 283
Lites, B.W., White, O.R.: 1973, High Altitude Observatory Research Memorandum 185, Boulder
Maltby, P., Avrett, E.H., Carlsson, M., Kjeldseth-Moe, O., Kurucz, R.L., Loeser, R.: 1986, Astrophys. J. 306, 284
Moore, C.E.: 1959, Nat. Bur. Stand. US Techn. Note 36, "A Multiplet Table of Astrophysical Interest", US Dept. of Commerce, Washington, Revised Edition
Rutten, R.J., Kostik, R.I.: 1982, Astron. Astrophys. 115, 104
Rutten, R.J., Van der Zalm, E.B.J.: 1984, Astron. Astrophys. Suppl. 55, 143
Rutten, R.J., Zwaan, C.: 1983, Astron. Astrophys. 117, 21
Vernazza, J.E., Avrett, E.H., Loeser, R.: 1976, Astrophys. J. Suppl. 30, 1
Vernazza, J.E., Avrett, E.H., Loeser, R.: 1981, Astrophys. J. Suppl. 45, 350
Warner, B.: 1969, Observatory 89, 11

SESSION 2

OBSERVATION OF FeII LINES
IN DIFFERENT ASTROPHYSICAL OBJECTS

FE II EMISSION LINE PROFILES IN THE ULTRAVIOLET SPECTRUM OF COOL, LUMINOUS STARS

Kenneth G. Carpenter
Center for Astrophysics and Space Astronomy
University of Colorado
Boulder, CO   80309-0391   USA

ABSTRACT. The differences between the outer atmospheres of the coronal and non-coronal cool stars and the importance of massive stellar winds to the latter group are summarized. The utility of Fe II as a probe of such winds is indicated and a brief review of previous observations of Fe II in these stars is given. The early results of a current IUE program to study the chromospheres and winds of the non-coronal, late-type stars are presented, including evidence of a strong dependence of the mid-UV Fe II profiles on stellar luminosity. In addition, the dependence of the Fe II profiles on intrinsic line strength in spectra of Alpha Ori and the discovery of a variation of the Fe II profiles with time in spectra of Gamma Cru are discussed.

1.   INTRODUCTION

The chromospheres of late-type, high-luminosity stars are best studied through ultraviolet spectroscopy. The approximately 10000K plasma found in these outer atmospheric layers produces numerous strong emission lines throughout the 1100 – 3200Å spectral region, where the competing photospheric continuum happens to be relatively weak and the emission can be easily seen. Various IUE studies (see e.g., Linsky, 1980) have shown that the outer atmospheres of cool (i.e., G0 and later) stars tend to fall into two classes. The coronal stars, which include all cool main-sequence stars, giants earlier than about K2, and supergiants earlier than about G8, show evidence in the UV of plasma at temperatures on the order of 100,000 degrees (in the UV) and of several million degrees (in the X-ray region), indicating the presence of transition regions and coronae in these stars. Stars later and more luminous than these limits show no evidence for any plasma hotter than about 20,000K, indicating the absence of transition regions and coronae in these stars. These non-coronal stars have much stronger winds, which presumably act as an important cooling agent and are certainly important to the long-term evolution of these stars.
   One of the potentially most important probes of these late-type winds are the numerous emission lines of Fe II which appear throughout

95

R. Viotti et al. (eds.), Physics of Formation of FeII Lines Outside LTE, 95–105.
© *1988 by D. Reidel Publishing Company.*

the UV and in particular in the 2200 - 3200Å spectral region. The
winds of these stars become stronger and more interesting as one goes
to later spectral types and higher luminosities. Unfortunately, these
high-luminosity late-K and M stars are among the astronomical targets
least well-studied with IUE, due to the paucity of targets that are
bright in the ultraviolet. I have, therefore, recently begun a study
of these stars with IUE using high resolution spectra of the 2200 -
3200Å region. The primary goal of this program is to extract
information on the stellar wind from a study of the observed
wavelengths, profiles, and intensities of the mid-UV Fe II emission
lines. The observed wavelengths and profiles can provide information
on the velocity structure of these winds. The relative intensities of
the Fe II lines can be used to estimate column densities and opacities
in the wind. These results, in combination with the observed
dependence of the profiles on luminosity, intrinsic line strength, and
time reported in this paper, may provide clues regarding the nature of
mass loss in this portion of the H-R diagram. In this paper, I
present and discuss some early results from this program. The reader
is referred to C. Jordan's paper later in this volume for a more
general discussion of cool star chromospheres and the modeling thereof.

2. FE II IN LATE-TYPE STARS

Although the vast majority of the Fe II chromospheric emission features
in the spectra of cool stars lie in the IUE spectral region, the first
observations of Fe II lines from these objects were obtained from the
ground. Herzberg (1948), Bidelman and Piper (1963), Boesgaard and
Boesgaard (1976), and others have observed in a number of M giants and
supergiants some of the approximately 17 lines of visual multiplets 1,
6, and 7 near 3200Å that are detectable from the ground. More detailed
observations and analysis of these lines in the spectrum of Alpha
Orionis are contained in Weymann (1962), Boesgaard and Magnan (1975),
and Boesgaard (1979).
      Van der Hucht, et al. (1979) obtained the earliest observations of
the mid-UV Fe II lines, using the Balloon-Borne Ultraviolet
Spectrograph (BUSS) in the 2750 - 3165Å region to detect two lines in
Alpha Boo (K2 III), one line in Alpha Tau (K5 III), and numerous lines
in Alpha Ori (M2 Iab).
      IUE has greatly expanded the range over which the UV Fe II lines
can be detected. Wing (1978) and Carpenter (1984a) have discussed IUE
observations of dozens of Fe II lines over the 2200 - 3230Å spectral
region in Gamma Cru (M3 III). A detailed study of the Fe II mid-UV
spectrum of Alpha Ori has been given by Carpenter (1984b), while the
entire 1150 - 2930Å spectrum (absorption and emission) of Alpha Boo
has been described in detail by Carpenter, Wing, and Stencel (1985). A
similar detailed summary of the full Gamma Cru spectrum is now in
preparation (Carpenter, Brown, Wing, and Stencel, 1986). An atlas of
UV spectra of late-type stars by Wing, Carpenter, and Wahlgren (1983)
presents a sampling of IUE high resolution spectra for the 2500 - 3200Å

region, which contains many of the important mid-UV Fe II lines, for a set of 13 cool stars.

Except for a few scattered lines of Mg II, C II, Si II, Al II, and Mg I, virtually all of the chromospheric emission in the mid-UV spectrum of late-type non-coronal stars is due to Fe II. In fact, although they are individually much weaker than the Mg II and Al II lines, the total radiative losses through the Fe II lines in Alpha Ori is on the order of 2/3 of the losses through the Mg II doublet, the primary chromospheric cooling channel (Carpenter 1984b). The Fe II radiative cooling is clearly very important to the energy balance of these chromospheres.

The major Fe II lines in the mid-UV fall into six groupings: around 2260Å (UV multiplets 4-5), 2370Å (UV 3, 35 and 36), 2450Å (UV 34, 128, 129, 209, 222, 299, 300 and 320), 2600Å (UV 1 and 64), and 2740Å (UV 32, 62, 63). There is also a pair of very strong Fe II lines near 2507Å, due to fluorescence by Lyman alpha. All but the shortest wavelength grouping can be seen in Figure 1, which shows a low resolution IUE spectrum of Rho Per (M4 II). These emission features are seen not only in normal K and M giants and supergiants, but also in spectra of carbon stars (Johnson and Luttermoser 1986) and in stars with a high dust/gas ratio in their outer layers (Stencel, Carpenter, and Hagen, 1986).

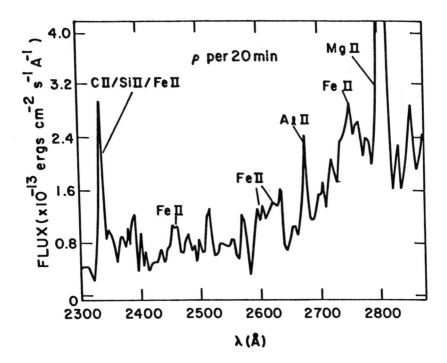

Figure 1. An IUE low-resolution spectrum of Rho Per (M4 II).

3. HIGH RESOLUTION FE II PROFILES

The high resolution IUE echelle mode allows us to obtain detailed
profile information on a large number of Fe II lines in the regime of
the long-wavelength cameras. An initial examination of the archival
and new observations which have been reduced at this time, has
revealed a dependence of the Fe II line profiles on several parameters.
The most dramatic dependence is on the surface gravity, i.e.
luminosity, of the star. However, the profiles are also seen to vary
with the intrinsic strength of the line in spectra of Alpha Ori and
spectra of Gamma Cru provide evidence for a significant change in its
Fe II profiles during the brief time that IUE has spent in orbit.

3.1. Dependence on Surface Gravity

The most important determiner of the strength and shape of the Fe II
line profiles appears to be the surface gravity of the star. Figure 2
shows the 2584 - 2602Å region of IUE spectra of four stars, all of
spectral class M2-M3, with luminosity classes ranging from a class III
giant to a class Ia-Iab supergiant. The "Fxn" notation above each
spectrum means that the fluxes for that star have been multiplied by
"n" for ease in viewing on the scale chosen for the plot. Note that
the profiles start as relatively narrow, single-peaked features in
Gamma Cru (M3 III) but grow progressively wider and develop
self-reversals with increasing luminosity. The width of the lines in
Gamma Cru is about 40 km/sec while in the supergiant Alpha Ori, the
widths approach 120 km/sec. The great breadth of these lines is
almost certainly due to opacity broadening, rather than to massive
turbulence in the chromospheres. The line centers become so opaque in
the more luminous stars, that the line photons cannot escape until
they are redistributed by scattering far out into the line wings. As
the lines begin to self-reverse in the bright giants, they show a
red-asymmetry (i.e. they have more flux to the red of the reversal
than to the blue). The lines continue to become broader and the
self-reversals deeper with increasing luminosity. In the highest
luminosity stars, the intrinsically strongest lines (e.g., at 2586Å in
Alpha Ori), turn over to a violet-asymmetry.
     The line profiles are very sensitive to small changes in the
luminosity of the star. As can be seen from Figure 2, it is very easy
to distinguish stars with Ia-Iab, Ib-IIa, II-III, and III luminosity
classes on the basis of the Fe II lines near 2600Å. These lines may
thus be a useful luminosity and gravity diagnostic for the late-type
non-coronal stars.

3.2. Dependence on Effective Temperature

In contrast to the case with surface gravity, the effective
temperature, or spectral type, of the star seems to have very little
effect on the Fe II line profiles. The contrast between the lines and
the photospheric continuum does weaken as one goes to earlier spectral
types, but the widths and general shapes of the lines appear to stay

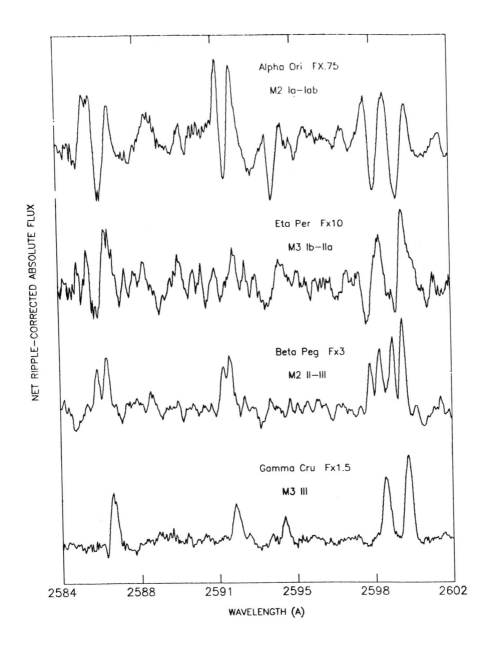

Figure 2.   Four high-resolution IUE spectra illustrating the dependence of the Fe II profiles on surface gravity.

relatively constant at a given luminosity class. Figure 3 shows the
2584 - 2602Å region of IUE spectra of four class III giants, ranging
in spectral type from K2 to M3. The lines are relatively narrow and
single-peaked in all four spectra. The apparent double-peaked feature
near 2591 in Beta And is due to a noise hit on the detector blueward
of the actual Fe II line.

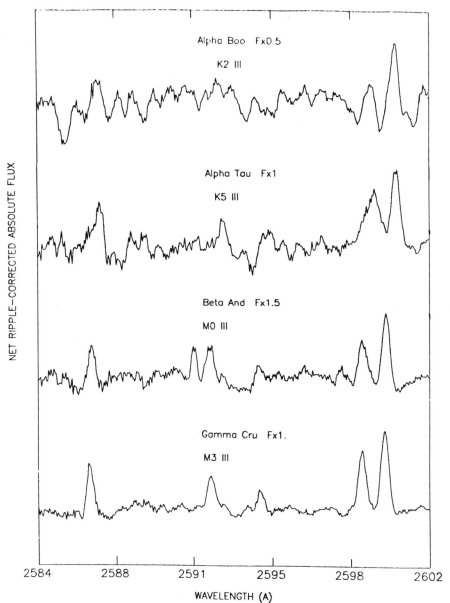

Figure 3.   IUE high resolution spectra of 4 late-type giants,
            illustrating the lack of a strong dependence of the Fe II
            profiles on effective temperature.

3.3. Dependence on Intrinsic Line Strength

Alpha Ori has the richest and strongest Fe II spectrum of any late-type star yet observed and the line profiles vary dramatically within a single spectrum of this star. The character of these profiles has been shown in Carpenter (1984b) to depend on the intrinsic strength of the transition producing the line. The intrinsically weakest lines in this spectrum are narrow, unreversed features very much like those seen in giant spectra. The intermediate strength lines are very broad, with strong self-reversals and show a red-asymmetry (i.e., more flux to the red of the reversal than to the violet), while the intrinsically strongest lines also are very broad with strong reversals, but show a violet-asymmetry. The only other star in which violet-asymmetric profiles have been seen in the mid-UV Fe II lines is in the M2 super-giant 119 Tau. In this star also, it is only the intrinsically strongest lines which are asymmetric to the violet. This correlation between profile asymmetry and intrinsic line strength is illustrated in Figure 4, which shows three lines of different intrinsic strength from UV multiplet 63 of Fe II.

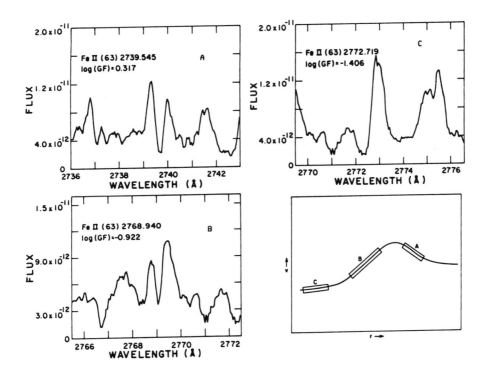

Figure 4. The dependence of Fe II profile asymmetry on intrinsic line strength in Alpha Ori and a proposed velocity curve for its outer atmosphere. Based on figures in Carpenter (1984b).

A possible explanation of this correlation is shown in the fourth panel in this figure, which shows a schematic velocity curve for the outer atmosphere of Alpha Ori. In this scenario, the intrinsically strongest lines, as in panel A, would be formed primarily in the outermost regions of the chromosphere in a region with a negative velocity gradient. The lines of moderate intrinsic strength are somewhat less optically thick and the escaping photons would originate from deeper in the chromosphere, primarily from region B in this picture, where there is a positive velocity gradient. The intrinsically weakest lines would allow photons to escape from deep within the chromosphere where there is little or no velocity gradient, as in region C.

The self-reversals and asymmetries are caused by repeated scatterings of line photons out of the optically thick line centers by atoms slightly higher in the chromosphere traveling at slightly different radial velocities. In a region with a negative velocity gradient, scattering atoms see slightly violet-shifted incoming photons and thus preferentially scatter photons out of the red side of the profile producing violet-asymmetric final profiles. In a region with a positive velocity gradient, the scattering atoms see slightly red-shifted incoming photons and thus preferentially scatter photons out of the blue side of the profile producing a red-asymmetric line.

3.4. Time Variability

Although the "normal" late K and M giants are not known for their stability, there has previously been no clear evidence for the variation of any of their Fe II line profiles with time. I am able to report here the first observational evidence of such variation in the form of IUE high resolution spectra of the M giant Gamma Crucis. Figure 5 shows the 2484 - 2602Å region of spectra taken over the period April 1978 to June 1986. The first two are taken with the LWR camera, the last two with the LWP camera. The early 1978 image has been reprocessed with current software to avoid the possibility of early processing errors. As the figure illustrates, sometime in the six-year period between April 1978 and April 1984, the character of the Fe II lines changed, with the lines altering from the single-peaked narrow profiles characteristic of late-type giants to double-peaked, significantly broader lines. The lines are still weaker than those in the M2 II-III star Beta Peg (see Figure 2), but they are significantly stronger than any of the pure class III giants shown in Figure 3. The latter spectra indicate that the change in the profiles has persisted and that the effect may be continuing to increase very slowly.

These data indicate that the opacity of the stellar wind has increased and suggest that mass has been added to this chromosphere either through a one-time ejection of mass or through a moderate increase in the star's mass loss rate. Further observations are planned to monitor this event and to help ascertain whether it is a part of a moderate, continuous increase in mass-loss rate, a one-time event, or part of a series of increases and decreases in chromospheric activity.

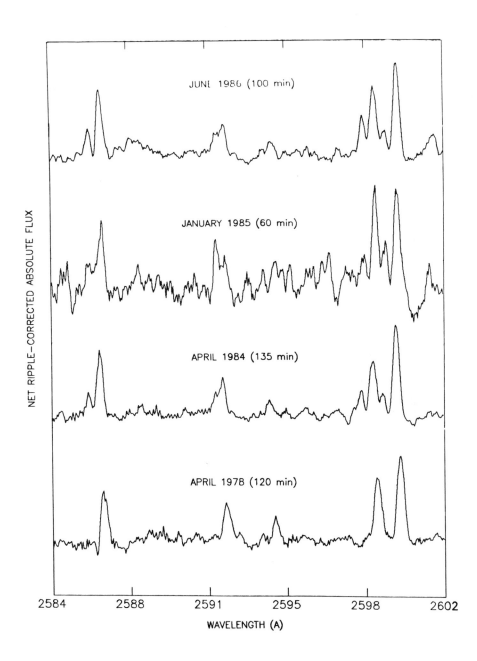

Figure 5. The variation of Fe II profiles with time in IUE spectra of Gamma Cru (M3 III).

4. SUMMARY

After a few introductory comments on the characteristics of late-type
chromospheres and a brief review of previous observations of Fe II in
late-type giant and supergiant stars, I have presented a sampling of
the data acquired with the IUE satellite during the initial stages of
a program to investigate the chromospheres and winds of late-type non-
coronal stars. The high resolution spectra obtained with IUE of the
2200 - 3230Å region contain a wealth of strong Fe II lines with a
variety of intrinsic strengths and profile characteristics. These
data show that the Fe II line profiles are very sensitive to the
luminosity (and thus surface gravity) of the star, but do not depend
strongly on effective temperature (i.e. spectral type) within the K2 -
M3 spectral range. Observations of Alpha Ori show that the Fe II
profiles in that star's spectrum are corelated with the intrinsic
strength of the line, while multiple observations of the M giant Gamma
Cru provide the first evidence of a significant variation in the Fe II
profiles in a "normal" late-type, high-luminosity star.
    Additional observations are currently being obtained and further
analysis of these data is now underway, including the determination of
the radial dependence of the wind velocity around each star and the
variation of the velocity field with spectral type and luminosity
class. Measures of relative Fe II fluxes will be used to determine
the opacity and hydrogen column density versus height in each
chromosphere.
    Finally, the profiles and strengths of the other lines (e.g., Mg
II, Al II, Mg I. and C II) will be used to further probe the
chromospheric winds of these cool, non-coronal stars.

This work is supported in part by grant NAG5-797 from the
National Aeronautics and Space Administration to the University of
Colorado.

5. REFERENCES

Bidelman, W. P. and Piper, D. M. 1963, Pub. A.S.P., 75, 389.

Boesgaard, A. M. 1979, Ap. J., 232, 485.

Boesgaard, A. M. and Boesgaard, H. 1976, Ap. J., 205, 448.

Boesgaard, A. M. and Magnan, C. 1975, Ap. J., 198, 369.

Carpenter, K. G. 1984a, in The Future of Ultraviolet Astronomy Based
    on Six Years of IUE Research, NASA Conference Publ. 2349, p. 450.

Carpenter, K. G. 1984b, Ap. J., 285, 181.

Carpenter, K. G., Brown, A. and Stencel, R. E. 1985, Ap. J., 289, 676.

Carpenter, K. G., Wing, R. F. and Stencel, R. E. 1985, Ap. J. Suppl., 57, 405.

Carpenter, K. G., Brown, A., Wing, R. F., and Stencel, R. E. 1986, in preparation.

Herzberg, G. 1948, Ap. J., 107, 94.

Johnson, H. and Luttermoser, D. G. 1986, preprint Indiana Astronomy Publication No. 86-103.

Linsky, J. L. 1980, in Annual Reviews of Astronomy and Astrophysics, vol. 18, 439.

Stencel, R. E., Carpenter, K. G., and Hagen, W. 1986, Ap. J., 308 (Sept. 1986), in press.

Van der Hucht, K. A., Stencel, R. E., Haisch, B. M., and Kondo, Y. 1979, Astr. Ap., 36, 377.

Weymann, R. 1962, Ap. J., 136, 844.

Wing, R. F. 1978, in Proceedings of the Fourth International Colloquium on Astrophysics, ed. M. Hack (Observatorio Astronomico Di Trieste), p. 683.

Wing, R. F., Carpenter, K. G., and Wahlgren, G. W., 1983, Perkins Obs. Spec. Publ. No. 1.

OBSERVATION OF FeII IN COOL VARIABLES

R. Viotti
Istituto Astrofisica Spaziale, CNR
Frascati, Italy

M. Friedjung
Institute d'Astrophysique, CNRS
Paris, France

1. INTRODUCTION

Many cool stars show different forms of variability associated with the presence of emission lines, which may
include some of FeII. There are many different categories
of cool variables with different behaviours and time
scales, and the origin of their variability - not to mention that of the formation of FeII emission lines - can be
quite different. We shall only briefly describe certain
aspects of what is a vast subject or rather a number of
vast subjects.
      The following categories of stars may be mentioned:
- cool main sequence - dMe, flare, BY Dra stars
- pre main sequence - T Tau stars, Herbig-Haro objects
- R CrB stars
- Mira variables
- binaries with a hot component - ζ Aur/VV Cep systems,
symbiotic stars, cataclysmic binaries.
      Some properties of FeII in cool stars are described in
the contributions of K. G. Carpenter and C. Jordan to this
volume , and shall not be repeated here.

2. MAIN SEQUENCE AND PRE-MAIN SEQUENCE STARS

Information about activity in cool dwarfs is reviewed in
the conference proceedings edited by Byrne and Rodono'
(1983). The presence of ultraviolet FeII emission (2585-
2631 A) is indicated by Johnson (1983) in that volume for
the star Woolley 644 AB. Baliunas et al. (1984) studied EQ
Peg B, which is a binary containing two UV Ceti type stars.
FeII emission was observed in the 2600-2650 region, and its
flux may have increased as much as 20% during a flare.

R. Viotti et al. (eds.), Physics of Formation of FeII Lines Outside LTE, 107–113.
© *1988 by D. Reidel Publishing Company.*

Groundbased observations of YZ CMi (Mochnacki and Schommer 1979) showed that FeII emission lines appeared during what seems to have been a spectroscopic flare. However, it may be noted that a study of flares on the BY Dra star AU Mic (Butler et al. 1986) showed no correlation between flaring of the ultraviolet FeII lines at 2620 and 2740 A and the optical U-band. Future study of FeII emission lines in these stars should give useful information about their chromospheres.

The pre-main sequence T Tau stars are reviewed by Bertout (1984). FeII emission lines can occur in optical and ultraviolet spectra. Forbidden lines including IFeIII are sometimes present, with radial velocities differeing from that of photospheric features. The interpretation of T Tau star observations has many uncertainties, and much work remains to be done. It may also be noted that Herbig-Haro objects, associated with pre-main sequence stars, have IFeIII emission lines. Some show a reflected T Tau type spectrum with FeII emission lines (Cohen et al. 1986).

3. R CORONAE BOREALIS STARS

These stars are not all cool; the 'prototype' R CrB has a late F spectrum. while MV Sgr has been a B spectral type. They seem to have a carbon overabundance and a hydrogen deficiency, while the most generally accepted interpretation of their variations as due to dust condensation, links these stars to cool variables. A review was given by Feast (1975).

'Chromospheric' FeII emission has been observed during fading, but seems then to disappear (Feast 1975). Alexander et al. (1972) studied the fading of RY Sgr; they attempted to physically interpret the absence of forbidden singly ionized iron lines. Payne-Gaposchkin (1963) studied a fading of R CrB, and analyzed equivalent widths of emission lines including those of FeII by a curve of growth method assuming LTE. The physical conditions of the 'chromosphere' did not appear to be very different from that of the photosphere seen at maximum. The present authors (Friedjung and Viotti 1976) studied the red and near-IR FeII emission lines of MV Sgr near maximum. The lines appeared to be formed in a region expanding with a velocity near 100 km/s, and having a surface area perpendicular to the line of sight of a disk with a radius of less than 1200 stellar radii. Finally, we can mention the work of Evans et al. (1985) on ultraviolet spectra of RY Sgr and MV Sgr using IUE. Emission lines were observed during extinction minima, but the low resolution mode of IUE used made line identification difficult. The absence or near absence of FeII claimed by the authors is doubtful and needs to be checked by spectral synthesis.

The study of FeII emission lines in spectra of stars where dust condensation may occur, should in future give important information about this process. In particular new studies using present semi-empirical methods of emission line analysis need to be carried out.

4. MIRAS

The appearence of FeII emisison lines in Mira variable stars was classically described by Merrill (1940). He mentioned the presence of bright spectral lines which appeared about halfway in the rise from minimum to maximum. FeII emission was stated as being visible near maximum, while observations of forbidden [FeII] emission lines near minimum was reported. Joy (1954) made a detailed study of Mira Ceti. He found that FeII emision lines had a surprisingly long lifetime, covering all the star's period, except for a few weeks following minimum light. He states that [FeII] lines were definitely seen before minimum. The FeII emission line radial velocity was found to decrease in a orbital cycle. The early work of Gorbatskii (1957, 1958) should also be mentioned in this connection. The theory of emission line formation was examined, and hydrogen emission after luminosity maximum considered to be due to recombination of hydrogen atoms which were previously ionized. The ionization (by Lyman continuum radiation) and excitation (by collisions with electrons and following recombination) of Fe+ were examined by Gorbatskii. He explained the disappearence of forbidden [FeII] lines near minimum by a fall in electron density following recombinations in the atmosphere.

Mira type variables are now thought to possess spherical shocks propagating outwards, associated with the production of line emission. Such a shock model is described by Fox et al. (1984, 1985). However in the latter of these papers calculation of predicted Balmer line fluxes leads to overestimates by factors of the order of 100; the authors think that the problem could be solved by assuming lower preshock densities and Lyman-alpha diffusion ahead of the shock front. Clearly one cannot yet expect a very good theory of FeII, if even that for the Balmer lines is uncertain. Semi-empirical methods of analysis of FeII may help after correction for overlying stellar atmospheric absorption.

5. BINARIES WITH A HOT COMPONENT

The cool component can be near the main sequence (cataclysmic binaries) or be a giant or supergiant (symbiotic stars, ζ Aur/VV Cep binaries), while the hot component may be a relatively normal hot star, as in the case of ζ Aur/VV Cep stars, a white dwarfs usually surrounded by an accretion disk and boundary layer (cataclysmic binaries), a white dwarf with an outer layer undergoing thermonuclear burning (perhaps some symbiotic stars), or a main sequence star surrounded by an accretion disk plus boundary layer (also for some types of symbiotic stars). The role of FeII in each case will be considered in turn.

In the case of a cataclysmic binary a star near the main sequence seems to fill its Roche Lobe and to transfer mass to the white dwarf companion, which usually appears to be surrounded by an accretion disk. Nova or dwarf nova eruptions can occur. In the review of Warner (1976) the presence of FeII emission lines is indicated in the spectra of the dwarf novae SS Aur and U Gem. Warner considered that the FeII emission lines of SS Aur were double; this suggests formation in cool, presumably outer regions, of the accretion disk.

In the case of a ζ Aur/VV Cep binary, a cool supergiant has a B type companion, the latter being thought to be immersed in the wind of the former (see Wright 1970). FeII emission was observed by Hagen et al. (1980) in the ultraviolet spectrum of VV Cep, and time variation was seen during egress from eclipse. The ultraviolet spectrum of a star of this class KQ Pup (Boss 1985) was observed by Altamore et al. (1982) who identified very many lines, most of which are of FeII, having P Cygni profiles with a velocity difference between emission and absorption components of about 40 km/s. The excitation potential of the FeII lines observed is up to 13 eV. Optical spectra of KQ Pup are described by Jaschek and Jaschek (1963); at one stage only forbidden [FeII] lines were detected. Several ionized iron emission lines were found by Swings (1969) in the near-UV. This star should prove in future to be a real mine of information about formation of FeII lines. This is also shortly illustrated in this volume by Altamore et al.

The theory of the formation of FeII lines in the wind of the cool components of ζ Aur/VV Cep binaries has been investigated in detail by the Hamburg group. Che et al. (1983) studied FeII lines of the UV multiplets 1 and 9 in the IUE spectra of ζ Aur, 32 Cyg and 31 Cyg. The former multiplet has the same upper term as UV multiplet 191, and the authors used branching ratio arguments to explain why scattered photons appeared in UV multiplet 191, so the lines of UV multiplet 4 appear only in absorption. The ionization equilibrium between Fe+ and Fe++ was estimated, and assuming a solar abundance theoretical calculations of equivalent widths and profiles of FeII UV multiplets 1 and 9 could lead to cool star mass loss rate estimates. Among

later studies of the Hamburg group, we can mention that of
Reimers and Schroeder (1983) on δ Sge. They suggest that
some broad FeII emission lines of UV multiplets 60-64 may
be formed in an accretion disk. Che-Bohnstengel (1984)
found a lower limit to the excitation temperature of 4200 K
for 32 Cyg from the relative population of the lower a4F
term of ultraviolet multiplets 43, 44, 45 (excitation potential of the order of 0.3 eV) to that of the less excited
lower 6D term of ultraviolet multiplets 1 and 8, using
theoretical line profiles and the level population theory
of Nussbaumer and Storey (1980). Schroeder (1985) considered curves of growth of what seem to be almost pure
absorption chromospheric lines of three binaries. One can
conclude that the method of the Hamburg group is fairly
rigorous, but this type of approach generally only relies
on a limited number of multiplets.

Of the binaries here considered, the symbiotic stars
are perhaps the most controversial. Many specialists considered them single stars, but the evidence for their being
generally binary is now very strong (Friedjung 1982). The
components strongly interact, but different processes dominate for different stars. S-type symbiotics have a normal
cool giant component and generally little sign of dust,
while D-type symbiotics have a Mira variable component with
dust. FeII emission lines are prominent in symbiotic spectra, but can in principle be produced in a number of different regions. For instance one can mention the D-type
symbiotic star RR Tel, whose UV spectrum rich in FeII
emission lines has been described by Penston et al.
(1983). The optical spectrum of this star was described by
Thackeray (1977); the FeII emission lines are among the
narrowest lines of its spectrum.

Among the possible regions of formation of FeII emission lines we can mention parts of the cool star's atmosphere iradiated by the hot component, as suggested by
Boyarchuck (1966) for AG Peg, or the cool component's wind,
or the outer parts of an accretion disk. For formation in
the cool star's atmosphere, one might expect optical thickness variation with phase, as the iradiated region is
viewed from different directions. Such an effect was su-
spected by one of us in an unpublished study (Friedjung
1967). Radial velocity studies can help to elucidate the
situation; Garcia (1986) pointed out that the FeII emission
lines of RS Oph had a radial velocity which tracked that of
the giant component. However, the profile is double peaked,
and Garcia suggested formation of the lines in a ring
rotating around the M star. Garcia mentioned double peaked
permitted FeII line profiles also for R Aqr, BX Mon, CH Cyg
and CL Sco; the question is whether they are formed in an
accretion disk around the hot component, a ring around the
cool giant, or perhaps a bipolar structure in the cool

star's wind.

It may be noted that Mira Ceti is not only the 'proto-type' of Mira variables, but may also be related to symbio-tic stars. It has a hot component, which may be surrounded by a disk. From a study of the high resolution ultraviolet (IUE) spectra of Mira B, Reimers and Cassatella (1985) found broad absorption lines of FeII which are broad with characteristic dish-shaped profiles. Multiplet UV 191 is present in emission. In general, it seems that emission line formation is more complex for symbiotic stars than for ζ Aur/VV Cep binaries.

6. CONCLUSIONS

Unlike in the case of hot stars, in the cool variables here discussed FeII emission lines probably arise in regions hotter than the cool star's photosphere. The observation of these lines in the optical and UV during different phases of their time variability should be of very much help to diagnose physical conditions in the outer atmospheres and circumstellar envelopes of these stars, but much work re-mains still to be done both observationally and theoreti-cally.

REFERENCES

Alexander, J.B., Andrews, P.J., Catchpole, R.M., Feast, M.W., Lloyd-Evans, T., Menzies, J.W., Wisse, P.N.J., Wisse, M.: 1972, Mon. Not. R. astr. Soc. 158, 305.

Altamore, A., Giangrande, A., Viotti, R.: 1982, Astron. Astrophys. Supplem. Ser. 49, 511.

Baliunas, S.L., Raymond, J.C.: 1984, Astrophys. J. 282, 728.

Bertout, C.: 1984, Rep. Prog. Phys. 47, 111.

Boyarchuk, A.A.: 1966, Astron. Zh. 43, 976.

Butler, C.J., Rodono', M., Linsky, J.L.: 1986, in "New Insights in Astrophysics", ESA SP-263, 229.

Byrne, P.B., Rodonò, M. (edsitors): 1983, "Activity in Red Dwarf Stars", Reidel, Dordrecht, The Netherlands.

Che, A., Hempe, K., Reimers, D.: 1983, Astron. Astrophys. 126, 225.

Che-Bohnestengel, A.: 1984, Astron. Astrophys. 138, 333.

Cohen, M., Dopita, M.A., Schwartz, R.D.: 1986, Astrophys.J. 307, L21.

Evans, A., Whittet, D.C.B., Davies, J.K., Kilkenny, D., Bode, M.F.: 1985, Mon. Not. R. astr. Soc. 217, 767.

Feast, M.W.: 1975, in "Variable Stars and Stellar Evolution", V.E. Sherwood, L. Plaut (eds.), Reidel, Dor-

drecht, Netherlands, p.129.
Fox, M.-W., Wood, P.R., Dopita, M.A.: 1984, Astrophys.J. 286, 337.
Fox, M.-W., Wood, P.R.: 1985, Astrophys. J. 297, 455.
Friedjung, M.: 1967, unpublished.
Friedjung, M.: 1982, in "The Nature of Symbiotic Stars", M. Friedjung, R. Viotti (eds.), Reidel, Dordrecht, p.253.
Friedjung, M., Viotti, R.: 1976, Astron. Astrophys. 53, 23.
Garcia, M.R.: 1986, Astron. J. 91, 1400.
Gorbatskii, V.G.: 1957, Astron. Zh. 34, 860.
Gorbatskii, V.G.: 1956, Astron. Zh. 35, 748.
Hagen, W., Black, J.H., Dupree, A.K.: 1980, Astrophys. J. 238, 203.
Jaschek, J., Jaschek, M.: 1963, Publ. astr. Soc. Pacific 75, 509.
Johnson, H.M.: 1983, in "Activity in Red Dwarf Stars", P.B. Byrne, M. Rodonó (eds.), Reidel, Dordrecht, p.109.
Joy, A.H.: 1954, Astrophys. J. Suppl. Ser. 1, 39.
Merrill, P.W.: 1940, "Spectra of Long Period Variable Stars", University of Chicago Press, Chicago, USA.
Mochnacki, S.W., Schommer, R.A.: 1979, Ap. J. 231, L77.
Nussbaumer, H, Storey, P.J.: 1980, Astr.Astrophys. 89, 308.
Payne-Gaposchkin, C.: 1963, Astrophys. J. 138, 320.
Penston, M. V., Benvenuti, P., Cassatella, A., Heck, A., Selvelli, P.L., Macchetto, F., Ponz, D., Jordan, C., Cramer, N., Rufener, F., Manfroid, J.: 1983, Mon. Not. R. astr. Soc. 202, 833.
Reimers, D., Schroeder, K. P.: 1983, Astron. Astrophys. 124, 241.
Reimers, D., Cassatella, A.: 1985, Astrophys. J. 297, 275.
Schroeder, K.P.: 1985, Astron. Astrophys. 147, 103.
Swings, J.P.: 1969, Astrophys. J. 155, 515.
Thackeray, A.D.: 1977, Mem. R. astr. Soc. 83, 1.
Warner, B.: 1976, in "Structure and Evolution of Close Binary Systems", P. Eggleton, S. Mitton, J. Whelan (eds.), Reidel, Dordrecht, Netherlands, p.85.
Wright, K.O.: 1970, in "Vistas in Astronomy", A. Beer (ed.), Pergamon Press, Oxford, England, Vol.12, p.147.

ULTRAVIOLET Fe II EMISSION LINES IN NOVAE[+]

A. Cassatella and R. Gonzalez Riestra[++]
IUE Observatory
European Space Agency
Apartado 54065, 28080 Madrid, Spain

ABSTRACT. Ultraviolet observations of classical Novae obtained with the International Ultraviolet Explorer since 1978, are used to illustrate the behaviour of the FeII lines during the decline from maximum, and to show the difficulty of line identifications.

1. INTRODUCTION

Since the International Ultraviolet Explorer was launched, nine Novae have been observed in the ultraviolet during their decline from maximum: V1668 Cyg (1978), V693 CrA (1981), V1370 Aql (1982), N Sgr 1982, N Ser 1983, N Mus 1983, PW Vul (1984a), Nova Vul 1984b, and RS Oph. Only in a few cases the objects could be followed in the decline phase until they reached the detection limit.

The analyses of these observations were generally aimed to obtain information on the chemical abundances in the ejecta (from spectra in the nebular phase), or on the dynamics and energetics of the ejection phenomenon (see Starrfield 1986, for a recent review). On the contrary, the problem of line formation, in particular of the FeII lines, is still open, also because of the difficulty of treating the line transfer problem in the expanding envelopes. Such investigations would be extremely interesting, also in view of the opportunity given by Novae to gain insight on the problem of line formation in other objects like, for example, active galactic nuclei, where analogous large mass loss phenomena occur, although on a much larger scale.

In this paper, we give some examples of the behavior of the FeII ultraviolet lines in Novae during the early post-maximum phases. In addition, we discuss some of the high resolution data available to show the difficulty of line identifications.

++Affiliated to the Astrophysics Division, Space Sciences Department.

+ Based on observations by the International Ultraviolet Explorer collected at the Villafranca Satellite Tracking Station of the European Space Agency or retrieved from the IUE Data Bank.

R. Viotti et al. (eds.), Physics of Formation of FeII Lines Outside LTE, 115–124.
© *1988 by D. Reidel Publishing Company.*

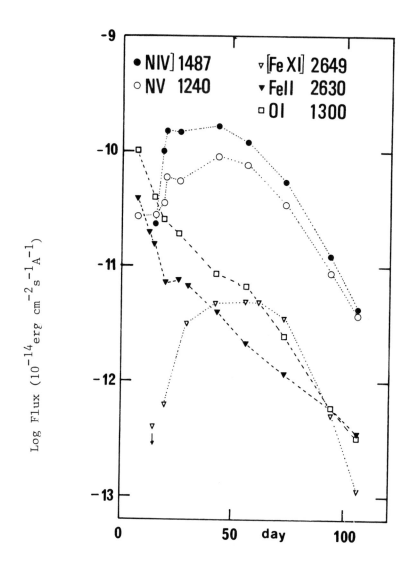

Fig. 1 Time variability of the emission lines in RS Oph during its
outburst in 1985. Abscissae represent the time after maximum, assumed
on Jan. 28, 1985. The figure shows that the lines of lower level of
ionization (OI 1300 A and FeII UV 1) peak very soon after the maximum,
while the high ionization line [FeXI] 2648.7 is maximum in a plateau
between day 42 and 62.

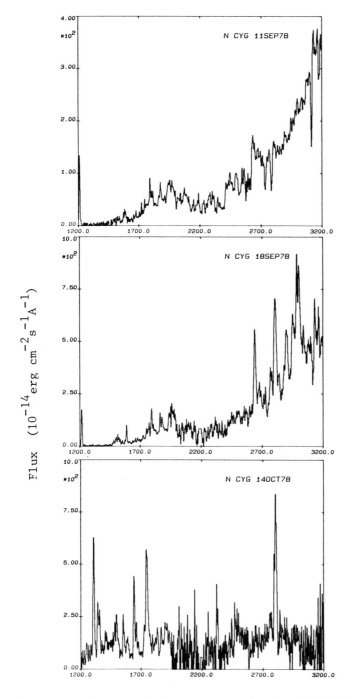

Fig. 2   Observed ultraviolet spectra of N Cyg 1978 in outburst. In the spectrum of Oct. 14, 1978 the nebular lines have become strong and a hot continuum emerges in the short wavelength region.

2. TIME VARIABILITY OF THE FeII LINES

N Cyg 1978 and RS OPH are the Novae most extensively observed by IUE, and therefore the most suitable for studying the time variability of the FeII lines.

The recurrent Nova RS Oph underwent an outburst in late January 1985 and was promptly followed in the ultraviolet with IUE (Cassatella et al. 1985; Snijders 1986). The cause for the recurrent outbursts of RS Oph (every 9 to 35 yrs since the first event was discovered in 1898) may be different from the thermonuclear runaway invoked for classical Novae (Livio, Truran and Webbink 1986). Also, it is possible that important differences exist, for example, in the total mass ejected and in the ejection velocity, compared to classical Novae. In any case, RS Oph offers an unique opportunity to study one important phenomenon: the interaction of the ejected matter with the stellar wind from the cool giant companion and with the surrounding circumsystem material, not yet dissipated at the time of the new outburst. The appearance of strong emission lines from very high ionization species observed shortly after the outburst both in the optical (Joy and Swings 1945; Rosino, Taffara and Pinto 1960) and in the ultraviolet (Cassatella et al. 1985), as well as the detection of a strong X-ray flux (Cordova et al. 1985) is a demonstration of the efficiency of such an interaction.

In Fig. 1 we show how the fluxes of the FeII emission lines from multiplet UV 1 (around 2600 A) vary with time during the 1985 outburst of RS Oph, compared with other strong emission lines of different ionization level such as OI 1300 A, N IV] 1487 A, NV 1240 A, and [FeXI] 2648.7 A. The figure shows clearly that the FeII lines peak in intensity very soon after the outburst, like the low ionization line OI 1300 A (and MgII 2800 A, not shown in the figure). Emission lines from higher ionization species, on the contrary, reach a maximum at later stages: NIV], for example, is maximum around day 35, NV around day 43, while the [FeXI] line is maximum in a plateau between day 42 and 62. The time of maximum is then correlated with the ionization potential of the line considered.

Apart from the UV 1 multiplet, other FeII lines are present, although fainter, in the post maximum spectra of RS Oph: the ones which could be identified with more confidence are those from multiplet UV 191 around 1786 A and those from UV 62 and 63.

Another suitable target for studying the time behaviour of the FeII lines is N Cyg 1978 (Cassatella et al. 1979; Stickland et al. 1979). During about one month after the maximum (near Sept. 12, 1978), the object showed strong FeII lines, especially from multiplet UV 1, which appeared with a P Cygni structure particularly prominent in the low resolution spectra of Sept. 11. In the subsequent days, the violet shifted absorption components of FeII UV 1 progressively fainted, while the emission components became stronger reaching a maximum probably near Sep. 23. Finally, in the spectra of middle October, the emission component also disappeared, and FeII UV 1 was only visible in absorption, probably produced by the zero voltage lines 2599.4 and 2585.9 A. These changes were accompanied by the appearance, at the later date, of nebular lines, and by a dramatic change of the UV energy distribution , as shown in Fig. 2.

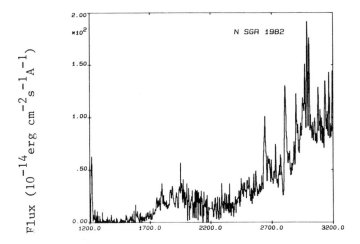

Fig. 3 Observed ultraviolet energy distribution of N Sgr 1982 on Oct.
14, 1982 (about four days after maximum). The spectrum is very similar
to that of N Cyg 1978 shortly after outburst (cf. Fig. 2), both in the
continuum energy distribution and in the emission lines, including
FeII.

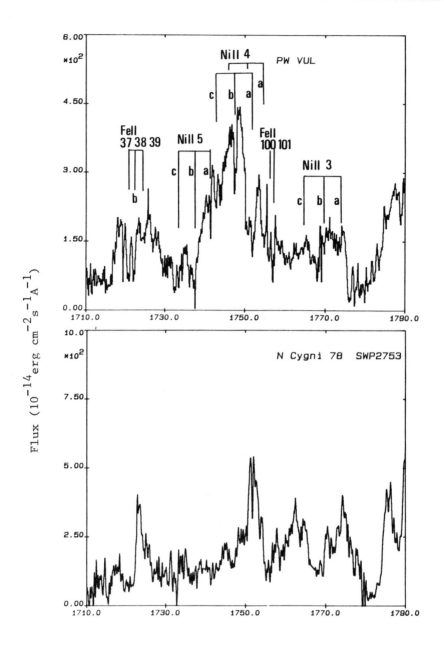

Fig. 4 High resolution spectra of N Cyg 1978 and N Vul 1984a in two
wavelength regions. The spectra of N Vul 1984a are crowded with absorption and emission lines from SiII, NiII and FeII. The absorption lines
have at least three separate components at different expansion velocities: at about 0 Km/s, -750 Km/s and -1550 Km/s. Particularily strong
is the emission-absorption structure from FeII UV191 and NiII UV4 and 5.

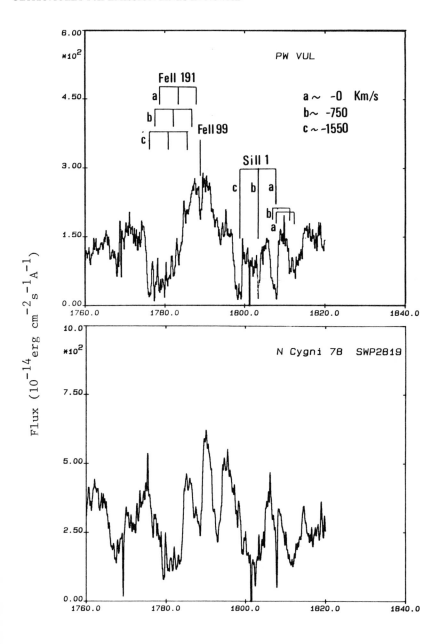

Fig. 4 (cont.)

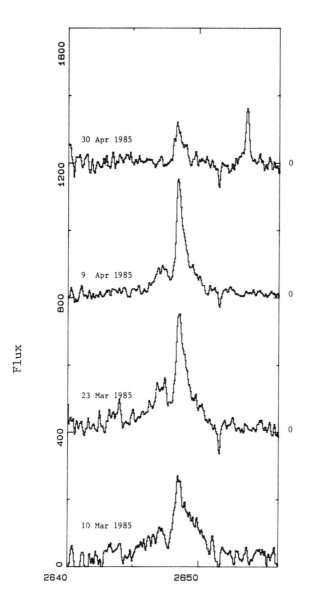

Fig. 5 High resolution data of the recurrent nova RS Oph (outburst near Jan. 28, 1985) showing the time evolution of the profile of [FeXI] 2549A.

Apart from RS Oph and N Cyg 1978, no other novae have been observed early enough after the outburst, or were monitored long enough to provide significant data on the behaviour of the FeII lines. In the first group we include N Mus 1983 (Krautter et al. 1984). In the latter group we include N Sgr 1982 (Drechsel, Wargau and Rahe 1984), whose monitoring was stopped just shortly after the outburst. The spectrum of N Sgr 1982 in outburst is very similar to that of N Cygni 1978 (Fig. 3), both in the continuum energy distribution and in the emission lines, including FeII.

3. THE PROBLEM OF LINE IDENTIFICATIONS

Only a few novae have been observed at high resolution with IUE. High resolution data are, of course, of primary importance in several respects and provide, for example, important information on the dynamics of the ejection, and on the line identifications.

The difficulty of line identifications in the UV spectra of Novae can be realized from the examples in Fig. 4, where we show two spectral regions from high resolution data of N Cyg 1978 and N Vul 1984a (PW Vul). The spectra of N Vul 1984a are crowded with absorption and emission lines from SiII, NiII and FeII. The absorption lines have at least three components at different expansion velocity: at about 0 Km/s, -750 Km/s and -1550 Km/s. Particularly strong is the emission-absorption structure from FeII UV 191 and NiII UV 4 and 5.

No detailed line identifications have been carried out for N Cyg 1978, but it is likely, considering the overall similarity of its spectrum with that of N Vul 1984a (see Fig. 4), that the observed emission-absorption features are due to the same species as in the latter object, although with different expansion velocities.

Another interesting case is that of the recurrent Nova RS Oph which was observed regularly at high resolution during the early outburst in 1985. These spectra were useful, among other things, to identify unambiguously the high ionization line [FeXI] 2648.7 A. The time evolution of the profile of this line is shown in Fig. 5. Such data are expected to provide important clues to the understanding of the dynamics and geometry of the ejection. The complex profile of [FeXI] is probably indicative of a non spherically-symmetric and perhaps discrete ejection process. It is worth recalling that RS Oph is the first astrophysical source in which so strong high ionization lines were detected in the ultraviolet, a part from the solar corona.

REFERENCES

Cassatella, A., Benvenuti, P., Clavel, J., Heck, A., Penston, M.V., Selvelli, P.L., Macchetto, F.: 1979, Astron. Astrophys. **74**, L18

Cassatella, A., Hassall, B.J.M., Harris, A., Snijders, M.A.J.: 1985, Recent Results on Cataclismic Variables, ESA SP-236, p. 281

Cordova, F.A., Mason, K.O., Bode, M.F., Barr, P.: 1985, IAU Circ. No. 4049

Drechsel, H., Wargau, W., Rahe, J.: 1984, Astrophys. Space Sci. **99**, 85

Krautter, J., Beuermann, K., Leitherer, C., Oliva, E., Morwood, A.F.M., Deul, E., Wargau, W., Klare, G., Kohoutek, L., van Paradijs, J., Wolf, B.: 1984, Astron. Astrophys. **137**, 307

Joy, A.H., Swings, P.: 1945, Astrophys. J. **102**, 353

Rosino, L., Taffara, S., Pinto, G.: 1960, Memorie Soc. Astron. Ital. **31**, 3

Snijders, M.A.J.: 1986, RS Ophuchi, ed. M.F. Bode, VNU Science Press, p. 51

Starrfield, S.: 1986, New insights in Astrophysics, University College London, ESA SP-263, p. 239

FE II in Luminous Hot Stars

F.-J. Zickgraf
Landessternwarte
Königstuhl
D-6900 Heidelberg
West Germany

ABSTRACT. The appearance of Fe II in the spectra of hot luminous stars of the Magellanic Clouds (Luminous Blue Variables and B[e]-Supergiants) is discussed.

1. INTRODUCTION

In the Magellanic Clouds (MC's) several hot and luminous supergiants, located in the HR-diagram close to the Humphreys-Davidson-limit (Humphreys and Davidson 1979) of known stellar luminosities, are found whose spectra (in the visual as well as in the satellite UV) show permitted and/or forbidden lines of singly ionized iron. These are the S Dor variables and the related P Cygni stars, summarized as Luminous Blue Variables (LBV's), and the B[e]-supergiants. The Fe II-lines proved to be a well suited diagnostic tool for the investigation of the peculiar and extreme stellar wind characteristics of these stars. In the following chapters some general properties and the appearance of Fe II in particular in the spectra of the LBV's (Section II) and B[e]-supergiants (Section III) will be discussed. Some concluding remarks are given in Section IV.

2. THE LUMINOUS BLUE VARIABLES

The S Dor variables (also called Hubble-Sandage variables e. g. in M 31 and M 33) of the LMC belong to the brightest stars in their galaxy. Spectroscopic observations of the proto-types S Dor and R 71 have been described by Thackeray (1974). A recent review on S Dor variables have been given by Wolf (1986). The S Dor variables have a high bolometric luminosity with $M_{bol} = -9^m \ldots -11^m.5$. Irregular photometric variations of the order of $\Delta m_V = 1 \ldots 2$ mag occur on time scales of years to decades. During minimum light the spectra

R. Viotti et al. (eds.), Physics of Formation of FeII Lines Outside LTE, 125–133.
© *1988 by D. Reidel Publishing Company.*

are characterized by [Fe II]-lines while at light maximum
strong Balmer emission lines dominate. The mass loss rates
during outbursts are typically 10^{-4} to 10^{-5} $M_\odot$/yr and approximately a factor of 100 lower during minimum light. The variability in visual light is regarded to be due to redistribution of radiation in the dense shells expelled during outbursts resulting in a redder colour in maximum than in minimum state. The bolometric luminosity, however, does not
change significantly between maximum and minimum (cf. Appenzeller and Wolf 1981, Wolf and Stahl 1982, Stahl et al.
1983a). Intrinsically the S Dor variables are OB-supergiants.
The proto-type S Dor is presently in a bright phase. Optical
spectra obtained during this phase (Stahl and Wolf 1982,
Leitherer et al. 1985) are characterized (apart from the
strong Balmer lines) by numerous Fe II- and (surprisingly)
[Fe II]-lines. The permitted Fe II-lines show strong P Cygni
profiles. The maximum velocity derived from their P Cygni
absorption components agrees with the expansion velocity obtained from the half widths of the forbidden lines of Fe II
showing that no further acceleration of the stellar wind
occurs between the inner and outer regions, where the permitted and forbidden Fe II-lines, respectively, originate
(Stahl and Wolf 1982). The profiles of the [Fe II]-lines
were found to be in agreement with predicted profiles of optically thin lines in an expanding, outward accelerated
wind (Leitherer et al. 1985). The UV-spectrum in the IUE-LWR
range also obtained during the bright phase (Leitherer et
al. 1985) was found to be completely crowded with lines of
singly ionized metals, mainly of Fe II. Lines of higher excitation potential show lower blue edge velocities than
lower excited lines. This finding was interpreted by Leitherer
et al. in terms of a temperature- and density field with
outward increasing velocity in agreement with the [Fe II]-
profile analysis. R 71 was studied during the present minimum phase by Wolf et al. (1981a). Spectroscopic observations
obtained during light maximum have been discussed by
Thackeray (1974). While in the optical spectrum during maximum absorption lines of Fe II (partly with P Cygni profiles)
were present, the minimum in the visual is dominated by numerous [Fe II]-lines. Permitted envelope lines are absent
or weak (e. g. Balmer lines). The minimum IUE satellite UV-
spectrum, however, is dominated by Fe II-absorption lines,
of which the stronger lines show P Cygni profiles. The lines
of Fe II-multiplet 1 were used by Wolf et al. (1981a) (together with a curve-of-growth analysis of Fe II and Fe III)
to derive the minimum mass loss rate $\dot{M}$ = 3 x 10^{-7} $M_\odot$/yr (see
above) by means of fitting theoretical profiles of Castor
and Lamers (1979) to the observed ones. The radius of the
[Fe II]-emitting region was estimated according to Viotti
(1976) to R[Fe II] > 100 R_*. The expansion velocity at this
distance was derived from the half width at zero intensity

of the [Fe II] -lines to 78 km s^{-1}. The UV-Fe II-absorption
lines, which originate closer to the photosphere, on the
other hand yielded a higher velocity of 127 km s^{-1}. This
indicates a decelerated outflow of matter. Support for this
model was found by Stahl and Wolf (1986) who observed
[N II] -lines, originating at even larger distance than the
[Fe II] -lines, which are still narrower than the latter.
Surprisingly R 71 was found to be an IRAS point-source
(Wolf, Zickgraf 1986). The presence of a cool dust shell
(T ≈ 140 K, R = 8000 R$_*$) could be the consequence of the decelerated stellar wind. R 127 was classified by Walborn
(1977, 1982) as OIaf/WN9-10. Stahl et al. (1983a) detected
an S Dor-type outburst which revealed R 127 to be the third
presently known S Dor variable of the LMC. The spectrum of
R 127 during outburst shows a very close resemblance to the
maximum spectrum of S Dor. In the optical range Fe II-lines
(the strongest with P Cygni profiles) were present during
the brightest phase. IUE high resolution spectra showed the
presence of numerous Fe II-lines with P Cygni profiles. The
absorption components are split into three distinct components, which appeared stable within 2 years of monitoring
(Stahl and Wolf 1985).

According to Lamers (1986) the P Cygni stars may be regarded as S Dor variables in a quiescent phase and possibly
both classes of stars are identical. The P Cyg star R 81 of
the LMC was described by Wolf et al. (1981b) as a close
counterpart of the galactic proto-type P Cygni itself. Surprisingly photometric monitoring of R 81 since 1982 revealed this star to be an eclipsing binary which a period
of 74.6 days (Stahl et al. 1986), offering the possibility
to determine directly the mass of a P Cygni star. An IUE
high-resolution spectrogram (taken outside eclipse) showed
the LWR-range to be dominated by strong Fe II-absorption
lines originating in the stellar wind of R 81 (Fig. 1). The
line profiles are split into three components and resemble
very closely the Fe II-line profiles of P Cygni (cf.
Cassatella et al. 1979). Optical spectra observed outside
eclipse did not show any trace of Fe II-lines. In a spectrogram obtained during the March 1986 eclipse, however,
Fe II-absorption lines with two components are clearly dis-
cernible. This result and the detection of variations of
the profiles of He I 4471 and Mg II 4481 with phase indicate the presence of a phase-dependent wind.

3. B[e] -supergiants of the Magellanic Clouds

Presently eight stars of this class are known in the MC's
(cf. Zickgraf et al. 1986). They are characterized by
strong Balmer emission lines (frequently with P Cygni profiles), emission lines of Fe II, [Fe II] , [O I] , etc. i. e.

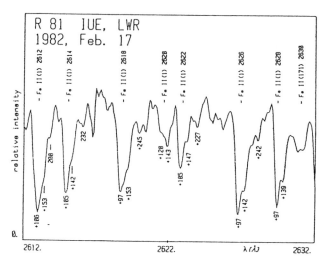

Figure 1. Section of a high dispersion IUE-spectrum showing the complex structure of the Fe II-lines of R 81 in the UV.

low excitation, and a strong IR-excess due to thermal radiation of circumstellar dust (T ≈ 1000 K). Unlike the S Dor Variables they appear rather stable both photometrically and spectroscopically. Their spectral types are around B0 to B3 except R 66 which is a B8-supergiant.

   R 126 of the LMC may be regarded as proto-type of the B[e]-supergiants. In its spectrum apart from the Balmer-lines narrow emission lines of Fe II and [Fe II] (FWHM ≈ 40 and 30 km s^{-1}, respectively) (Fig. 2) in the optical and long-wavelength IUE range are contrasted by broad (~2000 km s^{-1}) UV resonance absorption lines of highly ionized species in the IUE-SWP-range. This hybrid spectrum was interpreted by Zickgraf et al. (1985) in terms of a two-component model with a slowly expanding cool disk-like equatorial wind and a hot fast line-driven polar wind. The emission lines are supposed to originate in the pole-on seen disk. Zickgraf et al. (1986) extended this interpretation to the entire class of B[e]-supergiants of the MCs. Fe II-line formation in circumstellar disks was independently suggested by Friedjung and Muratorio (1980) in the case of S 22 and by Muratorio and Friedjung (1986) for several further B[e]-supergiants. Figs. 3 and 4 show further Fe II-line profiles of B[e]-supergiants. In contrast to R 126 the permitted Fe II-lines of R 50 of the SMC show a double-peak structure with a slightly blue shifted central absorption (Fig. 3). The [Fe II]-lines on the other hand only exhibit a singly emission peak, which furthermore is narrower than the permitted lines. The different profiles were interpreted by Zickgraf et al. (1986) within the two-component mo-

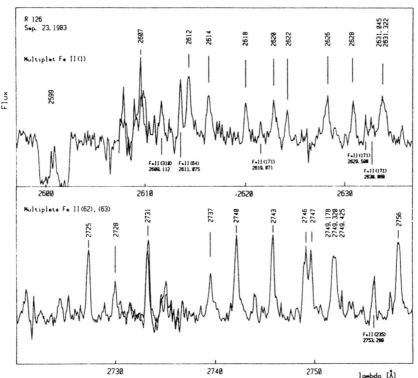

Figure 2.   Two sections of the IUE-LWP-spectrum of R 126
which are dominated by narrow emission-lines of the Fe II
multiplets No. 1, 62, 63. Note the complex structure of the
Fe II(1) 2599 interstellar feature with galactic, LMC and
halo components. Some further lines of higher Fe II multiplets are also indicated.

del assuming differential rotation of the slowly expanding
equatorial disk, which is seen edge-on. Fe II-line profiles
of R 66 are shown in Fig. 4 a,b. They consist of a narrow
emission component unshifted with respect to the systemic
velocity and a clearly separated blue-shifted (v_{max} = 300
km s^{-1}) absorption feature. This profile led Stahl et al.
(1983b) to the conclusion of a decelerated stellar wind.
However, as discussed by Zickgraf et al. (1986) and
Muratorio and Friedjung (1986) the profiles can also be interpreted within the (wind + disk)-model. In this model the
absorption components originate in the high-velocity wind
component of the B8-supergiant and the emission lines are
formed in the equatorial disk. The high velocity component
is particularly well visible in the UV where numerous strong
blue-shifted Fe II-absorption lines were found by Stahl et
al. (1983b).
    The puzzling line-spectrum of R 4 of the SMC was dis-

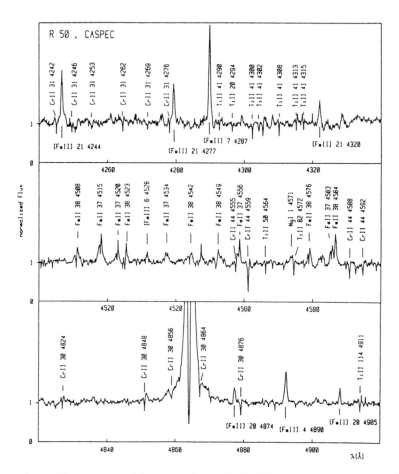

Figure 3.   Three sections of a CASPEC-spectrogram of R 50.
The permitted Fe II-lines show a double peak structure with
a central absorption feature which is slightly blue-shifted
(-7 km s^{-1}) relative to the single and narrower [Fe II]-
lines. Note also the numerous and very sharp "shell"-type
absorption lines of Ti II and Cr II.

cussed by Zickgraf and Wolf (1986). This particular star,
although fitting well to the general schemes of the B[e]-
supergiants, exhibits Fe II-emission lines which are blue-
shifted with respect to the broader [Fe II]-lines and other
emission-lines. Furthermore "shell"-type narrow absorption
lines of Ti II and Cr II are present which are red-shifted
relative to (presumably) photospheric absorption lines and
all emission-lines except [O I]. A binary system with spatially separated emission and absorption regions could provide an explanation for the complex line-spectrum of R 4.

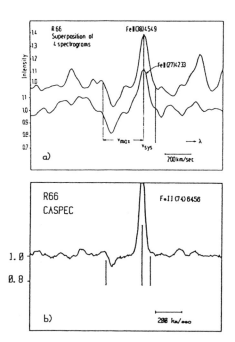

Figure 4. Profiles of Fe II-lines of R 66. The vertical lines in the upper and lower panel indicate the stellar systemic velocity (v_{sys}) and the extreme red and blue edges of the line profiles, respectively. V_{max} is the adopted maximum wind velocity of 300 km s^{-1}.

4. CONCLUSIONS

At present various possible mechanisms which may drive the mass outflow of S Dor variables are intensively discussed (e. g. Stothers and Chin 1983, Maeder 1983, de Jager 1984). Mechanisms suggested are e. g. radiative pressure exceeding the Eddington limit, and turbulent pressure exceeding the de Jager limit, i. e. $g_{grav} + g_{turb} = 0$. According to Appenzeller (1986) a radiatively driven wind is expected to be unstable if the radiative acceleration g_{rad} decreases less rapidly than $\sim T_{eff}^{4}$. In particular opacity variations in the copious Fe II absorption lines present in the UV spectra of the S Dor variables may play an important role in generating wind instabilities. Likewise radiation pressure due to the large number of optically thin metallic lines e. g. of Fe II in the Balmer continuum was suggested by Lamers (1986) to drive the mass loss in P Cygni stars.

The analysis of the Fe II- and [Fe II] -lines of the B [e] -supergiants gave evidence for circumstellar disks around these class of stars. A consequence of this geometry could be the presence of dust formed in the outskirts of the

disks. Like in the classical Be-stars stellar rotation may
lead to an enhanced slow mass loss from the equatorial region. Like in S Dor variables and P Cygni stars the combined opacity of merging UV-lines results in a very small
effective acceleration, which can be compensated even by a
centrifugal acceleration much lower than the gravitational
term.

REFERENCES

Appenzeller, I., Wolf, B.: 1981, in "The Most Massive Stars",
    ESO Workshop, eds. S. D'Odorico, D. Baade, and K. Kjär,
    p. 131
Appenzeller, I.: 1986, in "Luminous Stars and Associations
    in Galaxies", eds. C. W. H. de Loore and A. Willis,
    p. 139
Cassatella, A., Beeckmans, F., Benvenuti, P., Clavel, J.,
    Heck, A., Lamers, H.J.G.L.M., Macchetto, F., Penston,
    M., Selvelli, P. L., Stickland, D.: 1979, Astron.
    Astrophys. 79, 223
Castor, J. I., Lamers, H.J.G.L.M.: 1979, Astrophys. J.
    Suppl. 39, 481
de Jager, C.: 1984, Astron. Astrophys. 138, 246
Friedjung, M., Muratorio, G.: 1980, Astron. Astrophys. 85,
    233
Humphreys, R., Davidson, K.: 1979, Astrophys. J. 232, 409
Lamers, H.J.G.L.M.: 1986, in "Luminous Stars and Associa-
    tions in Galaxies", eds. C. W. H. de Loore and A.
    Willis, p. 157
Leitherer, C., Appenzeller, I., Klare, G., Lamers, H.J.G.L.M.
    Stahl, O., Waters, L.B.F.M., Wolf, B.: 1985, Astron.
    Astrophys. 153, 168
Maeder, A.: 1983, Astron. Astrophys. 120, 113
Muratorio, G., Friedjung, M.: 1986, Monthly Notices Roy.
    Astron. Soc. (in press)
Stahl, O., Wolf, B.: 1982, Astron. Astrophys. 110, 272
Stahl, O., Wolf, B., Klare, G., Cassatella, A., Krautter,J.,
    Ferrari-Toniolo, M.: 1983a, Astron. Astrophys. 127, 49
Stahl, O., Wolf, B., Zickgraf, F.-J., Bastian, U., de Groot,
    M.J.H., Leitherer, C.: 1983b, Astron. Astrophys. 120,
    287
Stahl, O., Wolf, B.: 1985, Astron. Astrophys. 154, 243
Stahl, O., Wolf, B., Zickgraf, F.-J.: 1986, "Instabilities
    in Luminous Early Type Stars", Proc. of a Workshop in
    honour of C. de Jager (eds. H.J.G.L.M. Lamers and C.
    de Loore, Reidel, Dordrecht)
Stahl, O., Wolf, B.: 1986, Astron. Astrophys. 158, 371
Stothers, R., Chin, C.: 1983, Astrophys. J. 264, 583
Thackeray, A. D.: 1974, Monthly Notices Roy. Astron. Soc.
    168, 221

Viotti, R.: 1976, Astrophys. J. 204, 293
Walborn, N. R.: 1977, Astrophys. J. 215, 53
Walborn, N. R.: 1982, Astrophys. J. 256, 452
Wolf, B., Appenzeller, I., Stahl, O.: 1981a, Astron.
    Astrophys. 103, 94
Wolf, B., Stahl, O., de Groot, M.J.H., Sterken, C.: 1981b,
    Astron. Astrophys. 99, 351
Wolf, B., Stahl, O.: 1982, Astron. Astrophys. 112, 111
Wolf, B.: 1986 in "Luminous Stars and Associations in Ga-
    laxies", eds. C.W.H. de Loore and A. Willis, p. 151
Wolf, B., Zickgraf, F.-J.: 1986, Astron. Astrophys. 164, 435
Zickgraf, F.-J., Wolf, B., Stahl, O., Leitherer, C., Klare,
    G.: 1985, Astron. Astrophys. 143, 421
Zickgraf, F.-J., Wolf, B., Stahl, O., Leitherer, C.,
    Appenzeller, I.: 1986, Astron. Astrophys. 163, 119
Zickgraf, F.-J., Wolf, B.: 1986, in "Circumstellar Matter"
    IAU Symp. 122 (in press)

FE II LINES IN HERBIG Ae STARS.

A. Talavera
I.U.E. Observatory.
European Space Agency.
P.O. Box 54065, 28080 Madrid, Spain.

(Affiliated to the Astrophysics Division.
Space Science Department.)

ABSTRACT. Herbig Ae are pre-main-sequence stars of intermediate mass and hotter than the TTau ones. They present a conspicuous Fe II spectrum. Its lines are variable, showing asymmetry, shortward shifted components and/or shortward shift of the whole line. Some lines are also present in emission. The time scale of variation of these characteristics goes from hours up to months or years. Some models are discussed.

1. Introduction.

Herbig Ae/Be stars are objects of intermediate mass (2-7 M⊙) in a pre-main-sequence stage of their evolution. Herbig (1960) published the first list containing 26 stars which were selected using the following criteria:
- spectral type A or earlier, with emission lines
- located in an obscured region
- associated with a reflecting nebula

Further work was done by Strom et al (1972), Cohen (1973), Cohen and Kuhi (1979), among others. These authors performed extensive studies of this group of stars specially from the observational point of view, leading to the conclusion that Herbig Ae/Be stars represent a prolongation of T Tauri stars towards higher effective temperature.

High resolution spectroscopy both in the visible and the ultraviolet regions of the electromagnetic spectrum has been done in the present decade by several groups of authors. All these observations allow us to define some of

135

R. Viotti et al. (eds.), Physics of Formation of FeII Lines Outside LTE, 135–142.

the most important characteristics of the spectrum of Herbig
Ae/Be stars:
- P Cygni profiles in the lines of Mg II and Hα
- emission lines of Ca II, Na I, He I, H I
- asymmetric profiles in the resonance lines of Ca II, Fe
II,...
- high degree of ionisation (lines of C IV, Si IV)
- variability of all these spectral features

All these characteristics point out the existence of
great activity in the envelope of these young stars.

Finkenzeller and Mundt (1984) published a list of 57
stars which they classified in three groups: double peak Hα
emission, single peak Hα emission, and P Cygni profile of
Hα.

In this contribution we shall try to summarise the work
done in the ultraviolet on the Herbig Ae group of stars with
special regard to the spectrum of Fe II as it is observed
with the I.U.E. satellite at high resolution.

2.   The UV spectrum of Herbig Ae stars.

Very few Herbig Ae stars are brigth enough to be
observed with IUE at high resolution. In table 1 we list of
stars used in this rewiew. They belong to the first and
third group of the classification scheme outlined by
Finkenzeller and Mundt (1984) which we have mentioned in
section 1.

We must say that very few systemetic data exist to
study for instance the variability of the Fe II lines. Only
AB Aur has been observed intensively to allow us such a
study. Praderie et al (1986) have performed a systematic
study of AB Aur using IUE.

The ultraviolet spectrum of Herbig Ae stars has been
described by The et al (1981), Talavera et al (1982), Catala
et al (1986a), among other authors. Their principal results
can be summarised as follows.

The Mg II resonance lines present a double peak
emission, or a P Cygni profile with a shortward wing
extending up to more than -500 km/s. The available data
indicate a correspondence between the Mg II and the Ho
profiles, however more simultaneous observations in these
wavelength regions are necessary to confirm this behaviour.

High ionisation lines such as the resonance lines of C IV and Si IV which are unusual in A-type stars are observed in Herbig Ae stars. These lines are observed in emission in HR 5999 (Tjin A Djie et al. 1982). The same lines show an asymmetric absorption profile in AB Aur (Catala and Talavera 1984) and in HD 250550. As we have mentioned before these stars are very faint to be observed with IUE at high resolution, specially in the region 1400-1600 A were these high ionisation lines are located.

Lines of singly ionised metals like Fe II,Cr II, Mn II are observed as well in this group of stars. These lines are generally observed in absorption. Their profiles are asymmetric or shortward shifted. Sometimes discrete shortward shifted absorption components have been observed for instance in AB Aur. We shall describe the Fe II spectrum in more detail in section 3.

3.   Fe II in Herbig Ae stars.

All the prominent lines of Fe II in the IUE wavelength range (1200-3200 A) are observed in Herbig Ae stars. We shall restrict our description to the multiplets lying in the "long wavelength" IUE domain (1900-3200 A).

In this region of the spectrum we can observe not only the resonance lines of Fe II (UV mult. 1,2,3), but also lines emerging from levels with excitation potentials as high as 7 eV (UV mult. 33, 35, 60, 61, 62, 63, 64, 78, 234,...).

These Fe II lines are observed in general in absorption and they show as principal characteristics, shortward displacement, asymmetric profile, discrete shortward shifted components and variability. Some lines are observed in emission in some stars.

Broad, symmetric, shortward shifted lines without any component located at the rest wavelendth are shown by HD 250550 and AB Aur. The velocity corresponding to this shift varies. We have observed in HD 250550 values like v=-290 km/s (1982), v=-200 km/s (1983), v=-110 km/s (1984-85). In AB Aur a value v=-50 km/s has been found by Praderie et al (1986) in some spectra of this star obtained in 1983.

Asymmetric lines with the deepest point of the profile located at the rest wavelength and a shortward wing reaching the continuum at velocities which vary depending of the star and the epoch of observations have been reported for AB Aur

(Praderie et al 1982, Vsw=-190 km/s) and HR 5999 (The et al 1985, Vsw=-135 km/s).

Discrete shortward shifted components appear sometimes in the most intense lines (both resonance lines and lines with excited lower levels). Praderie et al (1986) report up to three components at velocities 0, -56 and -156 km/s in all the spectra of a series of observations of AB Aur performed during 40 consecutive hours in October 1982. However in another monitoring of the same star by the same authors during five consecutive days in November 1984, only two components at velocities of 0 and -100 km/s appear in some of the spectra.

Components at 0 and -45 km/s have been observed in HR 5999 in September 83 in some lines of Fe II (The et al. 1985).

A common characteristic of all the described features is their variability. As we have seen, AB Aur has shown all these features at different epochs. The shortward displacement of the symmetric lines of HD 250550 varies in all the available spectra. The variations are observed in the velocity, the intensity and the shape of the lines and they appear in a long term basis (months or years) and also in a short term basis (days or hours).

4.   Emission lines of Fe II in Herbig Ae stars.

Emission lines of Fe II have been observed only in AB Aur (Praderie et al. 1986). Neither star of table 1 shows these emission lines in the UV spectra available up to now, except may be BD+46° 3471. This star is very faint to be observed with IUE and the existing spectra are too weak to ascertain the existence of Fe II in emission, although it seems that some lines are present indeed.

The Fe II emission lines observed in AB Aur belong to the multiplets listed in table 2. It is important to notice that none of the transitions ending in the less excited levels (eg. the resonance lines) do show emission, but the characteristics mentioned in section 3. Note as well that some of the multiplets in emission are intercombination transitions. We should remark that these lines, which appear in emission in AB Aur are absorption lines in all the other Herbig Ae stars studied up to now.

In their series of observations in October 1982 , Praderie et al. (1986) report no variability in the emission peaks although some fluctuations were measured in the equivalent width of the emission. No variation is reported by these authors in their observations of November 1984.

The emission lines of Fe II in AB Aur vary in a long term basis (Talavera, in preparation). The emission is more intense for instance in the spectra of October 1982 than in those of November 1984.

5.   A model of the envelope of Herbig Ae stars.

We shall refer in this section to AB Aur since it is the Herbig Ae star which have been studied more intensively. The first model for the line formation in the atmosphere of this star was presented by Catala et al. (1984). These authors proposed a semiempirical model composed of a deep chromosphere plus a stellar wind as the origin of the Mg II P Cygni profiles. This model was able to explaining the high ionisation found by Catala and Talavera (1984) in this star.

The observations of Praderie et al (1986) already mentioned, show that the variability observed in the Mg II profiles is rotationally modulated. Catala et al (1986b) have found variability in the Ca II K line which is also modulated by the rotation of AB Aur. The observed period is 45 hours in the Mg II lines and 32 hours in the Ca II K line. Taking into account the different depths of formation of these lines, these two periods point out the existence of differential rotation in AB Aur. These periods are compatible with the V sin i of this star.

In these works, Praderie, Catala and their coauthors propose a model to explain the variability observed in the Mg II and Ca II K lines in AB Aur . They suggest a non axisymmetric model in which fast and slow streams of material emerge from the stellar surface. Due to the rotation of the star, these streams will merge at a certain distance, and a so-called corotating interaction region (CIR) will be formed.

That model is inspired in observations of the sun and it was proposed for other stars by Mullan (1984) and Underhill and Faye (1984).

The CIR model for AB Aur is supported by the variability observed in the Mg II and Ca II K lines. However as Praderie et al (1986) mention and our own analysis of the available data shows, the variations observed in the Fe II lines are erratic, or in other words the Fe II lines are not rotationally modulated. This fact suggest that the Fe II lines are formed farther out than the Mg II and Ca II lines, in the deccelerating part of the wind. Probably the Fe+ ions are accumulated in some kind of shell. The formation of these shells is compatible with the erosion of the CIR's into pressure waves as it is proposed by Praderie and coauthors.

However it remains to be explained the "strange" behaviour of some multiplets of Fe II which appear in emission in AB Aur. If Fe+ is concentrated in shells, then the lines of Fe II should present the classical P Cygni profile which is not observed in AB Aur nor in any other Herbig Ae star.

It is well known, and this is one of the reasons for the celebration of this colloquium, that the physics of Fe II is very complex in any celestial object. Much more work is necessary, at least on the Herbig Ae stars, to understand how this ion is formed and which is its role in the structure of the outer layers and in the evolution of these pre-main-sequence stars.

NOTE.

Part of this work is based on data from the International Ultraviolet Explorer collected and/or dearchived at the Villafranca Satellite Tracking Station of the European Space Agency.

Table 1.

Star	Spec. Type	m(pg)	Hα
AB Aur	B9e	7.2-8.4	P Cyg
HD 250550	B9eq	9.7	P Cyg
BD+46 3471	A0e+s	10.1	P Cyg
HR 5999	A7III-IVe	6.4	Emiss.(double)

(Data taken from Finkenzeller and Mundt, 1984)

Table 2.

Fe II MULTIPLETS OBSERVED IN EMISSION.

Mult. (UV)	Lower Term	Upper Term	
60	a ^{4}D	z ^{6}F^0	*
61	a ^{4}D	z ^{6}P^0	*
78	a ^{4}P	z ^{4}P^0	
216	b ^{2}P	z ^{4}G^0	*
217	b ^{2}P	z ^{2}D^0	
234	b ^{2}H	z ^{2}G^0	
277	b ^{2}G	z ^{2}F^0	
282	b ^{2}G	y ^{2}H^0	*

* Intercombination transition

REFERENCES.

Catala C, Kunasz P, Praderie F, 1984, Astron.&Astroph., 134, 402

Catala C, Talavera A, 1984, Astron.&Astroph., 140, 421

Catala C, Czarny J, Felenbok P, Praderie F, 1986a, Astron.&Astroph., 154, 103

Catala C, Felenbok P, Czarny J, Talavera A, Merchant-Boesgaard A, 1986b, Astroph.J., 308, 791

Cohen M, 1973, MNRAS, 161, 105

Cohen M, Kuhi LV, Astroph.J.Sup.Ser., 41, 743

Finkenzeller U, Mundt R, 1984, Astron.&Astroph.Sup., 55, 109

Herbig GH, 1960, Astroph.J.Sup.Ser., 4, 337

Mullan DJ, 1984, Astroph.J., 283, 303

Praderie F, Talavera A, Felenbok P, Czarny J, Boesgaard AM, 1982, Astroph.J., 254, 658

Praderie F, Simon T, Catala C, Merchant-Boesgaard A, 1986, Astroph.J., 303, 311

Strom SE, Strom KM, Yost J, Carrasco L, Grasdalen G, 1972, Astroph.J., 173, 353

Talavera A, Catala C, Crivellari L, Czarny J, Felenbok P, Prade rie F, Proc.3rd IUE European Conf., ESA-SP176, p99

The PS, et al (13 authors), 1981, Astron.&Astroph.Sup., 44, 451

The PS, Tjin A Djie HRE, Brown A, Catala C, Doazan V, Linsky JL, Me we R, Praderie F, Talavera A, Zwaan C, 1985, Irish Astron. J. , 17-2, 79

Tjin A Djie HRE, The PS, Hack M, Selvelli PL, 1982, Astron.&Astroph. , 106, 98

Underhill AB, Faye RP, 1984, Astroph.J., 280, 712

Fe II LINE PROFILES VARIABILITY IN THE HERBIG Be STAR Z CMa

E. Covino(1), L. Terranegra(1), A. Vittone,(1), F.Giovannelli(2),
C. Rossi(3).
1 - Osservatorio Astronomico di Capodimonte, Napoli-Italia.
2 - Istituto di Astrofisica Spaziale, CNR, Frascati-Italia.
3 - Istituto dell'Osservatorio Astronomico, Università "La
    Sapienza", Roma-Italia.

ABSTRACT. Within the framework of an optical and infrared spectrophotome
tric monitoring on a sample of PMS stars, we present some preliminary re
sults on the behaviour of the Herbig Be star Z CMa.
   A strong variability in H and Fe II line profiles is detected.
The optical spectroscopic changes are connected with IR brightness varia
tions.

1. INTRODUCTION

The Herbig Be star Z CMa is the most interesting object of the
very young CMa R1 association. The spectral type of Z CMa is peculiar,
showing at the same time characteristics of both early (B5-B8) and late
(F5) spectral types (Covino et al., 1984).
   The emission spectrum generally shows strong H, Fe II and Ca II
lines with P Cyg profiles.
   A more detailed description of the spectrum of Z CMa can be
found in the following papers: Merrill (1927),Swings and Struve (1940,
1942), Herbig (1960), Strom et al. (1972).
   According to Strom et al. (1972) Z CMa is a pre-main sequence
(PMS) object of about 5-6 $M_{\odot}$ and with an extimate age of 1.2 x 10^6 yr.

2. OBSERVATIONS

Within the framework of a spectrophotometric monitoring on a
sample of PMS stars, simultaneous optical and infrared observations of
Z CMa were carried out on April 16-17, 1984; March 17-18, 1985 and April
8-11, 1985 at the European Southern Observatory (ESO) La Silla, Chile.
   Several spectra in the range $\lambda\lambda$ 3950-6800 Å with a reciprocal
dispersion of 59 Å/mm were obtained with a Boller & Chivens spectrograph
mounted at the Cassegrain focus of the ESO 1.52 m telescope.
   An Image Dissector Scanner (IDS) was used with a double 8 arcsec aperture decker.

R. Viotti et al. (eds.), Physics of Formation of FeII Lines Outside LTE, 143–148.
© *1988 by D. Reidel Publishing Company.*

An high resolution ($\lambda/\Delta\lambda \cong 10^4$) spectrum in the range $\lambda\lambda$ 5700-6750 Å was secured with the Cassegrain ESO echelle spectrograph (CASPEC) mounted at the ESO 3.6 m telescope.

Infrared observations in J,H,K,L,M photometric system were made with a photometer using an InSb photovoltaic detector cooled with liquid nitrogen at 77 K°. The observations were carried through a 15 arcsec diaphram at the focus of the ESO 1 m telescope.

3. RESULTS

Our ESO observations of Z CMa show a strong H and Fe II line profiles variability. In the spectrum got in 1984 (Fig.1), the H_β line

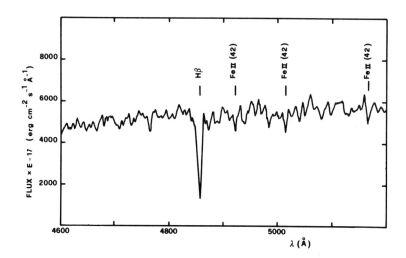

Figure 1. 14 April 1984 ESO 1.52 m + IDS spectrum

has a weak P Cyg profile while Fe II (42) lines and in general all Fe II features are seen in absorption. The spectrum is very similar to the one observed in 1983, when the star had approximately the same 9.3 visual magnitude (Covino et al., 1984).

On the contrary the spectrum taken in 1985 (Fig.2) shows very strong P Cyg profiles at H_β, Fe II (42) and probably at Fe II (49) lines. In the CASPEC spectrum (Fig.3) other Fe II lines are present only in emission.

The absorption components of H and Fe II lines are fairly broad and blue-shifted of about -300 km s^{-1}.

The 1985 spectrum of Z CMa is very similar to the one described by Merrill (1927) who first noticed that the P Cyg profiles were not present at all Fe II lines. Swings and Struve (1942) suggested that the Fe II lines which have a^{6}S as lower level show strong absorption components, whereas those with different lower levels are present only in emission.

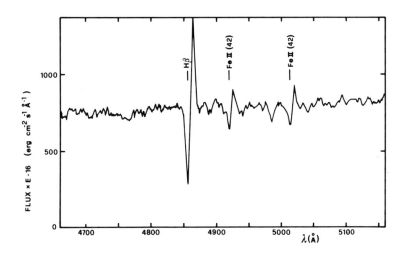

Figure 2. 16 March 1985 ESO 1.52 m + IDS spectrum

        In Table 1 we report all Fe II lines observed in Z CMa. We can
not confirm the suggestion made by Swings and Struve since Fe II (40) lines, which have a ^{6}S as lower level, show no evidence of P Cyg components.
        It is however evident the simultaneous appearance of strong P
Cyg profiles at H$_\beta$ and Fe II (42) lines.
        From our simultaneous IR photometric observations reported in
Table 2, we note that the line variability in the spectrum of Z CMa is
connected with the photometric variability of the star. The photometric
variability is more evident in U,B,V bands (Herbst, 1986).

TABLE 2

ESO 1m INFRARED PHOTOMETRY

Date	J	H	K	L	M	J.D. (2440000+)
16-18/4/1984	6.17±.03	4.90±.02	3.72±.03	1.68±.03	0.8±.3	5807.73120
17-18/3/1985	5.84±.03	4.63±.03	3.49±.01	1.55±.03	0.8±.2	6142.22960
8-10/4/1985	5.82±.03	4.69±.03	3.57±.03	1.62±.04	0.7±.3	5165.72849

TABLE 1

Fe II emission lines in ZCMa

Multiplet	Transitions	Lab. Wav.	Remarks
(28) $b^4P - z^4F°$	$2\frac{1}{2} - 3\frac{1}{2}$	4178.85	
(27) $b^4P - z^4D°$	$2\frac{1}{2} - 3\frac{1}{2}$	4233.17	P Cygni?
(38) $b^4F - z^4D$	$1\frac{1}{2} - \frac{1}{2}$	4508.28	
	$2\frac{1}{2} - 1\frac{1}{2}$	4522.63	
	$3\frac{1}{2} - 2\frac{1}{2}$	4549.47	
	$4\frac{1}{2} - 3\frac{1}{2}$	4583.83	
(37) $b^4F - z^4F°$	$2\frac{1}{2} - 2\frac{1}{2}$	4515.34	
	$3\frac{1}{2} - 3\frac{1}{2}$	4555.89	
	$4\frac{1}{2} - 4\frac{1}{2}$	4629.34	
(42) $a^6S - z^6P°$	$2\frac{1}{2} - 2\frac{1}{2}$	4923.92	P Cygni
	$2\frac{1}{2} - 2\frac{1}{2}$	5018.43	P Cygni
	$2\frac{1}{2} - 3\frac{1}{2}$	5169.03	P Cygni
(49) $a^4G - z^4P°$	$2\frac{1}{2} - 1\frac{1}{2}$	5197.57	P Cygni?
	$3\frac{1}{2} - 2\frac{1}{2}$	5234.62	P Cygni?
	$4\frac{1}{2} - 3\frac{1}{2}$	5275.99	P Cygni?
	$5\frac{1}{2} - 4\frac{1}{2}$	5316.61	P Cygni?
(46) $a^4G - z^4F°$	$5\frac{1}{2} - 4\frac{1}{2}$	5991.38	
	$4\frac{1}{2} - 3\frac{1}{2}$	6084.11	
(74) $b^4D - z^4P°$	$3\frac{1}{2} - 2\frac{1}{2}$	6456.38	
	$2\frac{1}{2} - 1\frac{1}{2}$	6247.56	
	$1\frac{1}{2} - \frac{1}{2}$	6147.74	blend
	$2\frac{1}{2} - 2\frac{1}{2}$	6416.91	
	$1\frac{1}{2} - 1\frac{1}{2}$	6238.38	
	$\frac{1}{2} - \frac{1}{2}$	6149.24	blend
(40) $a^6S - z^6D°$	$2\frac{1}{2} - 3\frac{1}{2}$	6516.05	
	$2\frac{1}{2} - 2\frac{1}{2}$	6432.65	
	$2\frac{1}{2} - 1\frac{1}{2}$	6369.45	

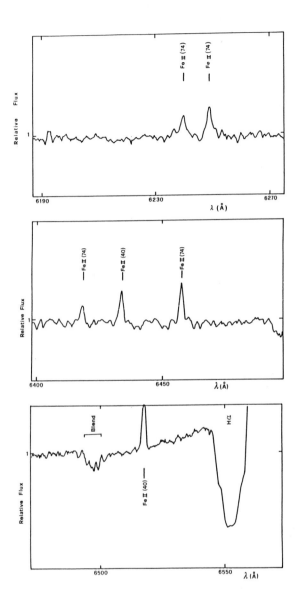

Figure 3. 11 April 1985 ESO 3.6 m + CASPEC spectrum

4. CONCLUSION

At the moment we cannot give final conclusions of the intriguing behaviour of Z CMa, since we must finish the analysis of the large amount of data we have acquired in optical and IR spectral ranges.

Anyway, we can give some precious information: i.e. the strong

H and Fe II line profiles variability connected with the photometric (U, B, V and IR) variability of the star and the complex behaviour of Fe II lines.

REFERENCES

Covino, E., Terranegra, L., Vittone, A., Russo, G.: 1984, Astron.J. 89, 1868.

Herbig, G.H.: 1960, Astrophys. J. Supp., 4, 337.
Herbst, W.: 1986, private communication.
Merrill, P.W.: 1927, Astrophys. J., 65, 291.
Strom, S.E., Strom, K.M., Yost, J., Carrasco, L., Grasdalen, G.: 1972 Astrophys. J., 173, 353.
Swings, P. and Struve, O.: 1940, Astrophys. J., 91, 426.
Swings, P. and Struve, O.: 1942, Astrophys. J., 97, 258.

A TRANSIENT FE II PHENOMENON IN THE OPTICAL SPECTRUM OF THE Be/X RAY BINARY SYSTEM HDE 245770/A0535+26.

D. de Martino, A. Vittone
Osservatorio Astronomico di Capodimonte, Napoli - Italia

ABSTRACT. Low-dispersion optical spectroscopic observations of the optical counterpart of the X-ray transient source A0535+26 were carried out in 1976-1978. Three spectra which show two rapid phase variations suffered by the Be star are presented. On 1977 February 25 the star lost its Be character and the onset of a new emission phase occurred in only 3 days. This episode has been characterized by several emission features mainly due to permitted and forbidden FeII lines. Radial velocity investigation shows that all FeII lines are red-shifted of about 200 km/sec, while permitted Fe II lines give us some lower values, namely about 50 km/sec. On 1977 March 3 all FeII features disappeared. The transient FeII phenomenon then suggests a rather explosive activity of HDE 245770 leading to the storage of a new circumstellar envelope.

1. INTRODUCTION

Be stars as optical counterparts of X-ray transient sources show different kinds of variability, both intrinsic and connected to orbital effects. In particular the most striking features of intrinsic spectroscopic variability of these objects are phase variations, i.e. the transformations of a Be spectrum into a "normal" B one and/or viceversa.
The 9th magnitude irregular variable star HDE 245770 is the optical counterpart of the X-ray transient source A0535+26 with an O9.7IIIe spectrum (Giangrande et al.,1980).A0535+26 X-ray source is probably windfed by its companion at the periastron passage. Different time scales variability of this object has been detected (de Martino et al., 1985; Giovannelli et al., 1985).
In this paper we present a transient FeII phenomenon connected to phase variations occurred in February 1977.

2. OBSERVATIONS

Spectroscopic observations of HDE 245770 were carried out at Asiago Observatory from 1976 to 1978 with 120 and 180 cm telescopes. The most significant spectrograms were got with prismatic dispersion of 40 Å mm^{-1} at $H\gamma$ on 103a − 0 emulsion.
Spectrograms were digitized on the PDS 1010 A microdensitometer using a

R. Viotti et al. (eds.), Physics of Formation of FeII Lines Outside LTE, 149–153.
© 1988 by D. Reidel Publishing Company.

scanning aperture of 10 μ width and 200 μ height and a sampling step of 10 μ. Data were processed by means of the ELSPEC software package for photographic spectrograms reduction.

3. RESULTS AND DISCUSSION.

Three spectra taken on 1977 February 2,25, 28 show that HDE 245770 suffe red phase variations.
In Fig. 1 we report the spectrum taken on the 2nd of February, which is

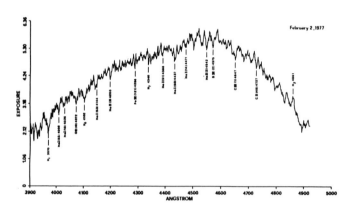

Figure 1. Spectrum of HDE 245770 prior the first phase transformation.

prior the first phase transformation. The B phase has been characterized by deep absorption features (Fig.2). Neverthless it is interesting to no te that the Balmer decrement is not observed by Hβ line, indicating that an emission contribute, even slight, from circumstellar matter must be present. HDE 245770 probably did not loose completely its envelope at this epoch.
Because of lack of data during the first transformation we cannot extima te the time scale of this phenomenon while the second phase variation oc curred rather rapidly, i.e. in only 3 days. Such kind of rapid variabili ty has also been observed in the spectrum of the Be star μ Cen (Peters, 1986). It is generally well known that spectra of Be stars are variable on time scales from days to decades and the total duration of each phase depends on the star and the epoch at which the variations take place (Doazan,1982). The spectroscopic behaviour of a Be star just prior the onset of an emission phase may reveal whether the associated activity is explosive or not, but the irregular behaviour of these objects is trouble sonme to the observer.
The Be phase (Fig.3) has been characterized by Hβ emission line and an Hγ

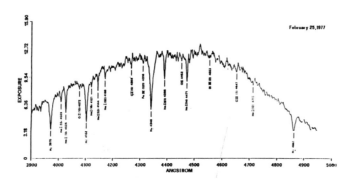

Figure 2. B phase

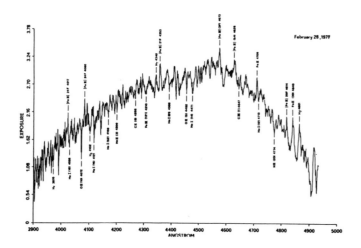

Figure 3. Be phase

emission reaching the continuum. This line may be considered as an activity indicator of the star (De Loore et al. 1984). HI lines are seen in
absorption from Hδ shortward. The most striking feature of the onset of
the emission phase is the presence of [Fe II] (λλ 4017. 4Å, 4080. 0Å,
4352. 8Å, 4814.6Å,), Fe II (λλ4627.9Å, 4708.9Å, 4840.0Å) and probably
[Fe III] (λ 4573.9A) lines. Although the apparence of these lines is not
uncommon in Be stars (Bep stars), HDE 245770 never showed these features

in its spectrum.

All [Fe II] lines are highly red-shifted, while Hβ and permitted Fe II lines show some lower radial velocity values, namely 94 km/sec and 50 km/
sec respectively. Table 1 shows radial velocity of permitted and forbid-

TABLE 1 : Fe II and [Fe II] Lines Radial Velocities

ELEMENT	λ (Å)	V_R (km s¹)
[Fe II] 24 F	4017.4	+ 196
[Fe II] 24 F	4080.0	+ 210
[Fe II] 21 F	4352.8	+ 206
[Fe II] 20 F	4814.6	+ 197
Fe II (54)	4627.9	+ 46
Fe II	4708.9	+ 52
Fe II (30)	4840.0	+ 50

den Fe II lines, whose standard deviations are 15 km/sec. In order to ac
count the presence of low excitation lines and their intensities, the en
velope in which they are formed must be very extended and subionized re=
spect to the photosphere. According to Be star model (Thomas, 1982) this
envelope is located well above the coronal and post-coronal transition
regions.

Because of lack of data in other spectral regions, expecially in the X-
rays, we cannot extimate mass flow parameters and whether A0535+26 X ray
source has been ignited.

Neverthless our data give us the evidence that a quiescent Be star can de
velope emission lines in few days. Furthermore it is interesting to note
that on 3 March 1977 HDE 245770 lost its Fe II features. This fact suggest that further material expelled by the Be star probably supplied the
envelope.

Taking into account the short time scale of this phenomenon, we can suppose that a rather explosive activity of HDE 245770, leading to the storage of a new circumstellar envelope, has been occurred.

REFERENCE :

de Loore, C., Giovannelli, F., van Dessel, E.L., Bartolini, C., Burger,
M., Ferrari-Toniolo, M., Giangrande, A., Guarnieri, A., Hellings, P.,
Hensberger, H., Persi, P., Piccioni, A., Van Diest, H.: 1984, Astron.

Astrophys., 141, 279.

de Martino, D., Vittone, A., Giovannelli, F., Ciatti, F., Margoni, R., Mammano, A., Bartolini, C., Guarnieri, A., Piccioni, A.: 1985, in Multifrequency Behaviour of Galactic Accreting Sources (ed. F.Giovannelli), 326.

Doazan, V.: 1982, in "B stars with and without emission lines", ed. A. Underhill and V. Doazan (NASA SP-456), 316.

Giangrande, A., Giovannelli, F., Bartolini, C., Guarnieri, A., Piccioni, A.: 1980, Astron.Astrophys. Suppl.Ser., 40, 289.

Giovannelli, F., Ferrari-Toniolo, M., Persi, P., Golynskaya, I.M., Kurt, V.G., Mizyakina, T.A., Shafer, E. Yu., Shamolin, V.M., Smirnoff, A.A., Sheffer, E.K., Zaytzeva, G.V., Van Dessel, E.L., Van Diest, H., Hensberge, H., Burger, M., de Loore, C., Bartolini, C., Guarnieri, A., Piccioni, A., Gnedin, Yu.N., Khozov, G.V., Larionov, V.M., Shakovskaya, N.I.: 1985 in Multifrequency Behaviour of Galactic Accreting Sources (ed. F.Giovannelli), 284.

Peters, J.G.: 1986, Astrophys.J., 301, L61.

Thomas, R.N.: 1982, in "B stars with and without emission lines", ed. A. Underhill and V. Doazan (NASA SP-456), 409.

FeII LINES IN THE PALOMAR BRIGHT QUASAR SURVEY

E. Joseph Wampler
European Southern Observatory
Karl-Schwarzschild-Str. 2
D-8046 Garching bei München, F.R.G.

Grandi (1981) and Netzer and Wills (1983) have shown that FeII lines are prominent in the spectra of quasars. Wills, Netzer and Wills (1985) have shown that FeII can be the single strongest contributor to the emission line spectrum. Because the FeII spectrum consists of thousands of lines there are many escape routes in the FeII spectrum for energy absorbed by the fairly abundant Fe^+ ion. It is becoming increasingly clear that the resonance lines of FeII are optically thick (Wills, Netzer and Wills, 1985). This contributes to the excitation of high lying levels through fluorescence pumping from the lower lying levels. Quasar spectra, unlike the spectra of galactic stars such as η Car, do not show forbidden lines in emission (Wampler and Oke, 1967; Wampler, 1985). Understanding the FeII excitation and radiative processes in quasars will lead to a better understanding of the energy balance in the broad line region and the abundance of metals in the nuclear engine.

The Palomar Bright Quasar Survey (BQS) is a useful collection of objects for studying the FeII phenomena. Since BQS quasars are bright ($m_B \gtrsim 16.5$ mag) it is possible to obtain the representative sample of reasonably high signal-to-noise spectra that is needed for statistical studies. Fig. 1 shows a selection of BQS quasar spectra obtained with the IDS spectrograph attached to the 3-meter Shane telescope of Lick Observatory. Wampler (1986) gives spectra of two additional BQS quasars and a complete set of the spectra obtained at Lick observatory is shown in Baldwin and Wampler (1987).

FeII lines are responsible for the broad features that are prominent in the wavelength interval 2200 Å < λ < 3500 Å. Among the quasars shown in Fig. 1 are two low redshift ones. PG 0923+201 has very strong FeII lines while PG 0052+251 has much weaker lines. Nevertheless even in the spectrum of PG 0052+251 one can see a feature near λ 3200 which is probably a blend of multiplets 1, 6 and 7 (Wampler, 1985). Thus PG 0052+251 likely also contains FeII emission in its spectrum but with an intensity about 0.05 to 0.1 the intensity of the FeII emission in the spectrum of PG 0923+201. In fact, it seems

155

R. Viotti et al. (eds.), Physics of Formation of FeII Lines Outside LTE, 155–159.

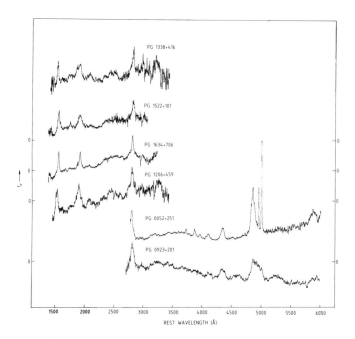

<u>Figure 1.</u>   A selection of BQS quasars. Strong FeII emission is
             seen in the spectral interval 2000 Å < λ < 3500 Å and
             near λ 5000 Å.

probable that all the BQS quasars would show FeII features if high
signal-to-noise spectra are obtained of them. Earlier Netzer et al.
(1985) showed that substantial FeII emission exists even in those AGNs
with weak FeII emission. Since in objects such as PG 0923+201 FeII
emission intensity exceeds that of all other broad emission lines it
is clear that even in a weak FeII object, such as PG 0052+251, FeII
emission plays an important role in the energy balance of the broad
line ionization region. In fact, no quasar known has been shown to
have FeII emission with an intensity relative to Ly-α less than about
0.02 the relative intensity of the strongest FeII emitters.

    There is some evidence that at least in a few quasar spectra
FeIII lines are also seen (Hartig and Baldwin, 1986; Wampler, 1986).
The strengths of FeIII emission relative to the FeII lines will likely
be an important future diagnostic tool to the understanding of these
spectra.

    The determination of accurate FeII line strengths is a difficult
problem. The FeII features are very broad. Not only does this blend
the forest of FeII lines together and confuse them with broad lines
from other ionic species but it is very difficult to determine the
continuum level. The gaps between the observed multiplets, particularly in the ultraviolet, are likely not continuum regions but rather

regions where FeII emission is relatively weak. It seems likely that
in the ultraviolet the features that are clearly seen are only local
peaks where particularly strong multiplets are seen lying over a
general forest of weaker lines. In this situation it is not possible
to describe the FeII spectrum or determine relative intensities of
ultraviolet to optical features without a rather detailed model of the
FeII emission. Wills, Netzer and Wills (1985) have recognized the
importance of this modelling for determining line strengths. There are
at least three important competing sources of continuum radiation in
the $\lambda\lambda$ 2000-4000 Å region: non-thermal continuum, thermal black body
continuum and Balmer emission continuum. The combined ultraviolet
continuum intensity is usually found by extrapolation from optical-
infrared wavelengths. Because both the supposed black body emission
and the Balmer continuum emission are strongly wavelength dependent
this procedure may lead to inaccurate estimates of the continuum
level.

One check on this procedure is to carefully study the spectra of
quasars in which MgII is seen in broad absorption. Two quasars,
PG 1700+518 (Wampler, 1986) and Q 1232+134 (He et al., 1984), show
strong broad MgII λ 2800 absorption that does not go to zero
intensity. It seems likely that the broad MgII absorption band in
these two objects is covering the non-thermal continuum as well as any
central 10^{15} cm diameter accretion disk that could be responsible for
thermal continuum. The depth of the MgII λ 2800 absorption feature in
PG 1700-518 is in good agreement with the extrapolated IR continuum, a
fact that lends support to this argument. There is a remaining uncertainty with respect to the Balmer continuum. In PG 1700+518 the
Balmer lines are relatively weak. Optically thin models give
Bac/H$\alpha \approx$ 2 (Osterbrock, 1974), while optically thick models give
Bac/Hα as high as 8 (Kwan, 1984). The models of Wills, Netzer and
Wills (1985 give 0.2 $\gtrsim$ Bac/H$\alpha \gtrsim$ 1. In PG 1700+518 the Balmer lines are
so weak relative to the total FeII lines that errors in the assessment
of the Balmer continuum causes relatively little error in the FeII
line strengths. In the 3000 Å < λ < 4000 Å region, the one most
affected by Balmer continuum emission, it is likely that the iron
emission in PG 1700+518 is at least twice the strength of the Balmer
continuum (Wampler, 1986). If the Balmer continuum is important one
should also see the Paschen continuum. Fig. 2 shows the HI Continuous
- Emission Coefficient γ_y (H,T) (in ergs cm^3 sec^{-1} Hz^{-1}) for two temperatures, T_e = 10^4°K and T_e = 2 × 10^4°K. It is clear that if there is
substantial Balmer continuum contribution at λ 2800 Å, there will also
be substantial Paschen continuum contribution at 6500 Å. It is then
not correct to ignore the hydrogen recombination continuum near Hα.

In the future, spectra with wide wavelength coverage will simpli-
fy the deconvolution of quasar continua into their component parts.
Polarization studies might be particularly valuable in this respect.
Miller and Schmidt (1985) have shown that it is possible to distinguish the non-thermal continuum radiation from the line radiation on
the basis of spectropolarimetry.

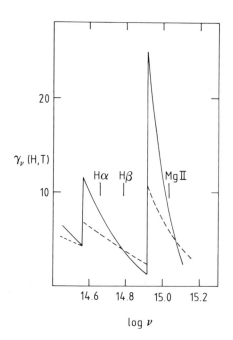

Figure 2.   The Hydrogen Continuous - Emission Coefficient γ_ν(H,T)
(in 10^{-40} erg cm^3 sec^{-1} Hz^{-1}) from Osterbrock (1974).
$T_e = 10^4$°K (solid line) and $T_e = 2 \times 10^4$°K (dashed
line).

    Perhaps 1/3 of the BQS quasars have FeII emission strengths sim-
ilar to that of PG 1700+518. For this object FeII emission is stronger
than all other emission lines in the UV-optical spectral region
combined. This great strength of the FeII lines may cause a problem
for explaining the emission process by radiative mechanisms. If, as
suggested by Hartig and Baldwin (1986), broad absorption line quasars
have stronger emission lines than quasars without broad absorption
features, some of the emission may be due to collisional processes
excited by a rapidly expanding jet. Using the models of Wills, Netzer
and Wills (1985) the derived Fe/H ratio is as much 50 times solar. But
Gaskell (1985) suggests that the strength of FeII is correlated with
line width. If so, the observed strengths of FeII might be correlated
with optical depth or with density. It might, therefore, be possible
to construct models that give the correct FeII line strengths without
invoking overabundant Fe/H ratios.

    I have indicated here some of the uncertainties of interpreting
these spectra. Clearly more theoretical work is needed as well as
better observational data; particularly needed are high signal-to-
noise spectra with very wide wavelength coverage.

REFERENCES

Baldwin, J.A., and Wampler, E.J.: 1987, in preparation.
Gaskell, C.M.: 1985, Ap.J. 291, 112.
Grandi, S.A.: 1981, Ap.J. 251, 451.
Hartig, G.F., and Baldiwn, J.A.: 1986, Ap.J. 302, 64.
He, X.-T., Cannon, R.D., Peacock, J.A., Smith, M.G., Oke, J.B.: 1984,
     M.N.R.A.S. 211, 443.
Miller, J.S., and Schmidt, G.D.: 1985, Ap.J. 290, 517.
Netzer, H., and Wills, B.J.: 1983, Ap.J. 275, 445.
Netzer, H., Wamsteker, W., Wills, B., and Wills, D.: 1985, Ap.J. 292,
     143.
Osterbrock, D.E.: 1974, Astrophysics of Gaseous Nebulae (W.H. Freeman
     and Co., San Francisco).
Wampler, E.J., and Oke, J.B.: 1967, Ap.J. 148, 695.
Wampler, E.J.: 1985, Ap.J. 296, 416.
Wampler, E.J.: 1986, Astron. Ap. 161, 223.
Wills, B.J., Netzer, H., and Wills, D.: 1985, Ap.J. 288, 94.

OBSERVATIONS OF FE II IN ACTIVE GALACTIC NUCLEI

Beverley J. Wills
McDonald Observatory and Department of Astronomy
University of Texas, RLM 15.308
Austin, Texas 78712
U.S.A.

ABSTRACT. A forest of thousands of Fe II lines competes with Lα as the strongest contributor to the broad emission line spectrum of QSOs, the nuclei of broad line radio galaxies and Seyfert 1 galaxies. The most prominent feature, always present in broad line spectra, is between 2000 and 3500 Å; the Fe II lines in the optical region, strongest near 4600 Å, vary widely in strength from one object to another. I will review developments, from the discovery and identification of Fe II lines in the optical spectra of the quasar 3C 273 and a few bright Seyfert 1 galaxies, to systematic studies of the optical Fe II lines in many Seyfert and radio galaxy nuclei, the recognition of UV emission (mainly 2000 to 3500 Å) in the resonance lines from about 5 eV, and finally to the identification of emission from high level transitions at 10 eV and above. I will discuss the use of recent model calculations in measuring the strength of the Fe II emission by matching synthetic and observed spectra, and present results on the strength of Fe II emission compared with other strong lines. Some statistical results are discussed, as well as some individual interesting objects whose spectra may elucidate or confuse our picture of Fe II production in the broad line region.

1. INTRODUCTION

Fig. 1 shows a typical spectrum of a low redshift active galactic nucleus (AGN). Note the presence of strong, broad permitted lines (FWHM ~ 2,000-20,000 km/s) superposed on a relatively smooth underlying continuum. Note also the presence of narrow

R. Viotti et al. (eds.), Physics of Formation of FeII Lines Outside LTE, 161–175.
© *1988 by D. Reidel Publishing Company.*

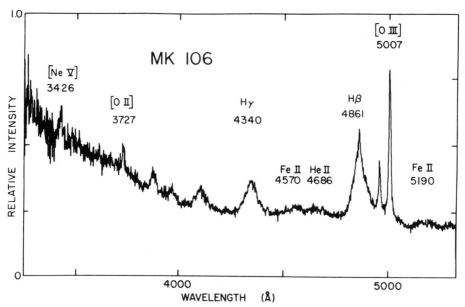

Figure 1: A typical low-redshift broad line spectrum (the Seyfert 1 Galaxy, Markarian 106. From the survey by Osterbrock (1977).

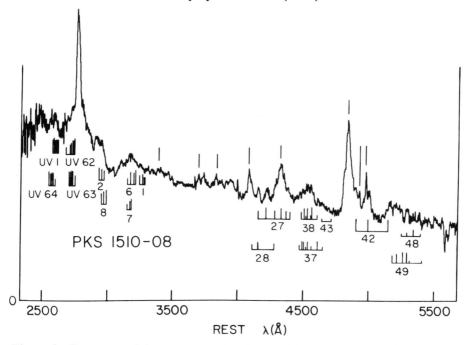

Figure 2: Spectrum of the low-redshift QSO, 1510-08 (z = 0.361), showing strong optical Fe II emission, the expected wavelengths of the strongest Fe II lines, and the apparently weak Fe II (UV) emission. From Phillips (1977).

forbidden lines (FWHM ~ 500 km/s) indicating that the narrow lines, both permitted and forbidden, arise in a low density gas, distinct from the broad line gas where densities are large enough to collisionally de-excite the levels giving rise to the forbidden lines. Models in which the gas is ionized by a central non-thermal continuum source suggest, for the broad and narrow line regions (BLR and NLR) typical densities n_e ~ 10^{9-10} cm^{-3} and 10^{4-5} cm^{-3}, with T_e ~ 10^4 K, and dimensions of $\lesssim$ 1 pc and up to many pc, respectively.

Fe II emission was first seen in extragalactic objects by Greenstein and Schmidt (1964) for the quasars 3C 48 and 3C 273. Features near 4450-4650 and 5100-5500 Å were re-observed by Wampler and Oke (1967) who suggested their identification with blended Fe II multiplets. Later, Sargent (1968) made the same identification in the Seyfert 1 galaxy, I Zw 1. From the absence of [Fe II], Wampler and Oke concluded that the emission region has n_e > 10^6 cm^{-3}, probably > 10^7 cm^{-3}. More Seyfert galaxies were observed and other Fe II emission features were recognized (*e.g.,* Fairall 1968, DeVeny and Lynds 1969, Adams and Weedman 1972, Baldwin 1975, Boksenberg and Netzer 1976, Oke and Shields 1976).

The first systematic spectroscopic survey of Seyfert galaxies by Osterbrock (1977) showed that Fe II was present in 90% of the Seyfert 1 galaxies, and in particular that the Fe II emission was associated only with the broad line objects. He also noted the similarity of the Fe II spectra from one object to another. Fig. 1 shows an example of a spectrum from his survey.

Phillips (1977, 1978a, b) discussed the properties of the Fe II emission in more detail. He derived a lower limit on n_e ~ 10^7 cm^{-3}, also from the absence of [Fe II]. He strengthened earlier conclusions concerning the similarity of the optical Fe II spectra by broadening the spectrum of I Zw 1 (in which the broad lines are quite narrow, ~1000 km/s) to match the widths of the broad Hβ line in other Seyfert 1 spectra. At the same time this supported the BLR origin of the Fe II. This was demonstrated more quantitatively by the match of synthetic Fe II and Balmer line spectra, broadened so that the width of Hβ was the same as that in the observed spectra. The widths of Fe II lines were within 25% of those of the broad Hβ line. He also searched for Fe II emission in a number of low redshift QSOs, finding it to be generally very weak.

Phillips discussed possible emission mechanisms --population of the 5 eV levels either by absorption of UV resonance photons from the strong non-thermal continuum or by collisions in the dense BL gas (n_e ~ 10^9 cm^{-3}). Either mechanism requires high optical depths in the UV resonance transitions to produce the observed strength of optical Fe II. As Wampler and Oke had suggested, the presence of absorption in the resonance lines between 2350 and 2750 Å would show that the excitation mechanism for population of the 5 eV levels was resonance fluorescence by photons from the strong non-thermal continuum. This requires observations in the UV, so Phillips observed two Fe II quasars near the atmospheric cut-off, down to rest wavelengths of 2370 Å (one is shown in Fig.2). He found no absorption; apparently there was only weak emission. Phillips preferred resonance fluorescence excitation because there appeared to be sufficient continuum photons, and predictions of the observed optical line ratios were better than for the collisional excitation mechanism (but only marginally -- the collision strengths and transition probabilities were too poorly known). However, the absence of Si II $\lambda5979$ was a clue that all was not well with the resonance fluorescence explanation. This line need not be present if Fe^+ is collisionally excited.

Meanwhile the development of photo-ionization models to explain the ratios of the strong, relatively isolated emission lines had led to a model in which the BLR covers only a small fraction of the QSO central ionizing continuum source ($\lesssim 0.1$) and Netzer (1980) and later others, argued that the clouds would therefore intercept too few photons for pure radiative excitation to work. Netzer showed that conditions for collisional excitation did exist in the framework of conventional photoionization models. The main argument against collisional excitation was the purported absence of strong resonance line emission near 2000-3000 Å.

The problem was resolved with the development of linear detectors for faint light levels (>15-17^m), the flux density calibration of spectra over a range of several thousand Ångstrom, and the observation of objects with z > 0.6, so that the appropriate rest wavelengths could be observed. We searched our available spectra, and showed that strong 2000-3000 Å emission in fact did exist (Wills, Netzer, Uomoto, and Wills 1980a). Some examples are shown in Fig. 3a and b (see also Grandi 1981, Puetter, Burbidge, Smith and Stein 1981). It is not surprising that these features had been overlooked on uncalibrated spectrograms; they are blends of thousands of lines, in a spectrum where lines are already broadened by several thousand km/s. This discovery proved that the collisional excitation mechanism was dominant (although it is still possible that in a very few broad line objects, radiative excitation is important). One of the main points that I wish to emphasise here is that Fe II emission is probably present in all broad-lined AGN, and that the UV resonance lines are strong even when the optical lines are not. Fig. 3 illustrates this. See also Gaskell (1981), Bergeron and Kunth (1984), Snijders *et al.* (1980), Wills (1983), and Netzer, Wamsteker, Wills and Wills (1985).

Standard models were able to explain the *apparent* strengths of the UV and optical Fe II features fairly well, but there were large discrepancies between the predicted and observed shapes of the spectra. For example, the observed UV Fe II feature (2200 - 2650 Å) is relatively smooth (see Fig. 1 of Wills, Netzer and Wills 1980b) compared with the dips and bumps in model spectra (Fig. 3 in Netzer and Wills 1983).

Wills *et al.* 1980b resolved this discrepancy by showing observational evidence for the excitation of levels far above 5 eV, perhaps beyond 11 eV. Further observational evidence for the population of high energy levels of Fe^+ is the presence of emission in the red wing of Ly α (UV 9?), other weak features near 1610-1670 Å, 1860 Å (Gaskell 1981, Uomoto 1981), and 2050 Å. The excitation of these levels arises because the Fe^+ atom has thousands of levels, and there are hundreds of coincidences in their energy separation. Thus transitions between low energy levels can pump higher levels. Models including this and other important processes are described by Netzer and Wills (1983), Wills, Netzer, and Wills (1985), and in the review by Netzer in these proceedings.

Here we will use these models in discussing the observed properties of Fe II emission.

2. THE STRENGTH OF THE FE II EMISSION

In order to measure the strength of the Fe II emission with any reliability we need a model. This is because in broad line spectra the actual continuum may lie far below the apparent continuum due to blending of thousands of broad Fe II lines. Also many strong lines are seriously blended with broad Fe II lines and modelling is the best way to remove their effects.

Our approach (Wills, Netzer, and Wills 1985) has been to synthesize spectra, including Fe II, Balmer lines and free-bound continuum, and other lines, convolving these with the profile of a relatively uncontaminated broad line such as Hα or, if that is unavailable, Hβ. The strengths of (relatively) isolated lines, such as low order Balmer and Mg II $\lambda2800$ are adjusted to agree with the observed spectrum after the latter has been corrected for Galactic (or possible internal) extinction. Fig. 4 gives one example of a model fitted to the observed spectrum of a quasar. We simultaneously fit optical, intermediate and UV Fe II blends, Balmer continuum and high order Balmer lines, and the underlying (non-stellar) continuum. Minor components of the fit include, *e.g.*, He II resonance fluorescence lines (3100-4000 Å). In general we assume a power law or almost power law continuum between about 2000 and 5000 Å. This appears to be quite a good approximation to most likely models of the underlying continuum[1]. Apart from the underlying non-stellar continuum, the major contribution to the flux density at the Balmer

---

[1] For some QSOs, the Fe II models usefully constrain the shape of the underlying continuum showing that it is not well fit by a standard power extrapolated from the IR plus a single temperature (20,000-30,000 K) black body, as is sometimes suggested. The actual optical-UV continuum is flatter than this.

edge (3646 Å) is from the Balmer continuum; the contribution from Fe I and Fe II emission is quite small here, so this is well determined. The shape of the Balmer continuum is quite model-dependent, determined mainly by the run of temperature and optical depth in the emitting region. This, together with observational calibration uncertainties, introduces most uncertainty in estimating Fe II and Balmer continuum strengths.

Table I shows typical relative strengths of QSO broad lines, including the Fe II. The most important results of the model fitting are:

(i) Numerous Fe II lines, Balmer continuum and high order Balmer lines form an apparent continuum in excess of the underlying central source continuum.

(ii) Fe II is the single largest contributor to the emission line spectrum. In particular I (Fe II)/I(Ly α) ~ 1-2, and there is clearly a problem with standard photoionisation models in explaining the large power emitted in low ionization compared with high ionization lines (Netzer 1985).

(iii) The observed ratio I(Fe II)/I(Mg II λ2798) ~ 10 is too large to be explained by any BLR models with normal abundances. Perhaps iron is overabundant compared with magnesium?

TABLE I

Typical relative strengths of the broad emission lines in quasars

L α	λ1216	10	Balmer continuum	8
C IV	λ1549	4.5	H ε	0.13
C III]	λ1909	1.5	H δ	0.20
C II]	λ2326	0.2	H γ	0.36
Mg II	λ2798	1.3	H β	1
Fe II	λλ2000-3000	7	H α	4
Fe II	λλ3000-3500	0.8		
Fe II	λλ3500-6000	0.7-2.0		

3. STATISTICAL RESULTS ON FE II LINES

Statistically, Fe II appears weaker in quasars with broader lines, in radio quasars compared with their narrower-lined radio-quiet counterparts, and in QSOs compared with narrower lined Seyfert 1 galaxies; it appears especially weak in the very broad lined radio galaxies and especially strong in QSOs with broad absorption lines. Perhaps part of this may be accounted for by blending of the broad lines. As can be seen in Fig. 4, because

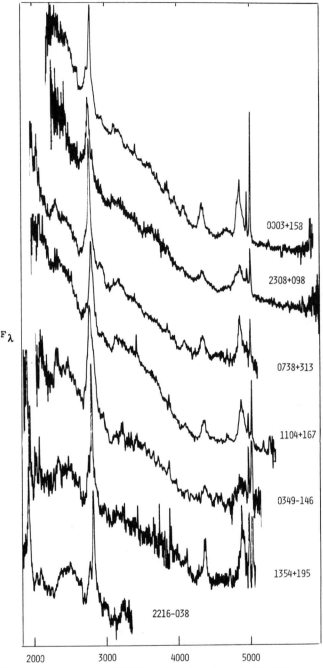

Figure 3a: Composite spectra of QSOs, showing that the Fe II feature between 2200 and 2650 Å is always present, but that the strength of Fe II$\lambda\lambda$4500-4680 is different in different objects. Note also the Fe II feature near 3200 Å. The ordinate is flux per unit wavelength on displaced scales, and the abscissa is rest wavelength in Å. Spectra are from University of Texas' McDonald Observatory, obtained by D. Wills and the author.

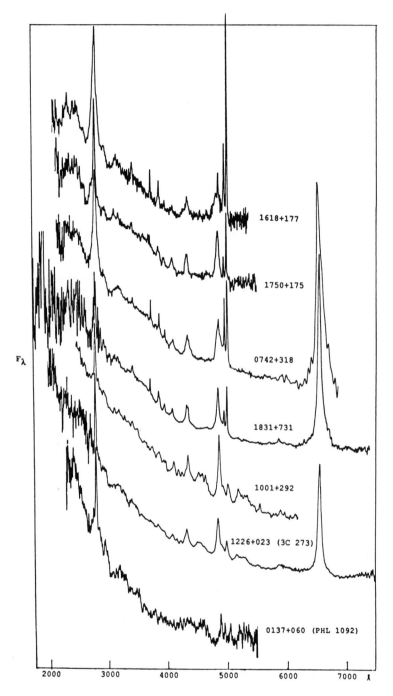

Figure 3b: As in Figure 3a, except that the spectra of 1831+731 and 3C 273 include UV (IUE) data. The high frequency structure in the narrow-lined spectrum of PHL 1092 between 3000 and 5200 Å is real and due to Fe II.

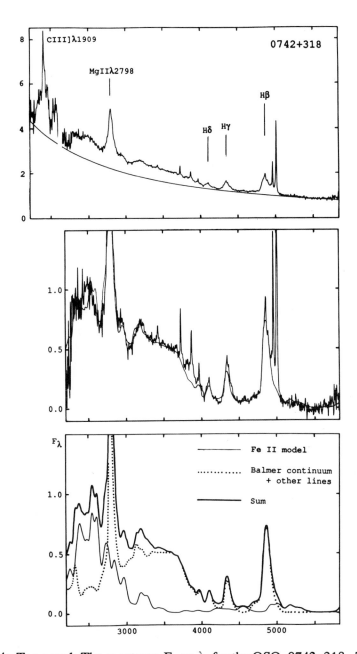

Figure 4: Top panel: The spectrum, F_λ vs.λ, for the QSO, 0742+318. The adopted power law continuum is shown. Second panel: the spectrum as above, but with power law subtracted, showing the complete fitted model for Balmer continuum and the broad lines. Third panel: the model spectrum, showing the two components - Fe II emission, and other broad lines plus the Balmer continuum. The narrow line component is not included in the model.

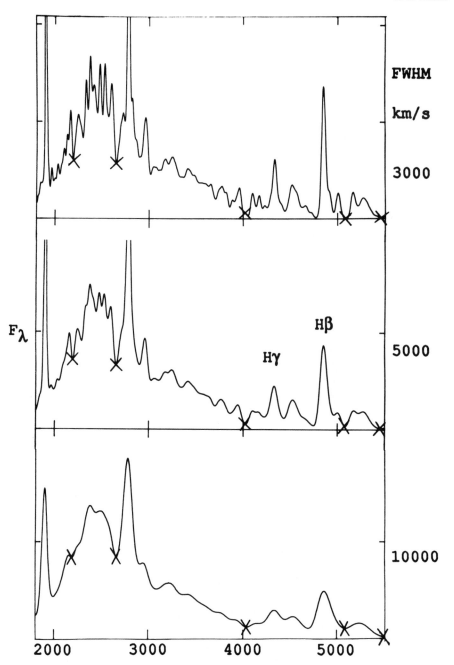

Figure 5: A synthetic spectrum of the QSO 3C 273, showing the effects of broadening a Gaussian line profile. Adopted "continuum" points are shown. Note the apparent decrease in strength of the Fe II $\lambda\lambda$2200-2650, 4500-4680 and 5100-5500 features with increasing line width.

of the blending caused by the broad line widths and the presence of so many emission
lines, the strength of the continuum can easily be overestimated, and consequently the
strength of the Fe II emission seriously under-estimated. We have attempted to measure
the dependence of this effect on line width.

In Fig. 5 we show a synthetic BLR spectrum that matches the observed spectrum
of 3C 273. We show this with line widths of 3000, 5000, and 10000 km/s (Gaussian
FWHM), and measure the Fe II features by using the continuum points shown. These
are standard continuum points in several published works. The Fe II $\lambda\lambda4500$-4680 and
Fe II $\lambda\lambda2200$-2670 features can be underestimated by factors of more than 3 or 4 at the
broadest line widths. The dependence of *observed* intensity ratios Fe II $\lambda\lambda2200$-
2670/Mg II $\lambda2798$ and Fe II $\lambda\lambda4500$-4680/Hβ as a function of FWHM, for various
classes of broad line AGN is shown in Fig. 6. The curves show the expected

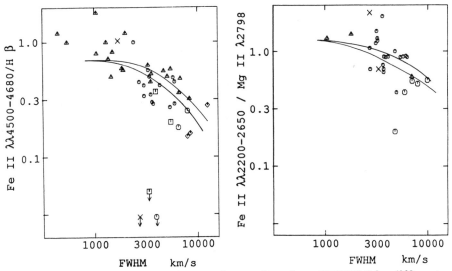

Figure 6: Measured (apparent) Fe II strengths as a function of FWHM for different
classes of broad line AGN, showing a comparison with the relations (smooth curves)
predicted from measurements like that shown in Figure 5, or from the broadening of the
observed spectrum of I Zw 1. This illustrates the need for modelling in determining true
line strengths. Symbols are: o (quasars with compact radio structure); O (quasars with
extended radio structure);× (radio-quiet QSOs);▵ (Seyfert 1 galaxies);◊ (broad line radio
galaxies); □ (quasars of uncertain radio structure).

dependence, based on measurements of spectra like those shown in Fig. 5 and on
broadened spectra of the Seyfert 1 galaxy, I Zw 1. We note:

(i) The data follow the predicted trend, and this can explain some observed differences
between different classes of AGN.

(ii) There is plenty of scatter. At least for Fe II λλ4500-4680 the scatter is real as determined by comparison of spectra with similar line widths but different apparent Fe II strengths (see Figures 6-8 in Osterbrock 1985).

(iii) Compared with the QSOs, the Seyfert galaxies seem to have statistically greater true Fe II λλ4500-4680/Hβ ratios that may depend on line width (see Gaskell 1985).

Boroson, Oke, and Persson (1985) find that QSOs with strong forbidden (narrow) emission lines in their surrounding nebulosity have weaker optical Fe II lines and broader permitted lines in the nuclear spectrum. Could the apparent Fe II strength simply be a function of line width? Boroson *et al.* use Fe II λλ5100-5500 as being less affected by line width than Fe II λλ4500-4680 but our results show the former is equally affected. This conclusion disagrees slightly with that of Boroson *et al.* (1985) perhaps because their curve does not extend reliably to line widths much beyond 5000 km/s. Boroson *et al.*'s correlation deserves further investigation.

Fe II can cause other problems because it is blended with many lines: *e.g.*, it makes Hβ look asymmetric. When corrected for blending, Hβ in Seyfert galaxies and low redshift quasars is quite symmetric in many cases, especially in the higher luminosity AGN (*e.g.*, Shuder 1984, de Robertis 1985).

In QSOs with very broad absorption line (BAL) spectra, Hartig and Baldwin (1986) find strong evidence for Fe II λ1781 emission, together with Al III and Fe III UV34 in the region of λ1909 for some BAL spectra. This feature is thought to be C III] in most non-BAL QSOs. The presence of Fe III and other differences between BAL and non-BAL QSOs indicate new ways of further investigating the conditions under which Fe II emission arises in QSOs.

Some statistical properties of Fe II emission in large samples of AGN, including the luminous QSOs, are now being studied (see Collin-Souffrin *et al.* 1986, and Joly, this colloquium).

4. SOME UNUSUAL FE II SPECTRA

PG1700+518: Wamoler (1986) derives enormous Fe II strength (Fe II/L α ~ 26) by deriving a very low continuum, based on the assumption that the strong, broad (BAL) MgII λ2798 absorption is due to material completely obscuring the continuum source. The resulting spectral distribution of blended Fe II emission is inconsistent with our simple models (Wills *et al.* 1985); these predict very little emission below λ = 2000 Å, and rather greater contrast in the Fe II features. Either there is an additional unexplained emission component or the Mg II cloud does not really cover the continuum source.

PHL 1092 (Bergeron and Kunth 1980, and Figure 3b): This relatively narrow-lined QSO is interesting because the spectrum shows especially strong Fe II, very weak Hβ, and also a lot of Fe I. There may also be Fe III lines.

NGC 1068 (Snijders, Netzer, and Boksenberg 1986): Weak, broad Fe II in the UV (IUE) spectrum was one of the first indications that an active broad line object was hidden in the nucleus. Other evidence is a broad line spectrum in the optical seen only in polarized light, which is normally overwhelmed by a very strong narrow line spectrum (Antonucci and Miller 1986).

3C 232: Grandi and Phillips (1978) first drew attention to the strong λ2950 emission, and Netzer (1980) was able to explain it as Fe II. We are still not really sure of its identification because of its great strength and the unusual shape of the optical emission near 4570 Å. There are quite strong features at the expected wavelengths of blended Fe I, and some Fe II features seen in many other QSOs are very weak in this spectrum.

QSO 0059-2735 : Hazard, McMahon, Webb and Morton (1986) have discovered a very unusual BAL QSO having very narrow Mg II and Fe II absorption at the emission line redshift. The Mg II line absorbs all the flux near 2798 Å. They suggest that this absorption is caused by emission line clouds in the line of sight and that further study of the absorption system could lead us to better understanding of the Fe II emitting gas. Other absorption lines represent a wide range in ionization, and some are identified with Fe III (UV34).

5. CONCLUDING REMARKS

- Fe II appears universally present in broad line AGN.
- Large variations are observed in the intensity ratio of optical to UV Fe II lines from object to object. This is almost certainly the result of different optical depths.
- The strength of Fe II emission equals or exceeds that of Lα, the strongest (relatively) isolated line in AGN broad line spectra. This, together with the great strength of Fe II emission compared with Mg II, is a serious problem in understanding QSOs' spectra.
- Modelling of the Fe II spectrum is needed to make correct measurements of the strengths of Fe II lines, and to investigate the strengths and velocity profiles of most other lines.
- Modelling the Fe II spectrum is also needed to better estimate the Balmer continuum and other continuum in the spectral region 2000 - 4000 Å.

In our own galaxy, Fe II emission is usually associated with dust, presumably because it indicates large optical depths of low-ionization material, *e.g.*, in T Tauri stars, Be stars, and Eta Carina. This is probably also true in AGN, where there is recent evidence for an emission bump near 3 μm suggesting hot (T ~ 1000 K) dust close to the BLR in many QSOs (Wills 1986).

I thank my collaborators in the study of Fe II emission: H. Netzer, D. Wills and W. Wamsteker, and D. Wills for comments on the manuscript. I also thank C. Hazard, S. Collin-Souffrin, M. Joly and D. Alloin, and E.J. Wampler for sending preprints. I thank the organizers of this meeting for their help and I thank them and the IAU for generous financial assistance. Our work on Fe II emission is supported by U.S. National Science Foundation (grant AST-8215477).

REFERENCES

Adams, T.F., and Weedman, D.W. 1972, *Ap.J.(Letters)*, **173**, L109.
Antonucci, R.R.J., and Miller, J.S. 1985, *Ap.J.*, **297**, 621.
Baldwin, J.A. 1975, *Ap.J.(Letters)*, **196**, L91.
Bergeron, J., and Kunth, D. 1980, *Astron.Ap.*, **85**, L11.
Bergeron, J., and Kunth, D. 1984, *M.N.R.A.S.*, **207**, 263.
Boksenberg, A., and Netzer, H. 1977, *Ap.J.*, **212**, 37.
Boroson, T.A., Persson, S.E., and Oke, J.B. 1985, *Ap.J.*, **293**, 120.
Collin-Souffrin, S., Joly, M., Pequignot, D., and Dumont, S. 1986, *Astron.Ap.*, in press.
de Robertis, M. 1985, *Ap.J.*, **289**, 67.
DeVeny, J.B., and Lynds, C.R. 1969, *P.A.S.P.*, **81**, 535.
Fairall, A.P. 1968, *P.A.S.P.*, **80**, 235.
Gaskell, C.M. 1981, Ph.D. thesis, University of California, Santa Cruz.
Gaskell, C.M. 1985, *Ap.J.*, **291**, 112.
Grandi, S.A. 1981, *Ap.J.*, **251**, 451.
Grandi, S.A., and Phillips, M.M. 1978, *Ap.J.*, **220**, 426.
Greenstein, J.L., and Schmidt, M. 1964, *Ap.J.*, **140**, 1.
Hartig, G.F., and Baldwin, J.A. 1986, *Ap.J.*, **302**, 64.
Hazard, C., McMahon, R., Webb, J., and Morton, D.C. 1986, preprint.
Netzer, H. 1980, *Ap.J.*, **236**, 406.
Netzer, H., and Wills, B.J. 1983, *Ap.J.*, **275**, 445.
Netzer, H., Wamsteker, W., Wills, B.J., and Wills, D. 1985, *Ap.J.*, **292**, 143.
Oke, J.B., and Shields, G.A. 1976, *Ap.J.* **207**, 713.
Osterbrock, D.E. 1977, *Ap.J.*, **215**, 733.

Osterbrock, D.E. 1985, in *Proc. Seventh Santa Cruz Astrophysics Workshop, Astrophysics of Active Galaxies and Quasi-Stellar Objects,* 16-27 July 1984, Santa Cruz, ed. J.S. Miller, University Science Books, CA.

Phillips, M.M. 1977, *Ap.J.* **215**, 746.

Phillips, M.M. 1978a, *Ap.J.Suppl.Ser.*, **38**, 187.

Phillips, M.M. 1978b, *Ap.J.*, **226**, 736.

Puetter, R.C., Burbidge, E.M., Smith, H.E., and Stein, W.A. 1982, *Ap.J.*, **257**, 487.

Sargent, W.L.W. 1968, *Ap.J.(Letters)*, **152**, L31.

Shuder, J.M. 1984, *Ap.J.* **280**, 491.

Snijders, M.A.J, Netzer, H., and Boksenberg, A. 1986, preprint.

Snijders, M.A.J., Boksenberg, A., Haskell, J.D.J., Fosbury, R.A.E. and Penston, M.V. 1980, *Proc. Second European IUE Conference,* April 1980, Tübingen, West Germany.

Uomoto, A.K. 1981, Ph.D. dissertation, University of Texas.

Wampler, E.J. 1986, *Astron.Ap.*, **161**, 223.

Wampler, E.J., and Oke, J.B. 1967, *Ap.J.*, **148**, 695.

Wills, B.J. 1983, *Proc. 24th Liège International Astrophysics Colloquium, "Quasars and Gravitational Lenses",* 21-24 June, 1983, Liège, ed. J.P.Swings, Institut d'Astrophysique, Université de Liège, p. 458.

Wills, B.J. 1986, *Proc. IAU Symposium No. 124, Observational Cosmology,* held 25-30 August, 1986, Beijing, China. eds. G.R. Burbidge, A. Hewitt and L.Z. Fang, and *Ap. J.*, submitted.

Wills, B.J., Netzer, H., Uomoto, A.K., and Wills, D. 1980a, *Ap.J.*, **237**, 319.

Wills, B.J., Netzer, H., and Wills, D. 1980b, *Ap.J.(Letters)*, **242**, L1.

Wills, B.J., Netzer, H., and Wills, D. 1985, *Ap.J.*, **288**, 94.

FeII LINES IN THE ULTRAVIOLET SPECTRA OF TYPE I SUPERNOVAE*

D. Branch, K.L. Venkatakrishna
University of Oklahoma, Norman, USA

ABSTRACT. A synthetic-spectrum technique applied previously to the interpretation of the optical spectra of Type I supernovae has been extended to the maximum-light near-ultraviolet spectrum. Spectral features in the interval 2750-3450 A are suggested to be complex blends of P Cygni profiles produced by nunerous FeII and CoII lines. This interpretation of the spectrum provides further evidence for radial composition mixing in classical Type I super-novae (Type Ia), and conflicts with suggestions that explosions of Wolf Rayet stars are responsible for the recently emphasized supernova subclass, Type Ib.

* The complete text of this article has appeared in The Astrophysical Journal Letters, Volume 306, L21 (1986).

R. Viotti et al. (eds.), Physics of Formation of FeII Lines Outside LTE, 177.

INFRARED LINES OF [FeII] FROM SUPERNOVA REMNANTS

E. Oliva
Osservatorio Astrofisico
di Arcetri
Largo E. Fermi 5
I- 50125 Firenze
Italy

A.F.M. Moorwood
I.J. Danziger
European Southern Observatory
Karl-Schwarzschild-Str. 2
D-8046 Garching bei München
F.R.G.

ABSTRACT. We present preliminary results of observations of IR [FeII]
lines in the direction of SNR. In all the spectra the [FeII] 1.644μm
line is by far the brightest feature in the observed spectrum and its
intensity is comparable to that of Hβ. Reddening and electron density
are estimated for the best measured objects. Implications on the Fe
abundance are briefly discussed.

1. OBSERVATIONS

The galactic remnants Kepler, RCW 103 and Puppis A plus N63 A and N49
in the LMC have been observed with the new ESO infrared spectrometer
IRSPEC (Moorwood et al., 1986) mounted on the ESO 3.6 m telescope
during November 1985 and June 1986. [FeII] emission was first searched
for in the brightest 1.644μm line. Other [FeII] lines of interest
(plus HI and H_2 lines in some cases) were then measured on the peak
position. Possible contamination of the 1.644μm line due to [SiI]
1.645μm appears to be negligible given the upper limits which we can
place on the [SiI] 1.607μm line.

2. RESULTS

The [FeII] (1.644μm) line was detected on all the observed remnants
with an intrinsic brightness comparable to Hβ. Observations of this
line can thus be a powerful means of detecting and studying highly
reddened SNR. RCW 103 and Kepler have been studied in most detail and
various observed line ratios of interest are summarized in table 1.
The [FeII] lines are by far the brightest features in our spectra and
only on a very bright spot in RCW 103 could we detect the Brγ hydrogen
recombination line (10 → 7, 2.165μm) at about 2% of the [FeII] 1.644μm
flux. This ratio is consistent with the value I(1.644μm)/Hβ ~ 1
estimated by comparing our data with calibrated optical spectra
(Leibowitz and Danziger, 1983; Danziger and Leibowitz, 1985).

R. Viotti et al. (eds.), Physics of Formation of FeII Lines Outside LTE, 179–181.

Table 1.   Observed Line Ratios

	RCW 103	Kepler
[FeII](1.644/1.599)	15	3.8
[FeII](1.644/1.265)	1.4	>.8
[FeII](4.89/1.64)	<.8	-
[FeII](1.64)/HI(2.17)	48	-
[FeII](1.64)/H$_2$S(1)(2.12)	17	>5

Useful information on the extinction can be obtained from the
I(1.644µm)/I(1.256µm) line intensity ratio which, according to the
latest computations, has an intrinsic value of .74 (Nussbaumer, this
conference). Measurements on RCW 103 give an optical extinction (A_v =
4.1±.7) in good agreement with that estimated from the Balmer
decrement (Leibowitz and Danziger, 1983).

The electron densities estimated from the I(1.644µm)/I(1.599µm)
line intensity ratios are: 3000 < n_e < 8000 for RCW 103 and n_e > 10000
for Kepler, adopting conservative criteria on the error estimates.
These are consistent with the values suggested by the [SII] 6717/6730
line ratios (Leibowitz and Danziger, 1983).

The gas temperature can, in principle, be deduced from the
I(4.889µm)/I(1.644µm) ratio. Unfortunately, the first line lies in a
spectral region where the strong atmospheric emission prevents good
S/N measurements. In the case of RCW 103 the measured upper limit on
this line ratio implies an electron temperature larger than 1600 K. A
more significant temperature determination can be obtained from the
I(0.862µm)/I(1.644µm) ratio. However, the first line is not accessible
to IR spectrometers and, at present, no reliable measurement of this
ratio exists.

3.   IMPLICATIONS ON Fe ABUNDANCE

The Fe^+/H^+ relative abundance can be estimated under the following
assumptions:

-   the hydrogen and [FeII] lines are emitted from the same region
-   the level population of hydrogen is that of a pure recombination
    spectrum (Menzel case B).

Under these assumptions we estimate

$$n(Fe^+)/n(H^+) \sim 5 \cdot 10^{-5}$$

from the [FeII](1.644)/Brγ ratio observed on RCW 103 suggesting, at first sight, an overabundance of iron in this remnant. However, the corrections one must apply to this number to deduce the Fe relative abundance can be very large, since stratification effects are probably significant. In particular, we note that in the region where FeII is formed, a large fraction of the hydrogen must be neutral and, therefore, not seen in the hydrogen recombination lines.

To our knowledge, the only theoretical model of a shock front including the [FeII] infrared lines is that of McKee, Chernoff and Hollenbach (1984) which, for a shock front propagating through a medium of density 100 cm^{-3} and where the iron abundance is 0.5 × solar, predicts a [FeII] 1.644μm line 4 times brighter than Hβ. Although this model is not necessarily representative of physical conditions inside a SNR, it suggests that the great strength of the IR [FeII] lines can be as sensitively dependent on the physical condition in the gas as on the abundance of iron.

REFERENCES

Leibowitz, E.M., Danziger, I.J.: 1983, M.N.R.A.S., **204**, 273.
Danziger, I.J., Leibowitz, E.M.: 1985, M.N.R.A.S., **216**, 365.
McKee, C.F., Chernoff, D.F., Hollenbach, D.J.: 1984, Galactic and
      Extragalactic IR Spectroscopy, Astrophysics and Space Science
      Library, **108**, 103, Kessler, M.F. and Phillips, J.P. eds.
      (D. Reidel Publ. Co.).
Moorwood, A.F.M. et al.: 1986, The Messenger, **44**, 19.

SESSION 3

THEORY OF LINE FORMATION. MODELS

THE NLTE FORMATION OF IRON LINES IN THE SOLAR PHOTOSPHERE

Robert J. Rutten
Sterrewacht "Sonnenborgh"
Zonnenburg 2
3512 NL Utrecht
The Netherlands

ABSTRACT. Solar iron lines provide extremely important diagnostics of the solar photosphere, including its deviations from LTE line formation in general. The NLTE effects in Fe I, Fe II and many other spectra are closely related because they are mainly determined by the amount of near-ultraviolet radiation present in the upper photosphere. The long-standing result that the ultraviolet is highly suprathermal, causing large overionization in minority species such as Fe I and overexcitation in majority species such as Fe II, is now challenged by Avrett on the basis of the ultraviolet blocking predicted from Kurucz's line compilations. Most of these blocking lines are from Fe I and Fe II, however, and since they set the temperature structure of the upper photosphere, the choice between LTE and NLTE Fe line formation can be self-fulfilling. The sun itself fulfills both choices all over its surface, according to Nordlund's simulation of the granulation, and the choice contributes also to Ayres' temperature bifurcation scenario. This issue is important in spatially-averaged atmospheric modeling as well as in modeling fine structure, for the sun and for other cool stars.

1. INTRODUCTION

1.1 IRON AS PROVIDER OF PHOTOSPHERIC DIAGNOSTICS

The literature on solar iron lines is quite large. It is roughly divided into three categories:

- stellar studies using solar lines to calibrate abundance determination and oscillator strengths;

- solar studies of (partially) resolved inhomogeneities, using a few specifically-chosen "magnetic" lines (large Landé factor g) or "velocity" lines (g=0);

- solar studies of unresolved inhomogeneities, using many iron lines

185

R. Viotti et al. (eds.), Physics of Formation of FeII Lines Outside LTE, 185–210.
© *1988 by D. Reidel Publishing Company.*

simultaneously to obtain spatially-averaged signatures of photospheric fine structure.

There are recent advances for each category. In the abundance work,
the solar iron spectrum remains the archetypical example of stellar
photospheric line formation. The advent of reliable Fe I oscillator
strengths (Blackwell et al. 1982 and references therein) represents
the elimination of an important source of error that has caused much
confusion in the past. Recent NLTE studies of this type are Steenbock
(1985) and Saxner (1984).

The second category, studying inhomogeneities, constitutes most
of the solar iron literature. It has now started to profit from the
space age (NRL HRTS spectrograph, e.g. Dere et al. 1984). Groundbased
observation generally suffers from much poorer resolution, with the
exception of some data sets taken at Sacramento Peak (e.g. Zwaan et
al. 1985). Basic fine-structure entities as granules and fluxtubes
will only be fully resolved, however, when a sufficiently large
telescope is put into orbit.

In the meantime, the third category of solar iron studies
presents the most important advances. Dravins (e.g. 1982) has
pioneered the use of precise wavelength shifts and profile bisectors
of optical iron lines as gauges of the spatially-averaged effects of
the solar granulation. Solanki and Stenflo (1984) have pioneered the
study of solar fluxtubes from the same iron lines by extracting
statistical properties from polarization spectra with very high
spectral resolution obtained with the Kitt Peak FTS employed as a
polarimeter. Since both of these techniques use spatially-averaged
data and resolve only the solar center-to-limb variation, they supply
promising diagnostics of stellar granules and stellar fluxtubes,
respectively; in both cases, their pioneers have started stellar
observations (Dravins and Lind 1984; Mathys and Stenflo 1986).

The reason why all these studies concentrate on iron lines is
simply that iron provides the overwhelming majority of the diagnostics
in the optical spectrum. In a compilation of unblended optical solar
lines (Rutten and Van der Zalm 1984), Fe I provides 354 lines out of
the total of 745, and Fe II is with 22 lines the only dominant stage
of ionization present with more than 10 lines. In particular, Harvey
(1973) and Sistla and Harvey (1970) list 18 good magnetic lines and 18
good velocity lines from Fe I respectively, more than all other
species provide together. In addition, the iron lines are not spoiled
by isotope and hyperfine splitting, in contrast to most other species;
they are truly clean.

While Fe I supplies most of the useful lines, the majority of the
iron atoms in the photosphere are ionized. One should therefore always
include Fe II lines in any analysis using Fe I lines, to obtain
complementary diagnostics and to avoid errors due to the large
sensitivity of the Fe I opacities to the ionization equilibrium. This
strategy has been adopted by Blackwell et al. (1980) in abundance
determination, by Dravins and Larsson (1984) in spatially-averaged
granulation analysis, and by Solanki and Stenflo (1985) in spatially-
averaged fluxtube polarimetry.

1.2 IRON AS A SHAPER OF THE PHOTOSPHERE

The enormous line densities of the Fe I and Fe II spectra not only result in their diagnostic prominence in the optical but also in severe line crowding in the ultraviolet. It seems likely that most of the "missing opacity" between 2000 and 3000 Å is due to iron (Rudkjøbing 1986; Kurucz, these proceedings), and thus iron is the major provider of the ultraviolet line blocking which affects the temperature structure of the upper photosphere. In addition, the combination of high abundance and rather low ionization energy makes iron the major contributor to the electron density in the temperature minimum region where hydrogen is neutral (see Vernazza et al. 1981, Fig. 47. Henceforth, I abbreviate this important and informative paper to VALIII). Finally, the Fe I photoionization edges from a^5D at 1575 Å and a^5F at 1768 Å contribute significantly to the continuous opacity near these wavelengths (VALIII, Fig. 36) and cause significant radiative cooling in the upper photosphere (VALIII, Fig. 49 [1]).

1.3 IRON AND NLTE DEPARTURES IN THE PHOTOSPHERE

In none of the three classes of solar iron research listed above has much attention been paid to departures from LTE. Usually, LTE is simply adopted. For many diagnostic applications, the question simply doesn't matter; for example, a g=0 velocity line will serve to measure Doppler shifts in helioseismology irrespective of its formation details. However, the details including departures from LTE in the excitation and ionization balances do matter in abundance deter- mination, in diagnostic usage where the height of formation or the geometry is important, and in studies that require evaluation of radiation losses to determine energy budgets.

This review is limited to the LTE-NLTE issue only, and only for the photosphere. The chromosphere, with its major and interesting NLTE problems (fluorescence, optically thin conditions etc.) is treated by Carole Jordan in these proceedings. Here I only discuss the general departures in the ionization and excitation balances, not details of specific lines. Even within these restrictions, there is good ground for a review at this time because the old (and heated) debate "LTE or NLTE" is flaring up again.

The old debate was never settled but has developed into an amazing schism between radiative transfer theoreticians and stellar abundance determiners. The first group, after completing the full theory of NLTE radiative transfer in the sixties and putting it into textbooks in the seventies (Jefferies 1968, Mihalas 1970, Athay 1972, Ivanov 1973), moved on to more complicated problems (partial frequency redistribution, expanding atmospheres, etc.). The second group simply continued Unsöld's trick of adopting LTE, sometimes "proving" its

---

[1] In VALIII the Fe II ground level is incorrectly adopted as parent term for ionization from a^5F. The actual parent is a^4F and the corresponding edges are near 1700 Å instead of 1768 Å.

validity from the good consistency of the results.

This split is most distinct in Fe I analyses. Lites' (1972) comprehensive study showed that ultraviolet overionization leads to large Fe I opacity deficits, which affect all strong and medium strong Fe I lines (section 2.3 below). On the other hand, Holweger (1967) built a very satisfactory model photosphere precisely from the medium-strong Fe I lines adopting LTE opacities. Although these two descriptions differ very much, both reproduce the observed iron lines quite well. Let me illustrate this split with two quotations. Athay wrote in 1972 (p. 181):

> "...somewhat to the surprise and consternation of those who have criticized the LTE analysis of Fraunhofer lines, Holweger does find, in fact, that a single model gives an acceptable representation of both the continuum and the lines..."

Holweger wrote in 1979:

> "...Among the problems that need further study are deviations from LTE. Unfortunately, these are easily arising in the computer if important collisional processes are neglected, or if radiative rates are not realistic. In cool stars, collisional excitation by hydrogen atoms is generally neglected; data on cross-sections are largely missing. However, in the Sun, hydrogen atoms outnumber the free electrons by a factor of 10000. The UV radiation field is complicated by a vast number of absorption lines..."

Holweger's model was succeeded by the very similar HOLMUL model (Holweger and Müller 1974) which is invariably the choice of LTE abundance determiners simply because it results in smaller spread than the elaborate semi-empirical NLTE models built in the meantime, most notably the HSRA (Gingerich et al. 1971) and the VALII (Vernazza et al. 1976) and VALIII models. These models are much cooler in the upper photosphere (Fig. 5), and it turns out that the LTE-or-NLTE formation issue is closely connected to the hot-or-cool upper photosphere issue (section 3.1 below).

There are two important recent papers concerning this issue, unfortunately both hidden in workshop proceedings. In the first, Avrett (1985) recovers Holweger's model by including ultraviolet line blocking in detail in a very comprehensive NLTE fit of the solar continua with a hot model (section 3.2 below). This indicates that photospheric overionization indeed arose artificially in the computer before, and that the actual solar photosphere does obey LTE. On the other hand, Nordlund (1984) shows from a very comprehensive simulation of the solar granulation that hot and cool photospheres are simultaneously present all over the sun (section 4.1 below). Nordlund finds that the spatial and temporal averaging over the granulation is so nonlinear due to the steep temperature sensitivity of the ionization equilibria that even the classical iron abundance determined from the standard models is quite wrong.

Thus, the NLTE-versus-LTE issue crops up again, now on the basis of comprehensive computer modeling. The issue is both controversial and important, being essential not only to the formation of Fe I but

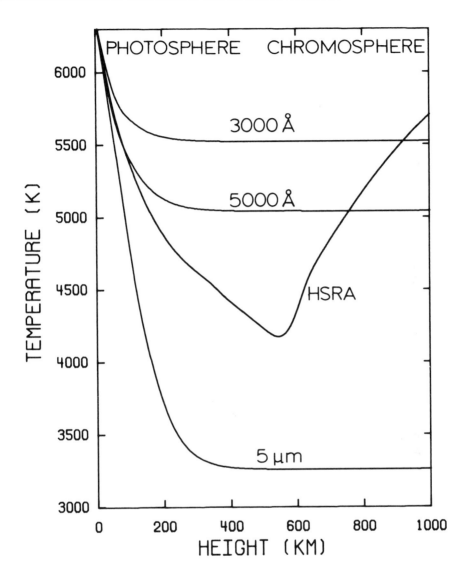

Fig. 1. Radiation fields in the solar atmosphere. The curve marked HSRA is the electron temperature of the HSRA model and represents the Planckian radiation field B_ν. The other curves represent the angle-averaged monochromatic continuous intensity $J_\nu(h)$ in the form of radiation temperatures, for the indicated wavelengths. The transition between photosphere and chromosphere is defined here as occurring at the temperature minimum.

also to the excitation balance in Fe II and the formation of lines of many other species. I will therefore mainly discuss the Fe I - Fe II ionization balance, and since Fe I lines are more sensitive to it than Fe II lines, I will pay much attention to Fe I line formation. I wish to apologize for this bad behaviour in a meeting devoted exclusively to Fe II!

2. NLTE IN PHOTOSPHERIC IRON LINES

2.1 NLTE MECHANISMS

Since this meeting brings together astrophysicists from widely different fields, it seems appropriate to outline the main points of photospheric NLTE departures even though detailed descriptions are available (e.g. Athay 1972, the VAL papers, Thomas 1983).

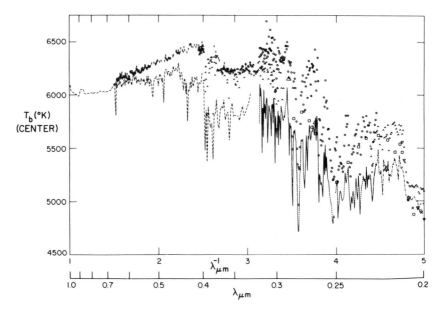

Fig. 2. Solar disk-center brightness temperatures against wavelength, from VALII. The brightness temperature corresponds with the observed disk-center monochromatic intensity. The upper values are for the highest continuum points between lines, the lower values are spectral averages including lines. From Vernazza et al. (1976), with permission from The Astrophysical Journal.

The reasons that most photospheric NLTE phenomena are set in the ultraviolet are that there the angle-averaged intensity J_ν exceeds the local Planck function B_ν, and that there the emergent radiation is

both hot and energetic. Figs. 1 and 2 illustrate these points following Avrett (1985). Fig. 1 shows J_ν and B_ν in the form of radiation
temperatures at three wavelengths for which the photon escape depth is
about the same: the disk-center Eddington-Barbier depth $\tau_\nu = 1$ is near
h = 0 km in each case. The surface value of J_ν is about 0.3 $B_\nu(\tau_\nu=1)$
at each wavelength, but because the temperature sensitivity of the
Planck function decreases with wavelength, J_ν drops below B_ν toward
the surface at long wavelengths whereas it exceeds B_ν at shorter
wavelengths. Note that this difference occurs here in LTE ($S_\nu \approx B_\nu$ at
$\tau_\nu = 1$; it also occurs in radiative equilibrium which requires only
the integral equality $\int \kappa_\nu S_\nu d\nu = \int \kappa_\nu J_\nu d\nu$ to hold at all heights). If
scattering contributes significantly to the continuous opacity, as it
does in the Fe I ionization edges below 2000 Å, J_ν decouples from B_ν
much deeper in the atmosphere than the depth where $\tau_\nu = 1$ (VALIII
Fig. 36).

Fig. 2 (taken from VALII) shows the observed brightness temperature at disk center. It is quite high in the near ultraviolet,
reaching 6500 K at 3000 Å between spectral lines. This graph is in
shape the reverse of the wavelength dependence of the monochromatic
opacity: a high brightness temperature corresponds to deep photon
escape. Note, however, that the average brightness temperature is
appreciably lower between 2000 and 4000 Å due to line crowding, and
that even the observed maxima are far below the continua computed from
model atmospheres if only the known bound-free and free-free opacities
are included. Nevertheless, the near-ultraviolet radiation escapes
rather deep, is therefore hot, and since it has a steep Planck
function, it has $J_\nu > B_\nu$ in the temperature minimum by about 1000 K
(Fig. 1).

It is also energetic, 4 eV per photon at 3000 Å. Which are the
most important 4 eV radiative transitions? For Fe I, photoionizations
from the intermediate levels; for Fe II, being the dominant stage of
ionization, the resonance lines. For both types of transition, the
photon destruction probability per extinction process ε is small, so
that the source function $S_\nu = (1 - \varepsilon) J_\nu + \varepsilon B_\nu$ follows J_ν rather
than B_ν, as shown in fig. 36 of VALIII for the a^5F polarization
edge. The ultraviolet excess $J_\nu - B_\nu$ therefore results in overionization from the intermediate Fe I levels at the temperature
minimum, and in overexcitation in the wings of the Fe II resonance
lines formed there. Note that scattering in the <u>cores</u> of strong lines
always results in $J_\nu < B_\nu$ even in the ultraviolet, because the
additional line opacity makes the gradient $dB_\nu/d\tau_\nu$ much shallower than
in the continuum. The cores of strong lines follow the textbook
example of a homogeneous layer which has $S_\nu = \varepsilon^{\frac{1}{2}} B_\nu$ at its surface.

2.2 NLTE DESCRIPTION

Another preamble which seems useful here although it is textbook
matter is to define the description of departures from LTE to be used
here. In fact, the definition of LTE itself isn't always clear; for
example, Bowers and Deeming (1984, p. 81) define weak, moderate and
strong versions. The definition used here is Ivanov's (1973, p. 6):

LTE requires that the matter is in thermodynamic equilibrium with the local kinetic temperature (the Maxwell, Boltzmann and Saha distributions are valid), while the radiation is not. It then <u>follows</u> that $S_\nu^\ell = B_\nu$ (ib., p. 46) while I_ν and J_ν can depart from $\overline{B_\nu}$ and the net flux from zero.

The usual way to measure departures from LTE is by specifying departure coefficients relative to the Saha-Boltzmann populations. But take care: their normalization differs among authors, and not all authors are aware that this can affect opacities. The original definition (Menzel and Cillié 1937) is usually followed (e.g. Jefferies 1968, eq. 6.3; VALIII eq. 15):

$$n_i = b_i \left(\frac{h^2}{2\pi\, m_e kT}\right)^{3/2} \frac{g_i}{2\, U^*_k}\, e^{(E_k - E_i)/kT}\, n_k n_e \tag{1}$$

or:

$$b_i = \frac{n_i/n_i^*}{n_k/n_k^*}$$

where n is the actual population and n^* the population computed from the Saha and Boltzmann distributions, respectively for a level i and the next higher ion k. This definition differs from what it is often taken to be:

$$\beta_i = \left(n_i/n_i^*\right) = \left(n_k/n_k^*\right) b_i \tag{2}$$

which was introduced by Wijbenga and Zwaan (1972). This coefficient relates the population departure of a level to the total amount of the element rather than the population of the next higher ionization stage. When the latter is out of LTE the two definitions can differ appreciably. For example, photospheric hydrogen has (VALIII p. 663):

$$\beta_1 = 1/b_k = 1 \quad \text{while} \quad b_1 = {}^1/\beta_k \approx \left(^B/_J\right)_{LyC} \approx 0.3$$

The departure coefficients are readily converted into corresponding temperatures (Wijbenga and Zwaan 1972). In particular, writing $\theta = 5040/T$, the excitation temperature T_{exc} for a line between lower level ℓ and upper level u is given by:

$$\theta_{exc} = \theta_e - (E_u - E_\ell)^{-1} \log(\beta_u/\beta_\ell) \tag{3}$$

It is the equivalent of the line source function; writing

$$\beta_u/\beta_\ell = 1 + \Delta\beta \quad \text{and} \quad h\nu/kT = \delta$$

one finds for the Wien approximation ($\delta > 1$):

$$S_\nu^\ell = B_\nu(T_{exc}) \approx (1 + \Delta\beta)\, B_\nu(T_e) = \left(^{\beta_u}/\beta_\ell\right) B_\nu(T_e) \tag{4}$$

and for the Rayleigh-Jeans approximation ($\delta < 1$):

$$S_\nu^\ell = B_\nu(T_{exc}) \approx (1 + \Delta\beta/\delta)\, B_\nu(T_e) \tag{5}$$

Finally, the lower-level departure coefficient affects the line extinction coefficient:

$$\kappa_\ell = \frac{\pi e^2}{mc} \frac{H(a,v)}{\sqrt{\pi}\Delta v_D} g_\ell f_{\ell u} \beta_\ell n_\ell^* [1 - (\beta_u/\beta_\ell)e^{-\delta}] \simeq \beta_\ell \kappa_\ell^* \qquad (6)$$

Many authors equate NLTE conceptually to $S_\nu \neq B_\nu$ but neglect the opacity dependence on β_ℓ (e.g. Cowley 1970, Chapt. 2-6).

2.3 PUBLISHED NLTE MODELING OF Fe I AND Fe II

Lites' outstanding thesis (Lites 1972, 1973; summary in Athay and Lites 1972; tables in Lites and White 1973) remains the only Fe I NLTE analysis published in detail so far. Fig. 3 summarizes its main results in the form of departure coefficients and excitation temperatures. The solid curves in Fig. 3a are for the a^3F and z^5G^o levels respectively, which are representative for low-lying levels and levels of intermediate excitation, respectively. Lites found that all such Fe I levels share the drop to $\log (\beta) = -0.7$ at $h = 500$ km, which is the location of the temperature minimum. This drop is the result of ultra-violet overionization, mainly from rather high levels (~ 4 eV) due to the hot radiation near 3000 Å, but shared out over all Fe I levels through the strong radiative and collisional coupling that the line- and level-rich Fe I term diagram provides.

The result of this shared drop in β for the optical Fe I lines is that their <u>excitation</u> is thermal ($\beta_u \approx \beta_\ell$), but that their <u>opacity</u>, which scales with β_ℓ, is reduced by about 0.7 dex (abundanese jargon for 10-log units, see Allen 1976) at the heights where their cores are formed. This is a large effect! The line cores are shallower than in LTE since they are formed deeper in the atmosphere. However, the effect is much smaller for equivalent widths since these are not very sensitive to the higher layers; the drop results typically in only 0.1 dex reduction in equivalent width (Holweger 1973, Rutten and Zwaan 1983).

Above the temperature minimum, the departure coefficients rise sharply because the ionizing radiation does not follow the chromospheric temperature rise. The 4 eV (z^5G^o) departure coefficients drop away from the ground-level coefficient due to the photon losses in the resonance lines; these are strong enough that excitation departures only occur above the temperature minimum.

Lites did not include higher levels in his modeling, except for some trial computations. These show the same pattern of $\beta_u/\beta_\ell < 1$ at the heights where photon losses in the strongest downward transitions from these high levels occur, i.e. where they become optically thin, which is below the temperature minimum for these lines which are actually weak in the solar spectrum due to their small Boltzmann factor. Thus, the strongest optical lines at high excitation have source function deficits as well as opacity deficits compared to LTE estimates. These deficits compete in their effect on the observed line depth. The drop in the line source function produces deeper-than-LTE profiles, while the drop in the line opacity produces shallower-than-

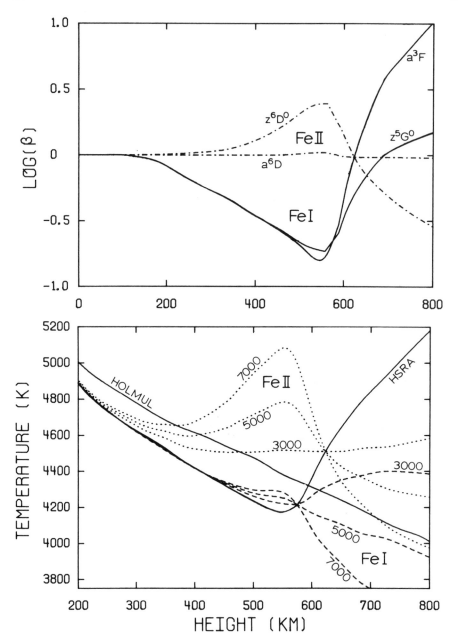

Fig. 3. Results of NLTE modeling of solar Fe I and Fe II, after Rutten
and Kostik (1982). Top: representative departure coefficients against
height, respectively for Fe I (solid) and Fe II (dot-dashed). The
even-parity curves represent typical lower-level departures (β_ℓ). and
the odd-parity curves typical upper-level departures (β_u).
Bottom: excitation temperatures against height, derived from the
departure coefficients in the top panel with eq. 3, for $\lambda = 3000$ Å,
5000 Å and 7000 Å and for Fe I (dashed) and Fe II (dotted),
respectively.

LTE profiles. The deficits may thus cancel, resulting in lines that look as if they were formed in LTE. This fortuitous cancellation of excitation and ionization departures is evident in Saxner's (1984) stellar model computations; it also seems confirmed by the actual photospheric high-excitation lines and their fitting by LTE abundance determiners (Rutten and Van der Zalm 1984). It is in conflict with the assertion (Jefferies 1968) that the high-excitation levels should have $\beta \approx 1$ through coupling to the reservoir of Fe II ground-level atoms, which have $\beta = 1$. If that were the case, the high-excitation Fe I lines would have $S_\nu^\ell > B_{\nu}$, since their lower level would still have the drop in β_ℓ shown by z^5G^o in Fig. 2a. The source function excess and the opacity deficit would then work together in producing lines much shallower than when computed in LTE. This is not seen; presumably, the collisional coupling to the continuum is not strong because the upper levels of the observable optical lines are yet about 1 eV from the continuum (Rutten and Zwaan 1983, Fig. 2), and the radiative coupling is small because most high levels ionize to excited Fe II parent terms.

The amount of collisional coupling between the levels isn't well known. Lites did not include neutral hydrogen collisions, but they should be taken into account as advocated by Holweger (see quote above). They were included by Steenbock (1985), but he has not yet published detailed rates. For small energy separations, these collisions must be as important as they are in producing collisional line broadening; and so one may expect that at any excitation energy all nearby levels will share their populations. This implies that a particular Fe I level will be in Boltzmann equilibrium with the low levels until the strongest downward transitions from the nearby levels become optically thin, further strengthening the general population sharing already found by Lites. The observed line-to-line differences should thus be small. This is indeed the case; although there is evidence for significant differences between specific multiplets (Blackwell et al. 1984), these are indeed very small, less than 0.1 dex. In any case, the basic NLTE effect, overionization by the suprathermal ultraviolet, is not sensitive to the amount of low-energy collisional coupling.

Let us now turn to Fe II. Fig. 3a specifies departure coefficients for the a^6D ground – level and the z^6D^o excited level, which is the upper level of the 13-line resonance multiplet near 2900 Å. The Fe II departure pattern is just the opposite of the Fe I pattern. The lower level population is in LTE; this holds for all the low levels in Fe II because they are all of even parity and only collisionally coupled, and Fe II is the dominant ionization stage. The upper level population peaks in the temperature minimum, and then drops in the chromosphere. The drop is simply due to photon losses in the resonance lines, but the peak has been controversial. Lites (1974) concluded that it is due to downward diffusion of photons out of the strong self-reversed emission cores he predicted for the resonance lines. Subsequently, Cram et al. (1980) concluded that the peak is due to the high value of J_ν near 2600 Å in the continuum (Fig. 2); it pumps the upper-level population via the wings of the resonance lines.

The presence of this peak is observationally required from the observed behaviour of the weak subordiante line at 3969.4 Å, which shares its z^6D^0 upper level with the resonance lines and which is located in the wing of the Ca II H line between the H core and H_ε. This line is seen in emission inside the solar limb, with extraordinary spatial intensity variations (Canfield and Stencel 1976; Rutten and Stencel 1980). It is even seen in emission at the center of the solar disk within small-scale structures (Cram et al. 1980). These emission features, which require $S_\nu > B_\nu$ locally, are probably photospheric in origin because their spatial distribution is not correlated with the chromospheric features seen in the adjacent core of the H-line. They are so clearly observable in this weak line because the added H-wing opacity pushes its height of formation to the peak in β_u and the corresponding peak in the line source function (Fig. 3b). Their spatial intensity variation is so large because they have ultraviolet temperature sensitivity although they are seen in the optical, just as in the familiar Zanstra mechanism in which the optical Balmer lines from planetary nebulae exhibit the ultraviolet radiation temperature of the central star.

Cram et al. (1980) showed that the same departure pattern holds for other subordinate lines sharing z^6D^0 upper levels (multiplets 1, 30 and 40). Similar patterns probably hold throughout the term diagram because Fe II typically has strong ultraviolet lines feeding odd-parity levels that are shared with optical transitions to the meta-stable even-parity levels. It follows that most optical Fe II lines will have suprathermal line source functions in the temperature minimum due to photospheric ultraviolet pumping. A direct consequence is that they should all turn into emission well inside the solar limb because the optical continua follow B_ν. This is indeed the case, as can be concluded from Pierce's (1968) compilation of solar limb emission lines. This list is not a list of the chromospheric spectrum, although its title says so – it is primarily a list of photospheric lines with suprathermal source functions near the temperature minimum (Rutten and Stencel 1980). Direct evidence from high-quality eclipse spectrograms has been given by Van Dessel (1975), who found that Fe II lines typically turn into emission already 1500 km inside the limb, while Fe I lines with their LTE source functions turn into emission much closer to the limb (Van Dessel 1974).

In conclusion, the overexcitation shown by the optical Fe II lines from the upper photosphere is set by the imbalance $J_\nu > B_\nu$ in the near ultraviolet. The Fe II overexcitation and the Fe I over-ionization are different manifestations of the same NLTE mechanism.

An important aspect of the equality of the upper-level population departures shared by the subordinate Fe II lines is that they result in enhanced effects toward longer wavelengths. This is illustrated by the excitation temperatures in Fig. 3b which are derived from the departure coefficients in Fig. 3a by eq. 3. The overexcitation set by J_ν below 3000 Å (which drops smoothly outward) results in a prominent peak in S_ν^ℓ at the temperature minimum for $\lambda = 7000$ Å. The peak is due to the decreasing temperature sensitivity of the Planck function. For example, in the Wien limit one can say that there is

source function equality according to eq. 4, but doubling S_ν by
$\beta_u/\beta_\ell = 2$ at the temperature minimum results sooner in emission for
lines formed there when the inward increase of the Planck function
toward the height of continuum formation is less steep. The effect is
even larger for the Rayleigh–Jeans limit due to the magnification
factor δ in eq. 5 (Mihalas 1978, p. 404). It implies that the best
observational test of Fe II NLTE is to study limb emission lines in
the infrared.

Optically thick emission is such a good NLTE diagnostic because
it is direct evidence of a suprathermal source function. In the case
of the Fe II resonance multiplet, however, the <u>absence</u> of intense
chromospheric emission in the line cores may be an important
diagnostic. This absence is surprising. At larger abundance than Mg
and with similar ionization energy, one would expect Fe II reversals
comparable to the prominent peaks in the Mg II h and k resonance lines
(Athay and Lites 1972). Although the profiles computed by Cram et al.
(1980) have negligible reversals, they were based on too small an iron
abundance and on the erroneous assumption of complete frequency redistribution. When monochromatic scattering is taken into account, the
line cores are coupled more closely to the chromospheric temperature
rise, just as in the case of the Mg II and Ca II lines. The absence of
chromospheric emission peaks thus remains unexplained; undoubtedly,
the complexity of the Fe II term diagram over the otherwise similar
Ca II term diagram plays a role.

3. IRON NLTE AND THE MEAN ATMOSPHERE

3.1 EMPIRICAL PLANE–PARALLEL MODELING FROM LINES

The NLTE results summarized above have large effects on optical Fe I
and Fe II lines formed near the temperature minimum. The stronger
photospheric Fe II lines have appreciable source function enhancements
(Fig. 3b); the optical Fe I lines, while having LTE source functions,
have large opacity deficits (Fig. 3a). How can it be that all these
lines can be reproduced quite well assuming LTE, as is the case in
numerous abundance-type analyses? This discrepancy was explained by
Rutten and Kostik (1982) by showing that the Holweger and HOLMUL
atmospheric models are "NLTE-masking": these departures from LTE are,
if present, implicitly compensated by the temperature structure of
these LTE models.

Fig. 4 schematically explains this masking by way of a thought
experiment for Fe I. Suppose that the NLTE modeling by Lites and by
Cram et al. using the HSRA is precisely correct. If one then neglects
these NLTE departures and constructs an empirical model from the
observed line-center brightness temperatures, a HOLMUL-like atmosphere
will automatically result. Neglecting the effect of β_ℓ on the
opacities (single arrows in Fig. 4) results in a spurious model height
scale, stretched where $\beta_\ell < 1$ and compressed where $\beta_\ell > 1$. Neglecting
the chromospheric source function departures (β_u/β_ℓ, double arrows)
requires that the model is a representative excitation temperature

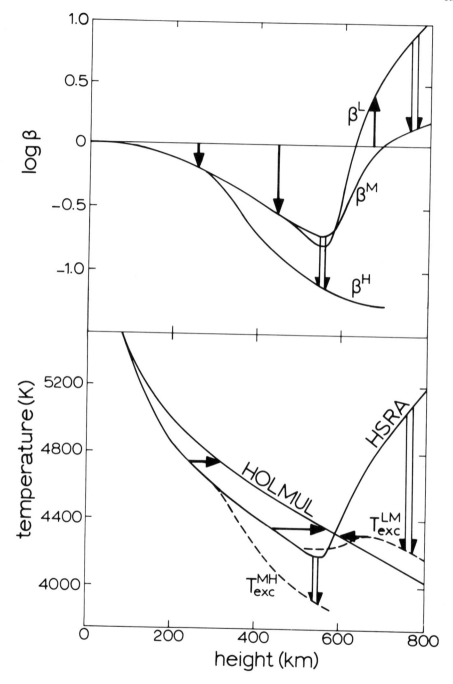

Fig.4. Schematic explanation of NLTE masking. The top panel shows
typical Fe I departure coefficients computed for the HSRA model,
respectively for low levels (L), intermediate levels (M) and high
levels (H). The corresponding excitation temperatures (eq. 3) are
shown in the bottom panel (dashed), together with the electron
temperatures of the HSRA and HOLMUL models (solid). The double arrows

in both panels demonstrate the effect of the NLTE departures on the line source functions, set by the ratio β^M/β^L for low-excitation lines (LM) and by the ratio β^H/β^M for high-excitation lines (MH), respectively. The single arrows in both panels indicate the NLTE shift in height of formation, set by β^L and β^M, respectively. For LM lines the combined effect of these arrows is that the HSRA and NLTE together produce the same line formation as HOLMUL and LTE together. The strongest MH lines, however, are deeper in the first case.

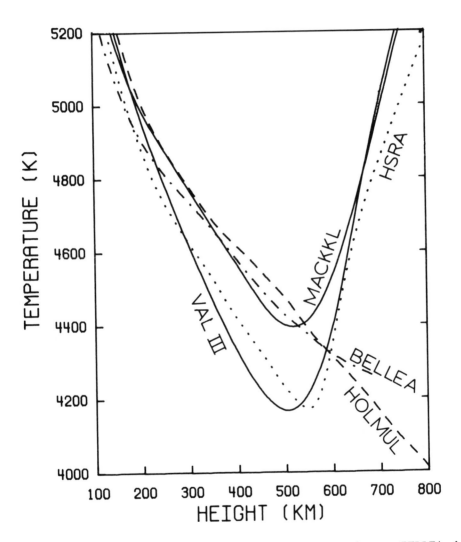

Fig. 5. Plane-parallel models of the solar atmosphere. BELLEA is a theoretical LTE radiative-equilibrium model, HOLMUL an empirical LTE model based on optical continua and lines, and HSRA, VALIII and MACKKL are semi-empirical models, based on NLTE fits of the solar continua throughout the spectrum.

instead of the kinetic temperature above the temperature minimum. The derived LTE model is hot in the upper photosphere to compensate for the opacity shifts and it follows the opacity-shifted Fe I excitation temperatures in the chromosphere. This NLTE masking works quite well for all optical Fe I lines. It also works for optical Fe II lines because their NLTE overexcitation is also set by the ultraviolet excess. A hot LTE model reproduces more or less the raised excitation temperatures shown in Fig. 3b, in this case without opacity shifts. It does not reproduce the observed emission features but one might argue that these are due to specific inhomogeneities anyhow.

Rutten and Kostik (1982) interpreted this explanation of the success of the HOLMUL model as a reconciliation with the HSRA which in roundabout fashion supports Lites' NLTE modeling. However, it is possible to reverse the sign and the arrows in this experiment, and to assume instead that the LTE modeling is precisely correct. One then concludes that the large NLTE departures found by Lites for Fe I and by Cram et al. for Fe II result artificially from their assumption of a too cool upper photosphere. If the actual upper photosphere is much hotter than the HSRA, the true ultraviolet overionization and over-excitation are small and the populations, Fe I opacities and Fe II source functions are close to LTE. The departures computed from the too cool model would be far too large, but just what is needed to reproduce the observed lines. The cool model would be "LTE-masking" in the photosphere.

In summary, we have an iron-line dilemma: one can model the observed Fe I and Fe II lines by either assuming (and believing) a HSRA-like cool model with a large excess of J_ν over B_ν in the ultraviolet, or by assuming (and believing) a HOLMUL-like hot model with a small ultraviolet excess and negligible departures from LTE up to the temperature minimum. The photospheric lines are well reproduced by each combination. Actual examples for many observed iron lines are given by Rutten and Kostik elsewhere in these proceedings.

3.2 EMPIRICAL PLANE-PARALLEL MODELING FROM CONTINUA

A similar quandary has recently appeared for the semi-empirical model atmospheres based on the observed solar continua. Traditionally, these had cool upper photospheres in order to fit the observed ultraviolet and infrared continua (HSRA, VALII, VALIII). However, Avrett (1985) has published results from a new model with a hot upper photosphere; the model itself has appeared very recently (Maltby et al. 1986, MACKKL model). It is shown in Fig. 5 together with the earlier models. The new model is similar in construction to the earlier VAL models. Its temperature-height relation was derived by fitting the observed continua with opacities computed from the statistical equilibrium and hydrostatic equilibrium equations taking full NLTE radiative transfer into account for all important opacity and electron sources. (The name of the computer code, PANDORA, seems aptly chosen).

The difference with the earlier models is that Kurucz's latest compilation with 17 million lines is used to predict the ultraviolet line blocking in a much more detailed manner than the opacity sampling functions employed before. The addition of all these lines results in increased effective ultraviolet opacity, thus in a higher height of formation of the emergent ultraviolet radiation. Fitting the observed ultraviolet brightness temperatures then requires a less steep temperature gradient, i.e. a hotter upper photosphere.

Avrett (1985) emphasizes that this new model unifies diverse diagnostics of the upper photosphere: the ultraviolet intensities, the spatially-averaged profiles of the inner wings of Ca II H and K, and the microwave continuum which shows the temperature minimum near 150 μm. They are all satisfactorily reproduced by the new model, in contrast to the VALII and VALIII models which predicted lower Ca II line-wing intensities than the empirical fit by Ayres and Linsky (1976) and lower microwave temperatures. The latter did agree with the older observations used in the VALII and VALIII modeling, but they are lower than the recent Swiss balloon data (Degiacomi et al. 1984). The computed ultraviolet intensities are about the same as before, and still fit the data already used by Lites (1972) to derive his ultraviolet radiation temperatures. The point of Avrett's addition of more line blocking is not that it results in lower ultraviolet intensities, but that the height of formation (photon escape) is now located higher in the atmosphere. This does affect the computed ultraviolet limb-darkening, which now agrees better with Samain's (1982) observations.

Thus, the difference with Lites' modeling is that J_ν departs from B_ν at about the same temperature but at larger height; the ultraviolet excess $J_\nu - B_\nu$ is therefore much smaller, and the iron excitation and ionization balances are much closer to LTE throughout the photosphere. The only remaining NLTE effect is the presence of photon losses in the most-probable lines at a given upper-level excitation energy. The only iron lines with NLTE source functions ($S_\nu^\ell < B_\nu$) in the photosphere are therefore the weak lines with large oscillator strengths at high excitation. All iron lines have LTE opacities throughout the photosphere. This holds for both Fe I and Fe II, and should hold also for H, Si, Mg, K, O and many other species that are sensitive to the ultraviolet radiation fields and that have NLTE imbalances when the HSRA is assumed.

With this new model Avrett more or less retrieves, proves and extends the LTE HOLMUL model. Fig. 5 shows that the photospheric parts of the MACKKL and HOLMUL models are quite similar. The combined assumptions of such a hot model, LTE opacities and LTE source functions therefore reproduce the optical lines and continua, the ultraviolet and infrared continua, and the Ca II line wings quite well. With enormous computational sophistication, plane-parallel modeling of the solar photosphere is back again to simple LTE!

One might conclude iron-ically that the whole NLTE issue has been a red herring, much ado about nothing, as far as photospheric line formation goes. I prefer to make two points instead. The first is that even if LTE modeling turns out to be not only an easy way out but in fact physically correct, the NLTE sophistication was necessary to prove it so. Obviously, LTE breaks down at some height in the atmosphere, and Avrett-like detailed modeling has to show where this breakdown occurs, in the photosphere or the chromosphere. In general, the only way a tractability assumption like LTE can be proven is by relaxing it and testing its validity in detail (see Rutten and Cram, 1981). The MACKKL model is clearly not the final truth yet, being empirical and having other tractability assumptions (like being plane-parallel with a specific microturbulence model).

Secondly, there is a hidden complexity. When taking all the ultraviolet blocking lines into account, it is necessary to specify both their height-dependent source functions and their height-dependent opacities. Initially, Avrett (private communication) assumed LTE for both, but when he found that these lines developed strong emission reversals in the chromosphere, he changed the line source functions by requiring them to follow a smooth transition between B_ν and J_ν in the upper photosphere, i.e. specifying them to become scattering lines with a source function without chromospheric rise, quite similar to the HOLMUL Planck function. The opacities are still added as derived from LTE. These two assumptions are ad-hoc and force a model fitted to the observed ultraviolet intensities at the computed height of formation to have a HOLMUL-like temperature-height relation. The upper photosphere is then quite hot, B_ν is close to J_ν, and the resulting Fe I and Fe II departures are small in the upper photosphere in agreement with the assumptions. The specified drop of S_ν below B_ν in the chromosphere is precisely what photon losses in strong lines should produce, and so the model is roughly self-consistent.

If on the contrary one would assume at the outset that the blocking lines behave as Fe I and Fe II lines (which constitute most of them) and that they follow the NLTE departure patterns shown in Fig. 3, then the opacities of the Fe I line cores would be smaller, resulting in deeper photon escape for the observed ultraviolet intensities, and the observed intensities would require corrections for Fe II excitation departures. A cooler model would result, and the large Fe I and Fe II NLTE departures predicted from it would agree with the assumed ones, again showing self-consistency.

In summary, the assumption of LTE or NLTE for the formation of the host of ultraviolet blocking lines is self-fulfilling: it results in empirical models which are respectively hot or cool, which produce the assumed line formation, and which reproduce the observations in either case. Just as with the optical iron lines, this is a dilemma that cannot be settled from the observed ultraviolet intensities although they are set by iron transitions. It is the new Swiss micro-wave data that really swing the balance to the MACKKL model, and therefore back to the classical LTE HOLMUL model, as the best representation of the solar photosphere within the constraints and assumptions of plane-parallel modeling.

3.3. RADIATIVE-EQUILIBRIUM MODELING

A third type of plane-parallel modeling is the one most often used in stellar physics: flux-constant modeling based on the assumptions of radiative equilibrium and LTE together. Solar examples are the solar controls of the model grids by Bell et al. (1976) and Kurucz (1979), and Kurucz's progress report in these proceedings.

Fig. 5 shows the Bell et al. (1976) solar model. It is close to the HOLMUL model in the upper photosphere but significantly cooler in deeper layers. Addition of yet more ultraviolet line-haze opacity may result in more backwarming and more and higher-located surface cooling, making it nearly identical to the HOLMUL model. It will therefore reproduce the observed optical iron lines assuming LTE quite well, just like the HOLMUL model.

The main difference with Avrett's modeling of the upper photosphere (the chromospheres are obviously different) is the latter's result that the optical H^- bound-free continuum has $J_\nu > B_\nu$ (Fig. 1) to such an extent that the total budget remains unbalanced, even for the new hot MACKKL model (Avrett 1985), thus requiring an unspecified energy sink. Although the empirical NLTE modeling and the radiative-equilibrium LTE modeling now produce about the same photosphere, the problem remains that a source of cooling rather than heating seems required just below the nonthermally heated chromosphere, at least in the context of plane-parallel modeling.

The main problem in radiative-equilibrium modeling is again to account for all the ultraviolet lines, and the problem of specifying their source functions and opacities arises again. Traditionally LTE is assumed for both in the construction of opacity distribution functions (e.g. Gustafsson et al. 1975, Kurucz 1979, Carbon 1979). Nordlund (1985b) has recently shown examples in which the line-haze source functions are changed from pure absorption to pure scattering. This results in a hotter upper photosphere because there is less surface cooling, just as if the ultraviolet lines were not present. The opacities of the haze lines are still assumed to be in LTE (Nordlund uses the term "scattering opacities" for opacity-binned averages over scattering source functions), but it is clear that departures from LTE in the ionization equilibria affect the location of the surface cooling and therefore affect the structuring of the upper layers, too.

4. IRON NLTE AND THE REAL SUN

4.1 GRANULATION

The real sun differs from the plane-parallel modeling discussed above in being highly inhomogeneous. The inhomogeneities are taken into account in the plane-parallel modeling by empirical adjustment of the micro- and macroturbulence parameters (often taken height-dependent and anisotropic as well) and also with collisional damping enhancement factors. Together these form a large set of ad-hoc fudge parameters. Although turbulence may be a reasonable description of motion fields as the 5-minute oscillation (which is certainly anisotropic), the scales of the latter are not well represented by the micro-macro separation (Carlsson and Scharmer 1985), and velocity broadening alone is clearly not a good physical description of the convective and magnetic small-scale structuring of the atmosphere (which are certainly height dependent).

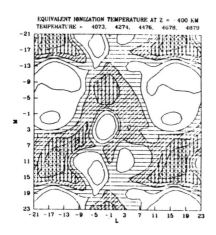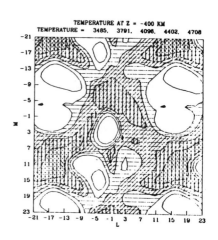

Fig. 6. A snapshot from Nordlund's numerical simulation of the solar granulation, taken from Nordlund (1984). The right-hand panel shows the distribution of the electron temperature in a horizontal plane at a height of 400 km. The temperatures are shown as shaded contours defined on top. The lefthand panel shows the corresponding iron ionization temperature. It is much higher than the electron temperature where the latter is low, due to ultraviolet overionization.

The major source of inhomogeneity in the deep photosphere is the granulation. Nordlund (1984) has computed typical iron line formation from his extensive time-dependent theoretical simulation of the solar granulation. The spatial and temporal variations in his results are very large; the temperature variation is much larger than the difference between the hot and cold mean models discussed above. This implies that the concomitant small and large departures from LTE are actually present side by side all over the sun. Fig. 6 (taken from Nordlund 1984) illustrates this. The righthand panel is a snapshot of the kinetic temperature distribution in a horizontal plane. The lefthand panel is the corresponding iron ionization temperature (i.e. the temperature that reproduces the computed NLTE ionizaton fraction from the Saha equation, cf. eq. 3). Its variation is smaller but still large, and its average is much higher in agreement with the over-ionization in Lites' (1972) plane-parallel modeling.

The resulting iron line profiles vary spatially and temporally with very large amplitudes. Nevertheless, the averaged observed Fe I profiles are reproduced very well. The computed line cores are deeper than the profiles in the Jungfraujoch Atlas (Delbouille et al. 1973), but this can be due to the neglect of NLTE overexcitation. Nordlund (1985a) finds that overexcitation indeed occurs even for Fe I, in contrast to plane-parallel modeling. It is due to the extra absorption of continuum photons by Doppler-shifted atoms. Another difference with plane-parallel modeling is that the Fe II opacities are also out of LTE. Comparison of Figs. 6.1, 6.6 and 6.7 in Nordlund (1984) shows that Fe I would be the dominant ionization stage in large fractions of the upper photosphere if LTE were valid. The overionization there leads to large relative overabundance of iron ions.

It is important to note that Nordlund requires no microturbulence, no macroturbulence other than a small contribution by the 5-minute oscillation, and no damping enhancement factors to fit the iron lines, and that he used no free parameters in the simulation itself. His fits are the first reproductions of Fraunhofer lines without any free parameters other than the iron abundance; this is a remarkable achievement.

It is also important to note that Nordlund obtains a very small iron abundance (A_{12} = 7.18). While it is larger than the abundance he finds for LTE ionization (A_{12} = 7.00), it is much smaller than the standard LTE or NLTE plane-parallel result (A_{12} = 7.6, e.g. Allen 1976, Rutten and Van der Zalm 1984). This difference is due to the nonlinear averaging caused by the steep temperature sensitivity of the ionization equilibrium, which typically scales as T^{10} (Nordlund 1984, eq. 6.4 but with the ratio reversed). This result indicates that comprehensive modeling of the real atmosphere can lead to marked changes from standard plane-parallel modeling. Taking the sun as an example for stellar abundance determination in general, it leads to questioning the numerical results of much of the latter.

However, also in this dynamical modeling there is the issue of the formation of the ultraviolet line haze. Nordlund (1984) computed the NLTE iron ionization equilibria with line-haze opacities that were scaled to reproduce the observed radiation temperatures at the iron ionization edges assuming the height-dependence of the H^- opacity. For high-excitation Fe II lines this height-dependence is reasonable in the upper photosphere, but it fails for high-excitation Fe I lines if the ionization is thermal and it fails also deeper down and for low excitation lines (Rutten and Van der Zalm 1984, Fig. 3). Nordlund argues that these opacities are not dominated by the iron ionization but this may be incorrect in the near ultraviolet where Fe I contributes heavily to the line haze and its intermediate levels contribute strongly to ionization imbalances.

Another problem is that in the simulation itself LTE was assumed for both the opacities and the source functions in the opacity-binned line-haze distribution functions. The resulting spatially-averaged temperatures are similar to the HSRA, with a cool upper photosphere. Nordlund (1984) describes an experiment in which he obtains instead a HOLMUL-like hot mean upper photosphere by applying height-dependent opacity enhancement factors in the near ultraviolet that bring the computed source functions closer to J_ν. The factors are just the inverse of the Fe I departure coefficients of Fig. 3a, and can be interpreted as representing the effect of stronger line blocking. The resulting change in the height of escape of the near-ultraviolet radiation affects the granular dynamics and leads to a hotter mean. Nordlund has not computed iron ionization equilibria in this experiment, but a much higher iron abundance would clearly be required to fit the observed Fe I lines.

In another experiment, Nordlund (1985b) shows that assuming pure scattering for the line haze source functions leads to <u>lower</u> temperatures in the upper photosphere than when LTE source functions are adopted. This experiment illustrates a major difference between the dynamic modeling and the plane-parallel radiative-equilibrium modeling discussed above. The granules in the simulation penetrate very far into the upper photosphere, and provide expansion cooling which would be very large if they rose adiabatically, but which is offset by strong radiative <u>heating</u> in the line haze. This heating by optically thin lines contributes to the high penetration; without it, the granules would be cooler and denser, and stop sooner.

Thus, the iron ionization and excitation equilibria are closely coupled together with the inhomogeneous temperature structure of the upper photosphere, even more closely than in the plane-parallel modeling because of the steep temperature sensitivity of the ionization balance. It seems prudent not to worry too much about Nordlund's small iron abundance before the role of the line haze has been further clarified, including its NLTE departures.

4.2 FLUXTUBES AND BIFURCATIONS

The chromosphere seen in Ca II K filtergrams is highly inhomogeneous. Bright Ca II emission coincides with locations of much magnetic field (e.g. Zwaan 1981), and the question one faces for the upper photosphere is to which extent the magnetic fluxtubes affect spatially-averaged modeling.

Ayres has proposed and given evidence that the low chromosphere is bifurcated into two components, a hot one which shows up in Ca II emission and presumably consists of fluxtubes, and a cool one which consists of the plasma between the tubes and which shows up in the infrared CO lines (Ayres et al. 1986 and references therein). The latter lines would actually trigger the bifurcation by providing large energy sinks wherever the temperature drops below the CO dissociation limit. Since the fluxtubes expand with height (e.g. Spruit 1981), the hot tubes would cover a negligible fraction of the surface in the photosphere.

In such a scenario, the strong Fe II lines will predominantly come from the hot fluxtubes and the strongest Fe I lines from the cool chromosphere in between. An attractive possibility arises to explain the absence of bright emission cores in the Fe II resonance lines from fluxtube geometry. Such slender structures may well be optically thin in lateral directions for the many weaker lines that share upper levels with the resonance lines. Lateral photon losses in subordinate lines result in lower source functions and lower opacities for the Fe II resonance lines than for the Mg II and Ca II resonance lines, because the latter have far fewer subordinate lines and metastable levels (Mg II has none, and indeed appears hotter in the empirical modeling by Ayres and Linsky 1976). The subordinate lines would be seen in emission at the limb, with large spatial intensity variation.

Below the temperature minimum, the cool component of the modeling by Ayres et al. (1986) equals the hot MACKKL model, in order to reproduce the latter with a negligible fluxtube filling factor. Photospheric iron lines should thus be only marginally affected by fluxtubes, as indeed shown by Stenflo and Lindegren (1977), except for their polarization (Solanki and Stenflo 1984, 1985). Nevertheless, iron line formation can be important in the bifurcation itself because, again, the line haze may contribute to it. Nordlund (1985b) obtains a similar bifurcation into stable hot and cool states simply from adopting a scattering ultraviolet line haze in plane-parallel radiative-equilibrium modeling, without invoking fluxtubes. Just as in Ayres' empirical modeling, the cool equilibrium state is caused by CO cooling in the infrared, and appears quite high in the atmosphere, above the location of the temperature minimum in the standard models. Thus, the line haze strikes again, and the iron ionization equilibria may well influence this thermal bifurcation and its location.

5. CONCLUSION

Current interpretation of solar iron line formation is amazingly multifarious. The same optical lines are used as gauges of the granulation and, independently, as spatially-averaged diagnostics of fluxtubes. Their formation may be very close to LTE or very far from it. The photosphere may be in radiative equilibrium up to the temperature minimum, or quite far from it. The inhomogeneities may be so important that interpreting spatially-averaged data is of little value even for abundance determination, or so unimportant that the classical turbulent fudge parameters provide a satisfactory description. The aspect perhaps most amazing is that each of these conflicting descriptions produces good fits to the observed lines. Clearly, fitting observations doesn't guarantee validity of the assumptions.

There are three main avenues to clarify these dilemmas. First and foremost, it is important in all descriptions to account in detail for the ultraviolet line haze, the more so because it largely consists of iron lines. All modeling of photospheric radiative transfer, whether plane-parallel or time-dependent 3-D, now depends on Johansson and Kurucz for further specification of the haze lines. These will have to be put into the computer with binning schemes that permit location-dependent NLTE departures in both the radiation fields and the populations; an example is given by Anderson (1985). Such simplification schemes are required because combining full 3-D radiative transfer for thousands of frequencies with full hydrodynamics and hydromagnetics as well is computationally prohibitive, even with the new efficient radiative transfer codes (Carlsson, these proceedings).

Second, the infrared iron lines as well as the infrared and microwave continua can provide important diagnostics, especially when observed from the extreme solar limb.

Third, observations with high spatial, temporal and spectral resolution supplied by space instruments will permit detailed evaluation of the role of inhomogeneities.

Finally, I would like to conclude this review with a philosophical speculation. The apparent success of classical plane-parallel LTE modeling of the spatially-averaged solar spectrum is in utter contrast to the extraordinary complexity that solar physicists deem necessary to explain solar granulation, magnetic fluxtubes etc., including intricate details of NLTE line formation for thousands of iron lines. It seems as if the classical modeling is going to be as adequate a first-order description for the spatially-averaged solar spectrum as Newtonian theory is for gravitation and Maxwell theory for electrodynamics, even though the "real" Sun requires a much deeper level of understanding, as in general relativity and quantum electrodynamcis. This deeper understanding will be profoundly different from the simple approximation, not only in its degree of complexity but also conceptually, but nevertheless, the latter may suffice for many applications.

ACKNOWLEDGEMENT. I thank Roberto Viotti for permitting me to discuss solar Fe I as well as Fe II, the Leidsch Kerkhoven-Bosscha Fonds for a travel grant, Jo Bruls, Petr Heinzel and Han Uitenbroek for comments, and Evelyn Cockburn for typing this paper.

REFERENCES

Allen, C.W.: 1976, "Astrophysical Quantities", William Clowes & Sons, London
Anderson, L.S.: 1985, in "Progress in Stellar Spectral Line Formation Theory", J. Beckman and L. Crivellari (eds.), NATO Advanced Science Institutes Series, Reidel Publ. Co., Dordrecht, The Netherlands
Athay, R.G.: 1972, "Radiation Transport in Spectral Lines", Reidel Publ. Co., Dordrecht, The Netherlands
Athay, R.G., Lites, B.W.: 1972, Astrophys. J. 176, 809
Avrett, E.H.: 1985, in "Chromospheric Diagnostics and Modeling", B.W. Lites (ed.), National Solar Observatory Conference, Sacramento Peak, Sunspot, p. 67
Ayres, T.R., Linsky, J.L.: 1976, Astrophys. J. 205, 874
Ayres, T.R., Testerman, L., Brault, J.W.: 1986, Astrophys. J. 304, 542
Blackwell, D.E., Shallis, M.J., Simmons, G.J.: 1980, Astron. Astrophys. 81, 340
Blackwell, D.E., Petford, A.D., Shallis, M.J., Simmons, G.J.: 1982, Mon. Not. R. Astron. Soc. 199, 43
Blackwell, D.E., Booth, A.J., Petford, A.D.: 1984, Astron. Astrophys. 132, 236
Bell, R.A., Eriksson, K., Gustafsson, B., Nordlund, Å.: 1976, Astron. Astrophys. Suppl. 23, 37
Bowers, R., Deeming, T.: 1984, "Astrophysics I", Jones and Bartlett Publishers Inc., Boston
Canfield, R.C., Stencel, R.E.: 1976, Astrophys. J. 209, 618
Carbon, D.F.: 1979, Ann. Rev. Astron. Astrophys. 17, 513

Carlsson, M., Scharmer, G.B.: 1985, in "Chromospheric Diagnostics and
        Modeling" B.W. Lites (ed.), National Solar Observatory
        Conference, Sacramento Peak, Sunspot, New Mexico
Cram, L.E., Rutten, R.J., Lites, B.W.: 1980, Astrophys. J. $\underline{241}$, 374
Cowley, C.R.: 1970, "The Theory of Stellar Spectra", Gordon and
        Breach, New York
Degiacomi, K., Kneubühl, F.K., Huguenin, D., Müller, E.A.: 1984,
        Internat. J. Infrared Millimeter Waves, $\underline{5}$, 643
Delbouille, L., Roland, G., Neven, L.: 1973, "Photometric Atlas of the
        Solar Spectrum from λ 3000 to λ 10000", Institut d'Astro-
        physique, Liège
Dere, K.P., Bartoe, J.-D.F., Brueckner, G.E.: 1984, Astrophys. J.
        $\underline{281}$, 870
Van Dessel, E.L.: 1974, Solar Phys. $\underline{38}$, 351
Van Dessel, E.L.: 1975, Solar Phys. $\underline{44}$, 13
Dravins, D.: 1982, Ann. Rev. Astron. Astrophys. $\underline{20}$, 61
Dravins, D., Larsson, B.: 1984, in "Small-scale dynamical processes
        in quiet stellar atmospheres", S.L. Keil (ed.), National
        Solar Observatory Conference, Sacramento Peak, Sunspot, New
        Mexico, p. 306
Dravins, D., Lind, J.: 1984, in "Small-scale dynamical processes in
        quiet stellar atmospheres", S.L. Keil (ed.), National Solar
        Observatory Conference, Sacramento Peak, Sunspot, New Mexico,
        p. 414
Gingerich, O., Noyes, R.W., Kalkofen, W., Cuny, Y.: 1971, Solar Phys.
        $\underline{18}$, 347
Gustafsson, B., Bell, R.A., Eriksson, K., Nordlund Å.: 1975, Astron.
        Astrophys. $\underline{42}$, 407
Harvey, J.W.: 1973, Solar Phys. $\underline{28}$, 9
Holweger, H.: 1967, Zeitschr. f. Astrophysik $\underline{65}$, 365
Holweger, H.: 1973, Solar Phys. $\underline{30}$, 35
Holweger, H., Müller E.A.: 1974, Solar Phys. $\underline{39}$, 19
Holweger, H.: 1979, in "Les elements et leurs isotopes dans
        l'Univers", 22me Liège Colloquium, Université de Liège,
        Belgium
Ivanov, V.V.: 1973, "Transfer of Radiation in Spectral Lines", English
        Edition, National Bureau of Standards, Special Publication
        385, U.S. Gov. Printing Office, Washington
Jefferies, J.T. 1968, "Spectral line formation", Blaisdell Publ. Co.,
        Waltham, Mass.
Kurucz, R.L.: 1979, Astrophys. J. Suppl. $\underline{40}$, 1
Lites, B.W.: 1972, "Observations and Analysis of the Solar Neutral
        Iron Spectrum", NCAR Cooperative Thesis no. 28, University of
        Colorado, Boulder
Lites, B.W.: 1973, Solar Phys. $\underline{30}$, 283
Lites, B.W.: 1974, Astron. Astrophys. $\underline{33}$, 363
Lites, B.W., White, O.R.: 1973, High Altitude Research Memorandum
        no. 185, Boulder
Maltby, P., Avrett, E.H., Carlsson, M., Kjeldseth-Moe, O., Kurucz,
        R.L., Loeser, R.: 1986, Astrophys. J. $\underline{306}$, 284
Mathys, G. and Stenflo, J.O.: 1986, Astron. Astrophys., in press

Menzel, D.H., Cillié, G.G., 1937: Astrophys. J. 85, 88

Mihalas, D.: 1970, "Stellar Atmospheres", 1st edition, Freeman and
        Co., San Francisco

Mihalas, D.: 1978, "Stellar Atmospheres", 2nd edition, Freeman and
        Co., San Francisco

Nordlund, A.: 1984, in "Small-scale dynamical processes in quiet
        stellar atmospheres", S.L. Keil (ed), National Solar Observa-
        tory Conference, Sacramento Peak, Sunspot, New Mexico, p. 181

Nordlund, A.: 1985a, in "Progress in Stellar Spectral Line Formation
        Theory", J.E. Beckman and L. Crivellari (eds.), NATO Advanced
        Science Institutes Series C 152, Reidel Publ. Co., Dordrecht,
        The Netherlands

Nordlund, A.: 1985b, in "Theoretical Probelems in High-resolution
        Solar Physics", H.U. Schmidt (ed.), Proceedings of the
        MPA/LPARL Workshop 1985, Max-Planck-Institut für Astrophysik,
        München

Pierce, A.K.: 1968, Astrophys. J. Suppl. 17, 1

Rudkjøbing, M.: 1986, Astron. Astrophys. 160, 132

Rutten, R.J., Cram, L.E.: 1981, in "The Sun as a Star", S.D. Jordan
        (ed.), CNRS-NASA Monograph Series on Nonthermal Phenomena in
        Stellar Atmospheres, NASA SP-450, part IV.

Rutten, R.J., Kostik, R.I.: 1982, Astron. Astrophys. 115, 104

Rutten, R.J., Stencel, R.E.: 1980, Astron. Astrophys. Suppl. 39, 415

Rutten, R.J., Van der Zalm, E.B.J.: 1984, Astron. Astrophys. Suppl.
        55, 143

Rutten, R.J., Zwaan, C.: 1983, Astron. Astrophys. 117, 21

Samain, D.: 1980, Astrophys. J. Suppl. 44, 273

Saxner, M.: 1984, "Some aspects of spectral lines in stellar atmo-
        spheres", thesis, Uppsala University

Sistla, G., Harvey, J.W.: 1970, Solar Phys. 12, 66

Solanki, S.K., Stenflo, J.O.: 1984, Astron. Astrophys. 140, 185

Solanki, S.K., Stenflo, J.O.: 1985, Astron. Astrophys. 148, 123

Spruit, H.C.: 1981, in "The Sun as a Star", S. Jordan (ed.), CNRS-NASA
        Monograph Series on Nonthermal Phenomena in Stellar Atmo-
        spheres, NASA-SP 450

Steenbock, W.: 1985, in "Cool stars with excesses of heavy elements"
        M. Jaschek, P.C. Keenan (eds.), Strasbourg Colloquium 1984,
        p. 231

Stenflo, J.O., Lindegren, L.: 1977, Astron. Astrophys. 59, 367

Thomas, R.N.: 1983, "Stellar Atmospheric Structural Patterns", NASA-
        CNRS Monograph Series on Non-thermal Phenomena in Stellar
        Atmospheres", NASA Special Publ. SP-471

Vernazza, J.E., Avrett, Loeser, R.: 1976: Astrophys. J. Suppl. 30, 1

Vernazza, J.E., Avrett, E.H., Loeser, R.: 1981, Astrophys. J. Suppl.
        45, 350

Wijbenga, J.W., Zwaan, C.: 1972, Solar Phys. 23, 265

Zwaan, C.: 1981, in "The Sun as a Star", S.D. Jordan (ed.), CNRS-NASA
        Monograph Series on Nonthermal Phenomena in Stellar
        Atmospheres, NASA SP-450, p. 167

Zwaan, C., Brants, J.J., Cram, L.E.: 1985, Solar Phys. 95, 3

STARK BROADENING OF THE Fe II LINES IN THE SOLAR AND STELLAR SPECTRA

Milan S. Dimitrijević
Astronomical Observatory
Volgina 7
11050 Beograd
Yugoslavia

ABSTRACT. Recently developped modified semiempirical approach (Dimitrijević and Konjević, 1980, 1986; Dimitrijević and Kršljanin, 1986) is applied to the Stark broadening calculation of Fe II lines in the Solar spectrum and the spectrum of Am 15 Vulpeculae.

1. INTRODUCTION

In stellar atmospheres calculations the collisional broadening parameters for a large number of various elements are required and they are often unavailable. Moreover, in O and B stars and white dwarfs atmospheres the Stark effect is the main pressure broadening mechanism. In A stars atmospheres its influence is very important and even in atmospheres of relatively cool stars as the Sun, where the line broadening caused by collisions with neutral perturbers is dominant, for higher number of spectral series the Stark effect may compete with neutral perturber interaction with emitter (Vince et al, 1985). A convenient method for Stark broadening calculations in astrophysics in the cases when more sophysticated calculations are avoided (e.g. lack of atomic data, complex spectra, large scale calculations or rough estimates) is the modified semiempirical approach (Dimitrijević and Konjević, 1980, 1981; 1986; Dimitrijević and Konjević, 1986). Tables of calculated Stark widths of prominent lines of some doubly- and triply-charged ions are given by Dimitrijević and Konjević (1981).

Recently, a detailed model atmosphere analysis of the Am 15 Vulpeculae star is carried out (Yo-ichi Takeda, 1984) in plasma conditions where the Stark broadening is not negligible. For a number of lines of more complex atoms, published Stark broadening data have not been found by the author of the mentioned article. Here are given the simple Stark broadening calculations of Fe II lines from the spectrum of Am 15 Vulpeculae and Solar spectrum in order to demonstrate the aplicability of the modified semiempirical method for such kind of the calculations in the case of more complex atoms and to provide also the Stark broadening data for some Fe II lines of

R. Viotti et al. (eds.), Physics of Formation of FeII Lines Outside LTE, 211–216.
© *1988 by D. Reidel Publishing Company.*

astrophysical interest.

2. THE MODIFIED SEMIEMPIRICAL APPROACH

In order to reduce the input set of atomic data and to extend the
applicability of semiempirical method published by Griem (1968) to
higher stages of ionization, Dimitrijević and Konjević (1980)
separated the transitions with $\Delta n = 0$ and introduced for them
different Gaunt factor. Transitions with $\Delta n \neq 0$ are summed separately.
half-half width w and shift d of ion spectral line broadening by Stark
effect become now

$$w + id = N \frac{4\pi}{3} \frac{\hbar^2}{m^2} \left(\frac{2m}{\pi kT}\right)^{1/2} \frac{\pi}{\sqrt{3}} \sum_{j=i,f} \{\vec{R}^2_{k,k+1} [\tilde{g}(x_{k,k+1}) +$$

$$+ i\epsilon_j \tilde{g}_{sh}(x_{k,k+1})] + \vec{R}^2_{k,k-1} [\tilde{g}(x_{k,k-1}) - i\epsilon_j \tilde{g}_{sh}(x_{k,k-1})]\} +$$

$$+ \sum_j (\vec{R}^2_{jj'})_{\Delta n \neq 0} [g(x_j) + i\epsilon_j g_{sh}(x_j)] - 2i\epsilon_j [\sum_{\Delta E_{jj'} < 0} (\vec{R}^2_{jj'})_{\Delta n \neq 0} g_{sh}(x_{jj'})] \tag{1}$$

Here $k = \ell_j$, i and f denote the initial and final levels $\mathcal{E}_j = +1$ if
$j = i$ and -1 if $j = f$ and $\vec{R}^2_{jj'} = |\langle j|\vec{r}|j'\rangle|^2$. Gaunt factors g, $\tilde{g}$, g_{sh}
and $\tilde{g}_{sh}$ are given as a function of x in the Table 1. Also
$x_{jj'} = 3kT/2\Delta E_{jj'}$, $x_j = 3kTn_j^{*3} /4Z^2 E_H$ where E_H is the hydrogen
ionization energy.

Table 1.

x	$\leqslant 1$	2	3	5	10	30	100
g	0.2	0.2	0.24	0.33	0.56	0.98	1.33
$\tilde{g}$	0.7	-1.1/Z+g					
g_{sh}	0.2	0.25	0.32	0.45	0.66	0.82	0.87
Z = 2 $\tilde{g}_{sh}$	0.35	0.40	0.47	0.58	0.70	0.82	0.87
Z = 3 $\tilde{g}_{sh}$	0.53	0.54	0.57	0.62	0.70	0.82	0.87
Z = 4 $\tilde{g}_{sh}$	0.62	0.63	0.63	0.65	0.70	0.82	0.87
Z 4 $\tilde{g}_{sh}$	0.88 - 1.1/Z+0.01+0.01x/Z;			$x \leqslant 100$			

$$\sum_{j'} (\vec{R}^2_{jj'})_{\Delta n \neq 0} \approx (\frac{3n^*_j}{2Z})^2 \frac{1}{9}(n^{*2}_j + 3\ell^2_j + 3\ell_j + 11) \tag{2}$$

and

$$R^2_{jj'} = (\frac{3n^{*2}_j}{2Z})^2 \frac{\ell_>}{2\ell+1}(n^2_\ell - \ell^2_>)\phi^2(n_{\ell-1}, n_\ell, \ell)$$

For all cases when $3kT/2\Delta E_{jj'} \lesssim 2$ one can use the values $g = g_{sh} = 0.2$ and $\tilde{g} = \tilde{g}_{sh} = 0.9 - 1.1/Z$. Furthermore, one can put in Eqs. 1 and 2 $\phi^\ell = 1$ what is a reasonable assumption for $\Delta n = 0$ (see e.g. Griem, 1974 p. 31) since the exact values of ϕ^ℓ usually range between 0.8 and 1. If one performs summation in Eq. (1) it is easy to obtain

$$w(\overset{o}{A}) = 0.2215 \cdot 10^{-8} \frac{\lambda^2(cm)N(cm^{-3})}{T^{1/2}} \sum_{j=i,f} [(\vec{R}^2_{jj'} + \frac{\bar{g}_{th} - g_{th}}{g_{th}} \cdot$$

$$\cdot (\frac{3n^*_j}{2Z})^2 (n^{*2}_j - \ell^2_j - \ell_j - 1)] \tag{3}$$

Since the contribution to the total line width of transitions with $\Delta n \neq 0$ does not exceed 25%, and it is compensated by assuming $\phi^\ell = 1$, we can neglect them and finally obtain

$$w(\overset{o}{A}) = 1.1076 \cdot 10^{-8} \frac{\lambda^2(cm)N(cm^{-3})}{T^{1/2}} (0.9 - \frac{1.1}{Z}) \sum_{j=i,f} (\frac{3n^*_j}{2Z})^2 \cdot$$

$$\cdot (n^{*2}_j - \ell^2_j - \ell_j - 1) \tag{4}$$

With analogous simplifications for the shift from Eq.(1) one may obtain

$$d(\overset{o}{A}) = 1.1076 \cdot 10^{-8} \frac{\lambda^2(cm)N(cm^{-3})}{T^{1/2}} (0.9 - \frac{1.1}{Z}) \sum_{j=i,f} (\frac{3n^*_j}{2Z})^2 \frac{\epsilon_j}{2\ell_j+1}$$

$$\{(\ell_j+1)|n^{*2}_j - (\ell_j+1)^2| - \ell_j(n^{*2}_j - \ell^2_j)\} \tag{5}$$

If all levels $\ell_{jf} \pm 1$ exist, an additional summation may be performed in Eq. (5) obtaining

$$d(\overset{\circ}{A}) = 1.1076 \cdot 10^{-8} \frac{\lambda^2 (cm) N (cm^{-3})}{T^{1/2}} (0.9 - \frac{1.1}{z}) \frac{9}{4z^2} \sum_{j=i,f} \frac{n_j^{*2} \varepsilon_j}{2\ell_j + 1} \cdot$$

$$\cdot (n_j^{*2} - 3\ell_j^2 - 3\ell_j - 1) \tag{6}$$

In order to test the modified semiempirical approach, selected experimental data for 36 multiplets (7 different ion species) of doubly-, and 7 multiplets (4 different ion species) of triply-charged ions were compared with linewidths calculated according to Eq. (1). The average values of the ratios of measured to calculated widths are as follows: for doubly-, 1.06 ± 0.31 and for triply-charged ions 0.91 ± 0.42 (Dimitrijević and Konjević, 1980).

Furthermore, in order to test the equations obtained for low temperature limit of modified semiempirical formula, comparison is made between the linewidth result from Eq. (1) and (4) and previously mentioned selected experimental results (Dimitrijević and Konjević, 1980), in the cases when Eq. (4) may be applied. The average ratio of experimental and calculated values from Eqs. (1) and (4) are 1.01 and 1.04 respectively (Dimitrijević and Konjević, 1986).

Recently, the modified semiempirical theory is applied to the most intensive lines of Ti II and Mn II observed in the Solar spectrum (Dimitrijević, 1982), in order to test the applicability of the approach to the case of more complex transitions and heavier elements. The obtained agreement between this approach and more sophysticated semiclassical calculations indicates that this method can be used for estimation of electron width for heavier elements.

3. RESULTS

In Table 2, electron impact full halfwidths (W_{MSE}) for 3 Solar multiplets ($a^4H - z^4F°$, $a^6D - z^6D°$, and $b^4F - z^4F°$) and 3 multiplets observed in the spectrum of Am 15 Vulpeculae (Yo-ichi Takeda, 1984) ($b^4P - z^4F°$, $b^4F - z^4D°$, and $b^4P - z^4D°$) are given for different electron temperatures (T) and for $N_e = 10^{17}$ cm^{-3}.

We hope that the presented method will be useful for Stark broadening calculations of Fe II lines especialy in the cases when we need an extensive set of data with good average accuracy and when the accuracy of each particular value is not so important.

Table 2. Electron impact full halfwidths (W_{MSE}) in Angströms for selected Solar lines and lines from the spectrum of Am 15 Vulpeculae (see the text) for different temperatures (T) and at the electron concentration 10^{17} cm^{-3} .

Ion	Transition (mult. No.) wawelength (Å)	T(K)	W_{MSE} (Å)
Fe II	$a^4H - z^4F^o$ (32) $\lambda = 4300.15$ Å	5000 10000 20000 40000	0.249 0.176 0.124 0.0880
Fe II	$a^6D - z^6D^o$ (UV 1) $\lambda = 2611.41$ Å	5000 10000 20000 40000	0.0856 0.0605 0.0428 0.0303
Fe II	$b^4F - z^4F^o$ (37) $\lambda = 4564.89$ Å	5000 10000 20000 40000	0.278 0.196 0.139 0.0982
Fe II	$b^4P - z^4F^o$ (28) $\lambda = 4293.11$ Å	5000 10000 20000 40000	0.250 0.177 0.125 0.0884

Fe II	$b^4F - z^4D^o$ (38) $\lambda = 4558.73 \overset{\circ}{A}$	5000	0.277
		10000	0.196
		20000	0.139
		40000	0.0995
Fe II	$b^4P - z^4D^o$ (27) $\lambda = 4558.73 \overset{\circ}{A}$	5000	0.281
		10000	0.199
		20000	0.141
		40000	0.0995

REFERENCES

Dimitrijević,M.S.: 1982, in Sun and Planetary Sistem eds. W. Fricke,
    G.Teleki, D.Reidel P.C., p. 101.

Dimitrijević,M.S., and Konjević, N.: 1980, J.Quant.Spectrosc.
    Radiative Transfer 24, 451.

Dimitrijević,M.S., and Konjević,N.: 1981, in Spectral Line Shapes,
    ed. B.Wende, W. de Gruyter, Berlin, p. 211.

Dimitrijević,M.S., and Konjević,N.: 1986, Astron.Astrophys. to be
    published.

Dimitrijević,M.S., and Kršljanin,V.: 1986, Astron.Astrophys. to be
    published.

Griem,H.R.: 1968, Phys.Rev. 165, 258.

Griem,H.R.: 1974, Spectral Line Broadening by Plasmas, Academic Press,
    New York.

Vince,I., Dimitrijević,M.S., and Kršljanin,V.: 1985, in Progress in
    Stellar Spectral Line Formation Theory, eds. J.E.Beckman,
    L.Crivelari, D.Reidel P.C., Boston, p. 373.

Yo-ichi Takeda: 1984, Publ.Astron.Soc.Japan 36, 149.

FE II LINES IN THE PRESENCE OF PHOTOSPHERIC OSCILLATIONS

C. Marmolino
Air Force Geophysics Laboratory
National Solar Observatory/Sacramento Peak
Sunspot, N.M. 88349 U.S.A.,
on leave from the Università di Napoli

G. Roberti
Dipartimento di Fisica   N.S.M.F.A.,
Università di Napoli .
Mostra d'Oltremare, Pad.20
80125 Napoli, Italia

G. Severino
Osservatorio Astronomico di
Capodimonte,
Via Moiariello, 16
80131 Napoli, Italia

1. INTRODUCTION

Line asymmetries and wavelength shifts can be produced by dynamical processes at work in the solar photosphere. Dravins, Nordlund and coworkers (ref.1,2,3) discussed the importance of the granulation in determining the C-shape of solar lines of FeI and FeII ions. Iron is a suitable atomic species to diagnose photospheric motions, since it has negligible asymmetries due to isotope composition and to pressure shifts and no hyperfine structure splitting. Marmolino, Roberti, Severino and coworkers studied the effects produced by photospheric oscillations (5 - min and short period acoustic waves) on the resonance line of KI at 7699 A (ref. 4,5,6). To extend this study to iron lines, in this paper we show the synthesis of the FeII 6516 line in the presence of granulation and 5 - min oscillation.

The granulation model is that by Nelson (ref.7) with the velocity amplitude increased by a factor 1.5, as in ref.6. The wave model is a monochromatic evanescent wave having a period of 300s and a velocity amplitude of 350 m/s at z=0, sligthly increasing with height. The line synthesis code allows for the use of non-LTE atomic level populations computed in the unperturbed model. However for the Fe II 6516 our LTE calculations with the VAL C atmosphere (without microturbulence), an Fe abundance of 2.51×10^{7} and the atomic data taken from ref.8, give a satisfactory comparison with the observed line profile from the Jungfrau Atlas (ref.9)

R. Viotti et al. (eds.), Physics of Formation of FeII Lines Outside LTE, 217–221.

(Fig.1).

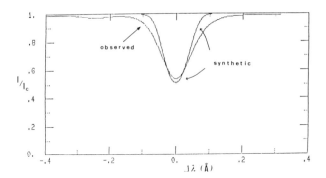

Figure 1. Synthetic vs.observed profile of the FeII 6516.081 line at disk center.

2. RESULTS

The asymmetry of the spatially resolved profiles in the presence of granulation is much stronger than that of the mean, unresolved profile (Fig.2). The bisector of the resolved profiles measures the vertical velocity gradients within the granulation, while the asymmetry of the mean profile is a combined effect of the different line shifts, different line strengths and different continuum intensities between the granular and intergranular components. This effect leads to a blueshift of the spatially averaged line profile corresponding to a velocity of 360 m/s in the center of gravity.

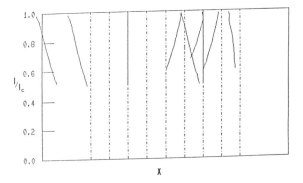

Figure 2. Bisector vs. horizontal coordinate in the presence of granulation. The last bisector refers to the spatially averaged profile. Note that the vertical dotted-dashed lines mark the positions of the unpertur bed bisectors and the separation of two successive lines is of 584 m/s in velocity scale.

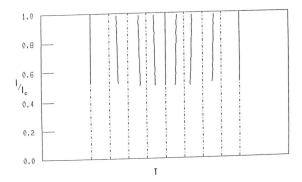

Figure 3. Bisector vs. time in the presence of 5-min oscillation. The last bisector refers to the temporally averaged profile.

The line bisector produced by the 5-min oscillation are plotted in Fig.3 as a function of the time. At each time the line bisector is roughly proportional to the vertical velocity vs. depth. The phase relations among the perturbations in the evanescent wave (dT and dP are 90 degree out of phase respect to v) ensure that when the velocity takes opposite values the thermodynamic structure is unchanged and the resulting profiles are mirror symmetric. Then the temporally averaged profiles has neither asymmetry nor shift.
    When both granulation and 5-min oscillation are present, the general behaviour of the spatially averaged line bisector vs. time is just the temporal fluctuation of the spatially mean C-shape due to the granulation (Fig.4). This agrees with the finding of Roca-Cortes et al. (ref. 10) for the KI 7699 line that the asymmetry is anticorrelated with the core shifts due to the 5-min oscillation.

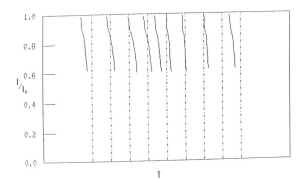

Figure 4. Bisector vs. time in the presence of the both granulation and 5-min oscillation. All bisectors are spatially averaged. The last bisector refers to the spatially and temporally averaged profile.

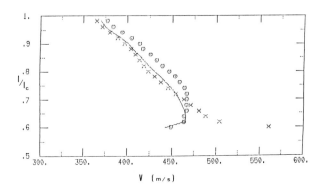

Figure 5. Residual intensity vs. amplitude of the velocity of fluctuations in presence of both granulation and 5-min oscillation. Crosses:
blue line flank; solid line: line bisector; circles: red line flank.

Examining the line profiles, we have found that, when the granulation is present, the oscillations do not equally affect the two line
flanks, as shown in Fig.5. The blue flank oscillates with a velocity am
plitude lower than the red flank does, down to I/Ic ~ 0.65. At I/Ic <
0.65 this situation reverses. The velocity amplitude of the bisector
points are approximatively the average of those of the flanks. These results seem to confirm the findings by Cavallini et al. (ref.11) for three
FeI lines. However there are two main differences between their observations and our line synthesis:
i) Cavallini et al. measured oscillations down to I/Ic = 0.4, while our
FeII 6516 line has a central depth of only I/Ic ~ 0.55;
ii) the velocity amplitude of the line bisector observed by Cavallini et
al. is of the order of 40 m/s, i.e. a factor 10 smaller than our calcula
tions and also than the values commonly accepted for the velocity amplitude of the 5-min oscillation.

REFERENCES

1. Dravins, D., Lindegren, L., Nordlund, Å.: 1981, Astron, Astrophys.,
   96, 345.

2. Nordlund, Å.: 1984, in S.L. Keil, ed: "Small Scale Dynamical Processes
   in Quiet Stellar Atmosphere", National Solar Observatory, Sacramento
   Peak, p. 181.

3. Dravins, D., Larsson, B., Nordlund, Å.: 1986, Astron.Astrophys, 158,
   83.

4. Marmolino, C., Roberti, G., Severino, G., Vazquez, M., Wöhl, H.: 1984,
   Proceed. 4th Europ. Meet. on Solar Phys., ESA SP - 220, p. 191.

5. Severino G., Roberti, G., Marmolino, C. and Gomez, M.T.: 1986, Solar

Phys. in press.

6. Marmolino, C., Roberti, G., Severino, G.: 1986, Solar Phys. submitted.

7. Nelson, G.D.: 1978, Solar Phys. 60, 5.

8. Cram, L.E., Rutten, R.J., Lites, B.W.: 1980, Astrophys. J. 241, 374.

9. Delbouille, L., Roland, G., Neven, L.: 1973, "Photometric Atlas of the Solar Spectrum from 3000 to 10000",Institut d'Astrophysique, Liège.

10. Roca-Cortes, T., Vazquez, M., Wöhl, H.: 1983, Solar Phys. 88, 1.

11. Cavallini, F., Ceppatelli, G., Righini, A., Alamanni, N.: 1985, in H.U. Schmidt ed: "Theoretical Problems in High Resolution Solar Physics", MPA/LPARL Workshop in München, p. 87.

ULTRAVIOLET FEII EMISSION FROM COOL STAR CHROMOSPHERES

C. Jordan
Department of Theoretical Physics
University of Oxford
1, Keble Road, Oxford OX1 3NP
U. K.

ABSTRACT.   A review is given of the presence and excitation of
ultraviolet FeII emission lines from the chromospheres of cool stars,
including   the   sun.   The   requirements   for   further   atomic   data,
particularly collision cross-sections, are stressed.   Some unresolved
issues of line identification and excitation mechanisms are discussed.

1.INTRODUCTION

FeII is the dominant ion giving rise to emission lines between 2200 Å
and 3000 Å in stars cooler than the sun and a major contributor in
spectra down to ~ 1500 Å.   The low lying terms of FeII have been known
for   many   years   and   given   sufficient   spectral   resolution   it   is
straightforward   to   identify   the   stronger   lines   present   in   stellar
spectra.   However,   the   extension   of   the   known   term   scheme   to   higher
levels by Johansson (1978a) has been important in understanding spectra
both in the uv and infra-red. (Johansson, 1977a, b, 1978b).   Here we
concentrate   on   the   uv   region   observable   with   the   International
Ultraviolet Explorer (IUE).
        The interpretation of FeII line emission is complicated, because a
number of different excitation processes are possibly taking place in
cool star chromospheres, for example, photo-excitation by the optical
continuum,   collisional   excitation,   transfer   of   photons   between
multiplets sharing a common upper level   and excitation by fluorescence
from accidental wavelength coincidences with strong lines.   Moreover,
the relative importance of these processes varies with stellar type.
        The lines of FeII have great potential for studying several aspects
of stellar chromospheres, i.e. the temperature and density structure,
variation   of   continuum   and   line   radiation   fields   with   height   and
relative velocities as a function of height.   At present much of this
potential   is   untapped,   not   only   because   of   the   complex   multi-level
radiative transfer calculations required, but also because many of the
atomic parameters, particularly collision cross-sections, are not known.
Analyses which are based only on oscillator strengths are feasible

R. Viotti et al. (eds.), Physics of Formation of FeII Lines Outside LTE, 223–234.
© *1988 by D. Reidel Publishing Company.*

because extensive calculations (Kurucz, 1981) and some measurements are available.

In Section 2 observations of cool stars, made with IUE, and of the sun, made from rockets and Spacelab 2 Shuttle flight, are reviewed, stressing recent line identifications. Excitation mechanisms and related diagnostic techniques are discussed in Section 3. In Section 4 some general conclusions and suggestions for future work are made.

2. OBSERVATIONS AND LINE IDENTIFICATIONS

2.1 The region 2000 Å - 3000 Å.

In stars such as the sun the photospheric continuum makes it difficult to observe emission lines above 2000 Å. However, by observing at the solar limb (e.g. Doschek et al., 1976) or during total eclipses the chromospheric emission line spectrum becomes observable and FeII is the dominant contributor between 2200 Å and 3000 Å. Observations of G-type main sequence stars from satellites are similarly difficult in this spectral range but studies of cooler dwarfs do not suggest that there are significant differences in the emitted line spectrum. A spectrum of the pre-main sequence star T Tauri (KO IV) obtained with IUE, shown in Figure 1, illustrates the multiplets present. (Brown et al., 1984). The schematic term diagram, (Figure 2), illustrates many of the transitions discussed below. The resonance multiplets (uv 1 to 5) are present, with uv 1 often being sufficiently strong to observe at high resolution with IUE. Transitions to the two lowest quartet levels, a^{4}F and a^{4}D, such as uv 32, 33, 35, 36 and uv 60 to 64, respectively, also occur.

The spectra of cool giants and supergiants have been discussed by Carpenter in these Proceedings so the more straightforward aspects are

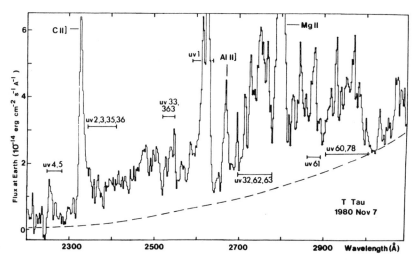

Figure 1   The spectrum of T Tauri (KO IV) in the wavelength region 2250 Å to 3000 Å, obtained with the IUE satellite, showing the location of multiplets of FeII. (From Brown, Ferraz and Jordan, 1984).

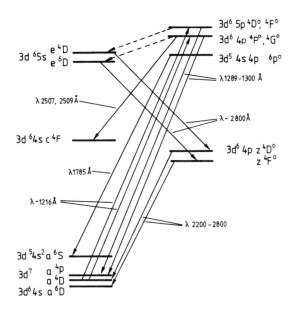

Figure 2   A partial, schematic term diagram for FeII, showing some of the transitions discussed in the text.

not reviewed here in detail.   The region from 2300 Å and 3000 Å was studied with the BUSS balloon payload (Van der Hucht et al., 1979) and in the early days of IUE (Wing, 1978, Carpenter and Wing, 1979).   A useful Atlas of high resolution spectra has been prepared by Wing, Carpenter and Wahlgren (1983) and some illustrations below are taken from that source.   These spectra show that all the strong multiplets visible in the T Tau spectrum, illustrated in Figure 1, are also strong in the cool giants and supergiants i.e. mults. uv 1 to 3; uv 32, 33; uv 35, 36; uv 60-64.   Lists of identified lines, their intensities and profile characteristics have been provided through the work of Carpenter (1984a,b).

Two wavelength regions of particular interest are illustrated in Figures 3, 4 and 5.   Figure 3, from Johansson and Jordan (1984), shows spectra of (a) α Tau (K5 III) and (b), β Gru (M5 III) in the region around 2500 Å.   Figure 4, from Wing et al. (1983) shows the same region in α Ori (M2 Iab).   These, and other spectra from Wing et al. (1983), show that mult. uv 33 is present and quite strong in K giants and early M giants, but that in the later M giants and M supergiants it is weaker and instead another set of lines becomes visible.   Two lines, at 2508.3 Å and 2506.8 Å have been identified as transitions in the multiplet c^4F-(b^{3}F) 4p $^4G^0$ and c^4F-5p $^6F^0$, from the $^4G^0$ and $^6F^0$ J = 9/2 levels, which are strongly mixed.   In β Gru the blend is dominated by the decay from $^6F^0_{9/2}$.   However, in α Ori (and RR Tel, Penston et al. 1983), the line from $^4G^0_{9/2}$ becomes stronger and is accompanied by a further line $c^4F_{9/2}$ - $^6F^0_{9/2}$ at 2504.9 Å.   Also the line at 2516.1 Å which may be either SiI(uv 1) or the $c^4F_{9/2}$ - $^4G^0_{7/2}$ transition becomes stronger, and SiI(uv

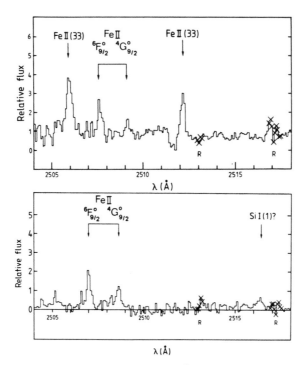

Figure 3 The region around 2504-2517 Å in (a) ∝ Tau (K5 III) and (b) β Gru (M3 III) showing mult. uv 33 and decays from levels. (the (b³F) 4p ⁴G°₉/₂ and 5p ⁶F°₉/₂, Revised from Johansson and Jordan, 1984).

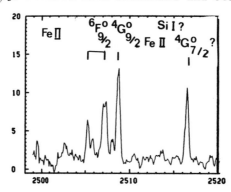

Figure 4 As for Figure 3, for ∝ Ori (M2 Iab) from the Atlas by Wing et al. (1983), with identifications added.

1) may also contribute at 2506.9 Å.

Figure 5 shows the regions between 2827 Å and 2902 Å in ∝ Ori (M2 Iab) and γ Cru (M3.4 III) from Carpenter et al. (1983). Apart from FeI (mult. 44), which is excited by MgII (Van der Hucht et al. 1979, Gahm, 1974), the remaining lines are due mainly to FeII, from multiplets uv 61, 399, 391 and possibly 380. Although uv 61 is strong in both stars,

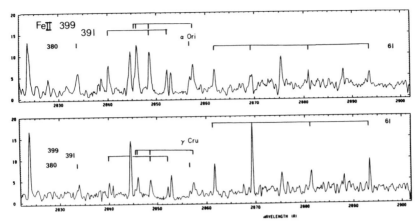

<u>Figure 5</u>   The region between 2822 Å and 2902 Å in α Ori (M2 Iab) and γ Cru (M3.4 III) showing transitions in multiplets uv 61, 380, 391 and 399. From Wing et al. (1983) with identifications added.

it can be seen that uv 399, 391 and 380 become relatively stronger from γ Cru to α Ori, i.e. as the stellar gravity decreases. The MgI line at 2852.1 Å may be blended with FeII uv 391.

2.2 The region between 1200 Å and 2000 Å.

Below 2000 Å emission lines of FeII are observable in the solar limb spectrum and below 1700 Å they become apparent in disk spectra. It is straightforward to identify most of the stronger lines using early tabulations (Moore, 1952) as transitions to a^6D (mult.uv 8), a^4F (mults. uv 38, 40-45) and weaker ones as transitions to a^4D (uv 68, 69) and a^4P (uv 85). The more extensive tabulation by Johansson (1981) includes further transitions to a^6D which are intersystem lines ($a^6D-z^2D^0$, $a^6D-y^4D^0$, $a^6D-y^4F^0$). These lines are most clearly seen at the solar limb. (See Figure 4 of Jordan et al., 1978). Similarly, further transitions to a^4F in mults. uv 39, 41, 44 and 46 and in $a^4F-z^2G^0$, $a^4F-z^2F^0$, listed by Johansson (1981), are present at the limb. A list of lines identified in the solar spectrum has recently been completed by Sandlin et al. (1986). The enhancement of the longer wavelength lines of FeII at the solar limb enabled Johansson (1977a) to first identify the octet terms $3d^54s(^7S)4d$ ^8D and $3d^54s(^7S)4p$ $^8P^0$, through transitions between them around 1750 Å in spectra obtained by Burton and Ridgeley (1970) and Doschek et al.,(1976). Decays from the $^8P^0$ term to the ground term, a^6D, were also found around 1890-1940 Å. This is just one example of how astrophysical spectra extend our knowledge of atomic term schemes.

    Transitions in FeII are difficult to resolve in IUE short-wavelength low resolution spectra but multiplets such as uv 38 and 39 probably contribute to blends observed around 1709-1730 Å and others to the background 'continuum' down to ~ 1550 Å. However, low resolution spectra of cool giants and supergiants show a feature around 1870 Å which has now been identified as FeII. (See Figure 2 in Johansson and Jordan, 1984).

High resolution spectra of a number of cool giants and supergiants have been obtained with IUE and those of β Gru (M5 III) have been particularly useful. Several stars show the presence of mult. uv 191 ($a^6S-x^6P^o$) at 1786.75, 1788.00 and 1785.27 Å , whose excitation mechanism is still a matter of debate. At shorter wavelengths weaker transitions have been observed in β Gru around 1288 – 1300 Å in multiplets classified by Johansson (1978). The strongest lines are at 1289.1 Å, 1293.6 Å and 1296.2 Å and are transitions between the a^4P and 5p $^4D^o$, 5p $^4F^o$ and 4p $^4S^o$ terms. (Johansson and Jordan, 1984). Their excitation is discussed in Section 3.2.1. These lines are present, but are very weak, in the solar spectrum. The spectrum of β Gru was used to locate a new $^4G^o$ term that gives rise to the feature at 1870 Å, which at high resolution is made up of two lines at 1869.53 Å and 1872.64 Å.

Three lines of moderate intensity and unusual spatial distribution appear in solar spectra obtained with the NRL's High Resolution Telescope and Spectrograph (see Figure 6 in Jordan and Judge 1984) and in the spectrum of β Gru (Judge, 1986). They are at 1347.03 ± 0.02, 1360.18 (±0.02) and 1366.40 (±0.02) Å, (Sandlin et al. 1986) and the latter two coincide with transitions in FeII. The line at 1360.18 Å could be $a^2G_{7/2}$ – $3d^6(b^3F)4p$ $^4G^o_{7/2}$ transition, which shares a common upper level with the line at 2516.0 Å, which is present, although weak in β Gru. This proposal is supported by the presence of an unidentified line at 1360.17 Å in the spectrum of RR Tel (Penston et al. 1983), where the line at 2516.0 Å is also detected. The line at 1366.40 Å could be the transition $b^4F_{7/2}$ – $3d^54s(^5P)4p$ $^4D^os_{5/2}$ but no other decays are observed to support this coincidence.

3. EXCITATION MECHANISMS AND DIAGNOSTIC TECHNIQUES

If the photospheric temperature and electron density are sufficiently high the even parity states up to ~ 38,000 cm^{-1}, when odd states begin, can achieve Boltzmann populations. Then photo-excitation in the optical continuum can excite upper levels which produce lines in emission at shorter wavelengths where the continuum is weaker. Since both optical and uv lines can also become optically thick, transferring photons, between multiplets, it is obvious that even in the absence of contributions from collisions very extensive calculations are required. Simultaneous modelling of the lines and continuum is required to put these processes on a quantitative basis. Viotti (1976) has discussed the relative importance of the various excitation and de-excitation processes. Because he was considering geometrically extended regions he worked in terms of the dilution factor, W, where

$$W = \tfrac{1}{2} [1-(1-(R_*/R_E)^2)^{\tfrac{1}{2}}]$$

(1)

and where R_* and R_E are the stellar radius and distance at which photo-excitation takes place. He constructed a diagram showing W against $N_e T_e^{-\frac{1}{2}}$, the latter quantity arising through the ratio of photo-excitation to collisional excitation. Three regions of parameter space occur; a 'radiative' regime where photoexcitation is more important than collisional excitation is populating excited levels; a collisional regime, where the reverse is the case and a 'nebular' regime where metastable levels decay radiatively. Although Viotti did not take into account the effects of high line opacity the approach is useful and an analogous diagram can be constructed for stellar chromospheres. Collin-Souffrin et al. (1979, 1980) did allow for the effects of high line opacities in the context of Seyfert galaxies and quasars, but included only levels up to about 5 eV. The conditions considered are similar to those in main sequence star chromospheres, i.e. $10^{10} \leqslant N_e \leqslant 10^{11}$ cm^{-3} and $7500 \leqslant T_e \leqslant 10,000$ K, but the line opacities are much larger ($\sim 10^5$ rather than 10^2-10^3).

Without making detailed calculations one can use Viotti's approach to examine the excitation regime of (a) metastable levels up to ~ 2.5 eV, (b) excited levels around 5 eV and (c) excited levels around 10 eV. Typical transition probabilities and collision strengths for forbidden lines are available from the work of Nussbaumer and Swings (1970) and Nussbaumer and Storey (1980) and one can conclude that at $T_e \sim 8000$ K, most of the metastable levels up to 2.5 eV have populations close to a Boltzmann distribution, provided $N_e \geqslant 2 \times 10^5$ cm^{-3}. This condition is met except perhaps in the most extended supergiants. The relative importance of collisional excitation and excitation, by photospheric radiation in populating excited levels around 5 eV and 10 eV can then be examined, making use of calculated transition probabilites and collision strengths (Kurucz, 1981; Nussbaumer et al., 1981). Although detailed calculations are required it appears that continuum photo-excitation of the 5 eV levels must be taken into account in all stars cooler than the sun. For levels around 10 eV , including radiative excitation out of metastable states up to ~ 2.5 eV, photo-excitation cannot be neglected but becomes less important than collisional excitation in stars cooler than about K0, in spite of the lower electron density in the giants and supergiants. By considering the situation of only collisional excitation one can see that this trend cannot explain why transitions from the 10 eV levels are comparable in strength with those from the 5 eV levels in the M giants and supergiants. However, there are no difficulties of principle in populating the 5 eV levels by the combination of photo-excitation and collisional excitation as appropriate. The 10 eV levels are considered further below.

3.1 Relative intensities of lines from a common upper level.

It is well known that when one multiplet of a set sharing a common upper level becomes optically thick, photons can be transferred to enhance the lines in multiplets of lower opacity. The ratio of the strengths of the different multiplets can then vary between their optically thin value and optically thick value – in some cases this can be a large factor. Where one multiplet remains optically thin the line ratios in the set can be used to determine the opacity in the remainder and thus the mass-column-density. Following the probability of escape method, which

has its origins in work by Zanstra (1949), the relative fluxes in two
lines can be expressed as

$$F_1/F_2 = \lambda_2\, q_1\, b_1/\, \lambda_1\, q_2\, b_2 \qquad (2)$$

where b is the branching ratio for the emission of a photon in a given
line and q is the probability, per emission, of a photon escaping in
that line. For a Doppler broadened line,

$$q = 1 - erf\,(\ln\tau_0)^{\frac{1}{2}} \qquad (3)$$

where $2\tau_0$ is the opacity through the whole region at line centre, and

$$\tau_0 = 6.0 \times 10^{-15}\lambda(\text{Å})f_{12}M^{\frac{1}{2}}\, \frac{N_e}{N_H}\Bigg\}_{\Delta h}\, \frac{N_i}{N_E}\, \frac{N_1}{N_i}\, \frac{N_H}{T_i^{\frac{1}{2}}}\; dh \qquad (4)$$

where $T_i^{\frac{1}{2}} = \Delta\lambda\, M_i^{\frac{1}{2}}/7 \times 10^{-7}\lambda$ $\qquad (5)$

and where M_i is the atomic weight.

Thus when one line is optically thin, i.e. $q_1 = 1.0$, and F_1/F_2 can
be measured, q_2 and $\tau_0(2)$ can be found. (Jordan, 1967). If only the
absolute intensity of a line is considered then one needs to know W, the
total fraction of photons created which eventually escape in a given
line, e.g.

$$W_1 = b_1 q_1 [1 - \sum_n b_n(1-q_n)]^{-1} \qquad (6)$$

This method has been applied to FeII in late-type stars (Brown et
al., 1981, 1984) and was also used to explain the variability of the
relative strength of mult. 42 in Seyfert galaxies (Jordan, 1979). It is
clear from the ratios of multiplets such as uv 1 and 32, uv 2, 33 and
60, uv 3 and 61, etc, that the sextet transitions (e.g. uv1, 2, 3) are
optically thick in the M giants and supergiants (e.g. β Gru and γ Cru,
illustrated by Wing et al. (1983). In α Tau and β Gru the ratio of uv 1
to uv 32 suggests optical depths of ~10^3. This line interlocking must
be taken into account when interpreting the asymmetries observed in FeII
line profiles (Carpenter, 1984b), but the FeII lines are potentially of
great value in studying the structure and dynamics of chromospheres.

3.2 Fluorescent Excitation

3.2.1 Excitation by H Lyα. Recognizing that fluorescent processes take
place in cool star chromospheres, eg the excitation of OI via H Lyβ, the
excitation of FeI (mult. 44) by MgII and excitation of H_2 by H Lyα,
(Jordan et al. 1978) one purpose of obtaining high resolution spectra
was to determine whether the blend with OI at low resolution was due to
FeII, pumped by H Lyα. (Brown et al. 1979). Although the SI (uv 9)
lines are stronger in α Tau (K5 III) the lines of FeII are clearly
visible in β Gru (M5 III). (See Johansson and Jordan, 1984). The

strongest lines at 1289.1 Å, 1293.6 Å and 1296.2 Å can be excited from $a^4D_{1/2}$ and $a^4D_{5/2}$ to the 5p $^4D^o_{3/2}$, 5p $^4F^o_{7/2}$ or $_{3/2}$ and 4p $^4S^o_{3/2}$ levels by radiation within ±1.5 Å of the H Lyα rest wavelength. In considering the effectiveness of this excitation one should be aware that because the oscillator strengths of the transitions are small they are susceptible to inaccuracies arising from the treatment of level mixing. For example, Johansson (1978a) has shown that because transitions between 4p $^4S^o_{3/2}$ and 5s $^4D_{5/2}$ and 6s $^4D_{5/2}$ occur the 4p $^4S^o_{5/2}$ level is mixed with at least one 5p level and one of the excitation routes at 1215.98 Å ($a^5D_{5/2}$ - 4p $^4S^o_{3/2}$) may have an oscillator strength larger than that calculated by Kurucz (1981). Thus estimates of the effectiveness of H Lyα pumping in Seyfert galaxies by Elitzur and Netzer (1985) may need revision when further calculations of the FeII oscillator strengths by Kurucz become available.

The transitions at 1869.53 Å and 1872.64 Å were also found to be excited by H Lyα from a^4G, to a new $^4G^o$ level around 13ev, the identification being made using the known intervals between the low FeII even levels (Johansson and Jordan, 1984).

The levels excited by H Lyα have other decay routes besides a^4P and a^4D, and strong lines are observed from them in the region between 2400 and 3000 Å. For example, the 4p $^4G^o_{9/2}$ and 5p $^6F^o_{9/2}$ levels decay to 4s c^4F causing the lines at 2508.3 Å and 2506.8 Å observed in β Gru, illustrated in Figure 3. The change in relation intensities between β Gru and α Ori can be understood in terms of an increased width for the H Lyα line. The other excited levels decay also through to e^4D and e^6D in transitions between 8000 Å and 9600 Å. It is these even levels that decay to z^4D^o and z^4F^o in the multiplets 380, 391 and 399 that are observed to be strong in β Gru, α Ori and the other M giants and supergiants (See Figure 5). In these stars all the strong transitions from levels above ~ 9.8 eV can be related to levels that can be reached by radiation within the H Lyα line.

The transitions at 1360.18 Å and 2516.0 Å from 4p $^4G_{7/2}$ can be excited by radiation in the far red wing of H Lyα around 1220.5 Å. These lines are not present in α Tau (See Figure 3) where there is little flux around 1220 Å but they appear in β Gru where the wing of H Lyα is still observable at 1220 Å (See Figure 3 of Johansson and Jordan, 1984). Similarly the line at 1366.40 Å from 4s4p $^4D^o_{5/2}$ can be excited from $a^4P_{3/2}$ by radiation at 1214.7 Å.

Another indication of selective excitation comes from the relative strengths of multiplets uv 399 and uv 63 in the M giants and supergiants. Using the observed flux ratio, oscillator strengths from Kurucz (1981) and allowing for uv 63 to have an opacity of ~ 10^2, one can show that the relative population of the e^4D and z^4D^o levels exceeds that expected from a Boltzmann distribution.

Modelling of stars such as β Gru (Judge, 1986) has only just reached the stage where a self-consistent treatment of H Lyα and the FeII excitation might be possible and this is an important area for future work.

The relative increase of the excitation of the 10 eV levels compared with the 5 eV levels between the K and M giants can be understood through general scaling laws proposed by Ayres (1979).

Whilst collisional excitation, depending on N_e, scales as $g_*^{1/2}$, the mass column density in the chromosphere, proportional to τ_0, scales as $g_*^{-1/2}$. Thus processes depending on a high line opacity and multiple scattering increase with respect to collisional processes according to g_*^{-1} and are thus more important in the M supergiants and giants. (See also Judge, 1986).

3.3 Multiplet 191.

The multiplet a^6S-3d^54s4p $^6P^o$ gives rise to three strong lines around 1785 Å which are observed in cool giants and supergiants. Observations at low resolution with IUE are difficult because of the position of a calibration pixel but the lines are definitely present in the high resolution spectra of β Gru and α Ori. It is not yet certain why the lines are so strong. Although photo-excitation in FeII uv 9 at 1260.5 Å by continuum radiation could be effective (Stalio and Selvelli, 1980; Viotti et al. 1980) there does not seem to be sufficient flux for this process in the M stars. A coincidence between uv 9 and SiII (uv 4) at 1260.4 Å has been noted by Nussbaumer (1977). This process is unlikely in M stars. SiII is present as an ion but emission in uv 4 is not expected to be strong because of the high excitation potential. Any emission in 1260.4 Å would be subject to interstellar absorption, but the component at 1265 Å should be visible. Engvold, et al.(1983) also favoured excitation in FeII uv 9. Collisional excitation directly from a^6S to the x^6P^o term was proposed by Nussbaumer et al. (1981). However, if the collision strengths of multiplets uv 1 and uv 191 are comparable (on the basis of the relative f-values), the Boltzmann factors in the populations of the lower levels and excitation rates would lead to a flux ratio of $\sim 10^3$ for individual lines. This is about an order of magnitude larger than observed in β Gru.

Recently Johansson and Hansen (1986), in these Proceedings, have suggested that di-electronic recombination from an unknown 4s4d ^6D term just above the ionization limit could decay through the x^6P^o term, a process that would favour only the uv 191 muliplet. The ion balance of FeII/FeIII needs a proper treatment before this suggestion can be checked in a quantitative way. We note that uv 191, besides other multiplets, is strong in the beam-foil source used by Dolby et al. (1979) to measure some FeII and FeIII level lifetimes. The mechanism for populating highly excited states in such sources is poorly understood, but uv 191 (upper excitation 9.8 eV) appears to be stronger than nearly multiplets such as uv 38 and 39 from levels around 7 eV. Finally, because of the close proximity of the upper level of uv 191 to e^4D and e^6D, which are known to be overpopulated with respect to lower levels, Penston et al. (1983) suggested that collisional processes might transfer population from e^4D and e^6D to x^6P^o.

4. CONCLUSIONS

The observations of FeII in the uv spectra of cool giants and supergiants with IUE have been of value in several respects, regarding the identification of energy levels and excitation processes. Given the richness of the FeII spectrum and the variety of processes occuring the lines have untapped potential for improving models of cool star chromospheres. Because of the importance of the radiation field, in the continuum and strong lines, particularly H Lyα, a full ab-initio treatment of the line excitation and ion balance is now required. A wide range of oscillator strengths and collision cross-sections are needed. Whilst there appears to be an ample source of excitation to the 5 eV levels through continuum photo-excitation and collisions, the high relative strength of lines from the 10 eV levels in the M giants and supergiants occurs through the importance of H Lyα excitation. The source of excitation of multiplet uv 191 yet clear and all the possible processes need to be modelled in more detail.

REFERENCES

Ayres, T.R. 1979, *Astrophys.J.* **228**, 509.
Brown, A. & Jordan, C. 1980, *Mon.Not.R.astr.Soc.***191**, 37P.
Brown, A., Jordan, C. & Wilson, R. 1979, Proc. of Symposium *'The First Year of IUE'*(Ed. A. Willis) UCL, p232.
Brown,A., Ferraz, M.C. de M. & Jordan, C. 1981, In Proc. Symp. *'The Universe at Ultraviolet Wavelengths''The First Two Years of IUE'* NASA, CP-2171, p897.
Brown,A., Ferraz, M.C. de M. & Jordan, C. 1984 *Mon. Not. R. astr. Soc.* 207, 831.
Burton, W.M. & Ridgeley, A., 1970, *Sol. Phys.* **14**, 3.
Carpenter, K.G. 1984a in *'The Future of Ultraviolet Astronomy: The First Six Years oF IUE'* NASA CP-2349 (Eds.J.M.Mead, R.D.Chapman & Y.Kondo), p450.
Carpenter, K.G. 1984b, *Astrophys.J.* **285**, 181.
Carpenter, K.G. & Wing, R.F. 1979, *Bull.Am.Astr.Soc.* **11**, 419.
Collin-Souffrin, S., Joly, M., Heidmann, N. & Dumont, S. 1979, *Astron. Astrophys.* **72**, 293.
Collin-Soufrin, S., Dumont, S., Heidmann, N. & Joly, M. 1980, *Astron. Astrophys.* 83, 190.
Dolby, J.S., McWhirter, R.W.P. & Sofield, C.J. 1979, *J. Phys. B.* **12**, 187.
Doschek, G.A., Feldman, U., VanHoosier, M.E. & Bartoe, J.-D.F. 1976, *Astrophys. J. Suppl.* **31**, 417.
Elitzur, M. & Netzer, H. 1985, *Astrophys.J.* **291**, 464.
Engvold, O., Jensen, F. & Moe, O.K. 1983, Paper presented at Nordic Astronomy Meeting, 15-17 August, University of Oslo.
Gahm, G.F. 1974, *Astron. Astrophys. Suppl.* 18, 259.
Johansson, S. 1977a, *Astrophys. J.* 212, 923.
Johansson, S. 1977b, *Mon.Not.R.astr.Soc.* 178, 17p.
Johansson, S. 1978a, *Physica Scripta*, 18, 217.
Johansson, S. 1978b, *Mon.Not.R.astr.Soc.* 184, 593.
Johansson, S. 1981, *'FeII Multiplet Table. A preliminary version for the region 1050-2200 Å'*. Internal Report, LRAP-8, Dept. of Physics, University of Lund.
Johansson, S. 1984, *Physica Scripta*, **T8**, 63.

Johansson, S. & Hansen, J.E. 1986, These Proceedings.

Johansson, S. & Jordan, C. 1984, *Mon.Not.R.astr.Soc.,* **210**, 239.

Jordan, C. 1967, *Sol. Phys.* **2**, 441.

Jordan, C. 1979, in *'Progress in Atomic Spectroscopy',* Part B
  (Ed. Hanle and Kleinpoppen) Plenum Press, 1453.

Jordan, C. and Judge, P., 1984, *Physica Scripta,* **T8**, 43.

Jordan, C., Brueckner, G.E., Bartoe, J.-D.F., Sandlin, G.D. &
  VanHoosier, M. E., 1978, Astrophys. J. **226**, 687.

Judge, P.G. 1986, *Mon.Not.R.astr.Soc.* In Press.

Kurucz, R.L., 1981, *Smithsonian Astrophysical Observatory, Special
  Report,* 390.

Moore, C. E., 1952, *NBS Circular No. 488* Section 2.

Nussbaumer, H. 1977, *Astron. Astrophys.* **58**, 291.

Nussbaumer, H. & Swings, J.P. 1970, *Astron. Astrophys.* **7**, 455.

Nussbaumer,H. & Storey, P.J. 1980, *Astron. Astrophys.* **89**, 308.

Nussbaumer, H., Pettini, M, & Storey, P.J. 1981, *Astron. Astrophys.*
  102, 351.

Penston, M.V., Benvenuti, P., Cassatella, A., Heck, A., Selvelli, P.L.,
  Macchetto, F., Ponz, D., Jordan, C., Cramer, N., Rufener, F.
  & Manfroid, J. 1983, *Mon.Not.R.astr.Soc.* **202**, 833.

Sandlin, G.D., Bartoe, J.-D.F., Brueckner, G.E., Tousey, R. & VanHoosier,
  M.E. 1986, *Astrophys. J. Suppl.* **61**, 801.

Stalio, R. & Selvelli, P.L. 1980, in *The Universe at Ultraviolet
  Wavelengths: The First Two Years of IUE'* (Ed. R.Chapman) NASA CP-2171,
  p201.

Van der Hucht, K.A., Stencel, R.E., Haisch, B.M. & Kondo, Y. 1979, *Astron.
  Astrophys. Suppl.* **36**, 377.

Viotti, R. 1976, *Mon.Not.R.astr.Soc.* **177**, 617.

Viotti, R. 1976, *Astrophys. J.* **204**, 293.

Viotti, R., Giangrande, A., Riciardi, O., Altamore, A., Cassatella, A.,
  Freidjung, M. & Muratoio, G. 1980, *Proc. Second. European IUE
  Conference,* p39.

Wing, R.F. 1978, in *Proc. of the Fourth International Colloquium on
  Astrophysics.* (Ed. M.Hack) p683.

Wing, R.F., Carpenter, K.G. & Wahlgren, G.M. 1983, *Perkins Obs. Spec.
  Publ. No.1* (Ohio Stae University, Ohio Wesleyan University).

Zanstra, H. 1949, *Bull. Astron. Inst. Neth.* **11**, 1.

ON THE EXCITATION OF THE FE II MULTIPLET UV 191 IN STELLAR SPECTRA

Sveneric Johansson
Department of Physics
University of Lund
LUND, Sweden

Jørgen E Hansen
Zeeman Laboratory
University of Amsterdam
AMSTERDAM, The Nederlands

ABSTRACT. Different excitation mechanisms that have been proposed as responsible for the strength of the Fe II multiplet UV 191 in stellar spectra are discussed. Dielectronic recombination is suggested as an alternative explanation. Some theoretical atomic data are presented.

1. INTRODUCTION

The Ultraviolet Fe II Multiplet 191 (henceforth UV 191) has turned out to be a conspicuous feature in many astrophysical spectra, particularly in spectra of cool stars (Engvold et al 1983). The multiplet consists of three lines around 1780 Å, which were observed early on in IUE spectra of binary systems (Stencel et al 1979). Different excitation mechanisms have been proposed for the high population rate of the upper term of the multiplet, x^6P^0 at 10 eV. In the eclipsing binaries VV Cep and 32 Cyg photoexcitation by continuum radiation from the hot companion has been suggested by Hagen et al (1980) and by Hempe and Reimers (1982). An accidental resonance between a Si II line and one line in UV 9 of Fe II, which shares the upper term with UV 191, led Viotti et al (1980) to think of selective photoexcitation in the UV 9 channel with subsequent fluorescence in UV 191. Collisional excitation in the UV 191 channel itself was introduced by Nussbaumer et al (1981) and collisional redistribution of overpopulated 5s levels was suggested by Penston et al (1983) as a possible excitation mechanism in RR Tel.

In this paper we will consider another excitation mechanism, dielectronic recombination. Recent studies of doubly-excited configurations in Fe II (Brage et al 1986) indicate that the upper x^6P^0 levels in UV 191 may be populated through cascading from highly excited levels, predicted to lie just above the ionization limit. However, we will begin by reviewing the mechanisms that already have been proposed to give a large population of the x^6P^0 term with a subsequent fluorescence in UV 191.

235

R. Viotti et al. (eds.), Physics of Formation of FeII Lines Outside LTE, 235–241.
© 1988 by D. Reidel Publishing Company.

2. DIFFERENT EXCITATION MECHANISMS

The two spectroscopic terms involved in UV 191, a^6S and x^6P^0, belong to the system of doubly excited configurations (see the paper on the atomic structure of Fe II in this volume by Johansson). The multiplet corresponds to the $3d^5(^6S)4s^2(^1S)^6S - 3d^5(^6S)4s4p(^1P)^6P^0$ transition, abbreviated to $a^6S - x^6P^0$ in most compilations and tables. Besides the selection rules for parity and J-value, the LS-selection rules for three "inner quantum numbers" - grand-parent term, intermediate term and final LS term - have a great influence on the strength of a transition. In UV 191 all these selection rules are obeyed. In Table 1 we give calculated oscillator strengths (emission) for the strongest component in the multiplets that are discussed in the present paper. We have also included some other transitions in order to give an idea of the branching ratios. As a measure of the significance of the intermediate term for "allowed" transitions we see a difference of two orders of magnitude in the gf values for UV 190 and 191, log gf= -1.41 and 0.87.

In Figures 1, 2 and 3 we have shown a partial term diagram of the spectroscopic terms that are of interest here, i.e. the ground term a^6D, the lower level of UV 191, a^6S, and the even parity terms of $3d^65s$, denoted e^6D and e^4D. The odd parity terms are $y^6P^0 = 3d^5(^6S)4s4p(^3P)^6P^0$, $x^6P^0 = 3d^5(^6S)4s4p(^1P)^6P^0$ and $w^6P = 3d^6(^5D)5p^6P^0$. The terms above the ionization limit belong to the doubly-excited configurations, $3d^54s4d$ and $3d^54p^2$. They are not experimentally known but predicted positions are indicated in the figures.

2.1. Photoexcitation

When detailed studies of high-resolution IUE spectra started the Fe II multiplet UV 191 became a challenging part of the analysis, as discussed and elucidated by Engvold et al (1983). These authors made the problem even more puzzling by reporting that one component of the multiplet is missing in the spectra of some cool stars. However, they could not offer any explanation for that observation.

Hagen et al (1980) noticed the conspicuous UV 191 triplet already in low-resolution IUE spectra of VV Cephei, an eclipsing binary, and suggested photoexcitation from the ground term to the upper term, x^6P^0, in the UV 9 Multiplet (see Fig.1). They also studied the evolution of the spectrum during and after an eclipse and noticed that the fluorescence lines dominated the spectrum during the eclipse. This fact implied that the pumping radiation came from the hot companion star.

Another explanation of the excitation of UV 191 was given by Viotti et al (1980) in their study of the emission line star Boss 1985. They noticed the close coincidence between one of the lines in UV 9 at 1260.4 Å and one component in multiplet UV 4 of Si II and suggested a selective excitation in this channel, similar to the Bowen mechanism involving O III. Since this kind of processes has been referred to by many different names, Kastner and Bhatia have introduced a special acronym, PAR, standing for "photoexcitation by accidental resonance", which emphasizes the significance of the chance coincidence in wavelength. The PAR process, proposed by Viotti et al for Boss 1985, can obviously not

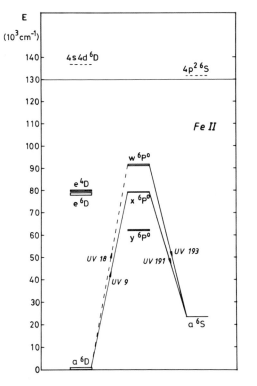

Fig. 1. *PHOTOEXCITATION*
          of the UV 191 multiplet

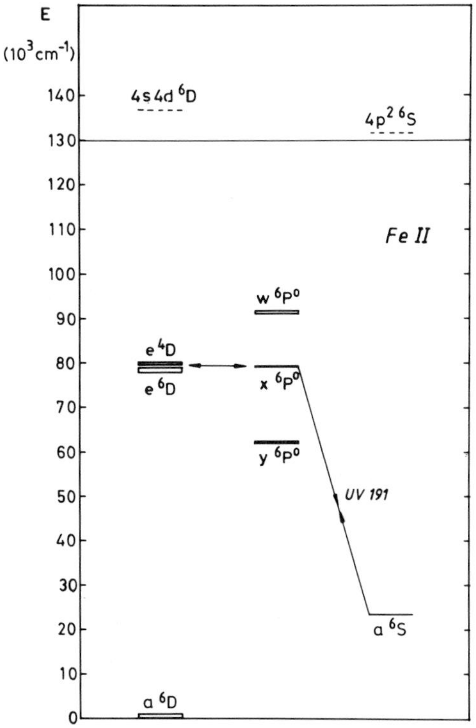

Fig. 2. *COLLISIONAL EXCITATION*
          of the UV 191 multiplet

account for the excitation of all the UV 191 lines, as was pointed out by Nussbaumer et al (1981).

The excitation scheme, discussed above for VV Cep, was confirmed by Hempe and Reimers (1982) in a study of the eclipsing binary 32 Cygni. The analysis of the observed spectra, taken during and at the egress of an eclipse, gave an unambiguous answer to the suggested photoexcitation by radiation from the hot companion star. In the IUE spectra taken at different times all Fe II lines behaved consistently, except for the lines in the UV 9 Multiplet. In the spectra of 32 Cyg there were also emission lines of the UV 193 multiplet present (see Fig. 1), which verified the photoexcitation mechanism. These lines correspond to transitions from the a^6S term to the w^6P^0 term of the $3d^65p$ configuration and are present in the laboratory spectrum of Fe II. They should therefore show up in emission if the w^6P^0 levels are populated. The w^6P^0 term has combinations to the ground term corresponding to fairly strong lines in UV Multiplet 18 and should be populated by continuum radiation with the same degree of probability as the x^6P^0 term. However, the absorption lines in UV 18 lie around 1100 Å, outside the range of IUE, so they can not be observed in order to confirm the photoexcitation of the UV 193 multiplet.

2.2 Collisional excitation

In a paper on sextet transitions in Fe II, Nussbaumer et al (1981) discussed the appearance of the UV 191 multiplet and proposed collisional excitation of metastable ions in the a^6S state as an alternative explanation to the population of the x^6P term. This means that the two terms in the UV 191 multiplet define a "closed system" (see Fig 2), where radiative transitions in UV 191 follow directly after collisional excitation. The branching ratio between UV 191 and UV 9 supports this scheme, if the a^6S state could be fed through collisional excitation from the ground state. However, this kind of mechanism should in principle generate a large number of multiplets in emission by collisional excitation from other low metastable levels.

In their study of RR Tel, Penston et al (1983) introduced collisional redistribution of the 5s terms e^6D and e^4D as a possible mechanism (see Fig 2). It is known from the $3d^6(^5D)4p - 3d^6(^5D)5s$ supermultiplets, i.e. UV multiplets 363, 373, 380, 391 and 399, that the 5s levels have a large population. In competition with radiative transitions to the 4p levels in these multiplets the 5s levels could be depopulated in energy transfer collisions with Fe^+ in the ground state, leaving the ions in a x^6P state. An examination of the energy region in the proximity of the 5s levels shows that additional odd states should be populated through collisional redistribution, unless there is some collision selection rule, which favours population of the sextet levels. It should be mentioned that the 5s states were thought to be populated through radiative recombination in RR Tel. Later on it has been shown (Johansson and Jordan, 1984) that the 5s levels might as well be populated through cascading from 5p levels, photoexcited by H Lyα.

2.3. Dielectronic recombination

We will here introduce another excitation mechanism, dielectronic recombination, that may be relevant for the excitation of the UV 191 multiplet. As mentioned in section 2.1 the upper term in the multiplet, x^6P^0, belongs to the $3d^54s4p$ configuration of the doubly-excited system. The coupling conditions in this configuration give high significance to two "inner quantum numbers" ascribed to the grand-parent term and to the intermediate term. As can be seen from the notation in Table 1 the x^6P^0 term has the intermediate term ^1P and the grand-parent, ^6S, i.e. the same grand-parent as the a^6S term. The strength of UV 191 compared to UV 9 may be ascribed to the LS-selection rule for the intermediate terms. The $a^6S - x^6P^0$ transition involves a $^1S-^1P$ transition for the intermediate term while the $a^6S-y^6P^0$ transition corresponds to the "intercombination" transition $^1S-^3P$.

The two levels above the ionization limit in Fig 3, $4s4d\ ^6D$ and $4p^2\ ^6S$, are not experimentally known, but their predicted positions (Brage et al 1986) are indicated by dashed lines. In Table 1 we can see that these terms are built on the same grand-parent term as x^6P^0 and they have ^1D and ^1S terms as intermediate terms. The calculated transition probabilities given in Table 1 imply that their fastest decay channels terminate on the x^6P^0 term. (For comparison we have in Table 1

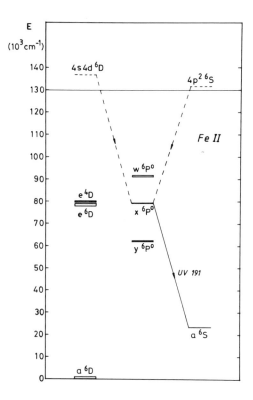

Fig. 3. Excitation of UV 191 by
DIELECTRONIC RECOMBINATION

inserted the transition probabilities for transitions to the y^6P^0 term, having an intermediate triplet term, and we see that the branching ratios for the transitions from the auto-ionizing states down to x^6P^0 and y^6P^0 are of the same order of magnitude as for the multiplets 190 and 191). Thus, if these two doubly-excited terms are populated they may either decay through auto- ionization or through radiative transitions to x^6P. Since they are supposed to interact with the $3d^6(^5D)\epsilon s$ and/or ϵd continuum they may have a high probability for being populated through dielectronic recombination.

In this paper we restrict our discussion to what consequences can be expected if these levels are populated. We have not done any calculations of rate coefficients for dielectronic recombination but propose this mechanism as a possibility in analogy with more well established cases in planetary nebulae (Harrington et al 1980, Storey 1981). There are at least two conditions that are critical for the dielectronic recombination process in the present case, the density of Fe^{++} in the stellar atmosphere and the actual position of the doubly-excited states in the continuum. In some spectra, where the UV 191 multiplet is strong, Fe III lines have been identified. However, the density of Fe^{++} ions in the ground state can not be deduced from IUE spectra by observation of resonance lines in Fe III, since all of them appear below the lower wavelength limit of IUE. The most probable Fe III multiplet is UV 34, $a^7S - z^7P$, around 1900 Å having an excitation potential for the lower term of 3.7 eV. The level values of the predicted terms were calculated by making a parametric fit to new level data in the doubly-excited system of Fe II (Brage et al 1986). The $4p^2\ ^6S$ level is predicted to lie only 0.2 eV and the $4s4d\ ^6D$ term 0.8 eV above the ionization limit. It is of course difficult to estimate the uncertainty in such predictions but we expect them to be less than 0.1-0.2 eV.

It has been noticed by Engvold et al (1983) that the intensity of UV 191 increases strongly with the effective temperature of cool stars, which should indicate that the excitation depends on the physical conditions in the stellar atmosphere itself. A higher temperature increases the population of the ground term of Fe III and should consequently increase the probability for dielectronic recombination for

the $4p^2$ ^6S and $4s4d$ ^6D terms. The subsequent cascading to x^6P should
thereby increase the intensity of UV 191.

*Table 1. Atomic data for those Fe II transitions, which are
relevant to the discussion of the excitation of the x^6P term.*

Multiplet Number[*]	Transition		Wavelength (Å)	log gf[+] em.
	Terms	Configurations		
UV 8	$a^6D-y^6P^0$	$d^6(^5D)4s -d^5(^6S)4s4p(^3P)$	1608–1639	−0.39 K
UV 9	$a^6D-x^6P^0$	$d^6(^5D)4s -d^5(^6S)4s4p(^1P)$	1260–1275	−0.60 K
UV 18	$a^6D-w^6P^0$	$d^6(^5D)4s -d^6(^5D)5p$	1096–1105	−0.29 K
UV 190	$a^6S-y^6P^0$	$d^5(^6S)4s^2-d^5(^6S)4s4p(^3P)$	2572–2586	−1.41
UV 191	$a^6S-x^6P^0$	$d^5(^6S)4s^2-d^5(^6S)4s4p(^1P)$	1785–1788	0.87
UV 193	$a^6S-w^6P^0$	$d^5(^6S)4s^2-d^6(^5D)5p$	1465–1495	−0.29
new	$e^6D-w^6P^0$	$d^6(^5D)5s -d^6(^5D)5p$	$\approx$ 7500	0.10 K
pred.	$y^6P^0-^6D$	$d^5(^6S)4s4p(^3P)-d^5(^6S)4s4d(^1D)$	$\approx$ 1400	−1.53
pred.	$x^6P^0-^6D$	$d^5(^6S)4s4p(^1P)-d^5(^6S)4s4d(^1D)$	$\approx$ 1900	1.10
pred.	$y^6P^0-^6S$	$d^5(^6S)4s4p(^3P)-d^5(^6S)4p^2(^1S)$	$\approx$ 1400	−1.47
pred.	$x^6P^0-^6S$	$d^5(^6S)4s4p(^1P)-d^5(^6S)4p^2(^1S)$	$\approx$ 1900	0.57

[*] The multiplet at 7500 Å is experimentally known but not included
   in the Revised Multiplet Table. "pred." means predicted.
[+] In this column is given the calculated log gf value (in emission)
   for the strongest component of the multiplet. A value marked with a
   K is taken from Kurucz (1981).

3. CONCLUSIONS

With the assumption that the doubly-excited states $4s4d$ ^6D and $4p^2$ ^6S
are located where we have predicted them, dielectronic recombination
followed by cascading to x^6P^0 is a plausible explanation of the strong
Fe II multiplet UV 191 in spectra of stars, having a suitable density of
Fe^{++} in their atmospheres. A possibility to check this excitation scheme
would be to look for unidentified features in stellar spectra in the
1700-1900 Å region with intensities correlated to the intensities of the
UV 191 lines and try to find the lines that would establish the unknown
levels above the ionization limit.

4. REFERENCES

Brage, T., Nilsson, A.E., Johansson, S., Baschek, B., and Adam, J. 1986,
          J. Phys. B. (in press).
Engvold, O., Jensen, E., and Kjeldseth-Moe, O. 1983, Inst. Theor.
          Astrophys. Oslo University Report No. 59, p. 65-75.
Hagen, W., Black, J. H., Dupree, A. K., and Holm, A. V. 1980,
          Astrophys. J. **238**, 203.

Harrington, J.P., Lutz, J.H., Seaton, M.J., and Stickland, D.J. 1980,
        Mon.Not.R.astr.Soc., **191**, 13.
Hempe, K., and Reimers, D. 1982, *Astron. Astrophys.* **107**, 36.
Johansson, S., and Jordan, C. 1984, *Mon.Not.R.astr.Soc.*, **210**, 239.
Kastner, S.O., and Bhatia, A.K. 1986, *Comments At. Mol. Phys.*, **18**, 39.
Kurucz, R.L. 1981, *Smithsonian Ap.Obs.Spec.Rept.,*No.390.
Nussbaumer, H., Pettini, M., and Storey, P.J. 1981, *Astron. Astrophys.*
        102, 351.
Penston, M.V., Benvenutti, P., Cassatella, A., Heck, A., Selvelli, P.,
        Macchetto, F., Ponz, D., Jordan, C., Cramer, N.,
        Rufener, F., and Manfroid, J. 1983, *Mon.Not.R.astr.Soc.*
        202, 833.
Persic, M., Hack, M., and Selvelli, P.L. 1984, *Astron. Astrophys.* **140**,
        317.
Stencel, R. E., Kondo, Y., Bernat, A.P., McCluskey, G.E. 1979,
        Astrophys, J., **233**, 621.
Storey, P.J. 1981, *Mon.Not.R.astr.Soc.*, **195**, 27P.
Viotti, R., Giangrande, A., Riccardia, O., Altamore, A., Cassatella, A.,
        Friedjung, M., and Muratorio, G. 1980,
        Proc. 2. Europ. IUE Conf. ESA SP-157, p. 229.

DOUBLY EXCITED FE II LINES IN THE A STAR 21 PEGASI

J. Adam, B. Baschek
Institute of Theoretical Astrophysics
University of Heidelberg
Im Neuenheimer Feld 561
D-6900 Heidelberg
Federal Republic of Germany

T. Brage, A.E. Nilsson, S. Johansson
Department of Physics, University of Lund
Sölvegatan 14
S-22362 Lund
Sweden

ABSTRACT. The majority of about 120 ultraviolet iron lines, among them strong lines around 1850 Å, which have recently been identified as transitions from doubly excited levels in Fe II during a new investigation of the hollow-cathode spectrum of iron, coincide with absorption lines in the high-resolution IUE spectrum of the sharp-lined A star 21 Pegasi. Model atmosphere calculations of the line strengths based upon Sadakane's (1981) parameters for 21 Peg and quantum-mechanically calculated f-values confirm the identification for about one third of the lines, located mostly in the region 1750-1900 Å. For the remaining coincidences the stellar line is probably blended with another contributor.

A new investigation of the hollow-cathode spectrum of Fe in the region 1300-3800 Å is in progress (Baschek and Johansson, 1986). In the course of this work some 120 laboratory lines in the IUE range have been identified as transitions from doubly excited states of the $(3d^5 4s4d + 3d^5 4p^2)$-configuration complex in Fe II (Brage et al., 1986). The results are applied to the spectrum of the sharp-lined B9.5V star 21 Pegasi = HR 8404 (V = 5.8 mag) which has been observed with IUE in high resolution by R. Wehrse in 1983 (SWP 21398, LWP 2173). Details of this work will be published elsewhere (Adam et al., 1986).

From an analysis of the optical spectrum of 21 Peg, Sadakane (1981) derived an effective temperature of 10 500 K, a surface gravity of $3 \cdot 10^3$ cm s^{-2}, a microturbulent velocity

R. Viotti et al. (eds.), Physics of Formation of FeII Lines Outside LTE, 243–245.

of 1 km s^{-1}, and a slight underabundance of Fe.

21 Peg exhibits a well-developed Fe II absorption line
spectrum; e.g. in the region $\lambda \leq$ 1900 Å all lines of Johansson's (1978) list with laboratory intensity $\geq$ 2 and many
fainter lines are present. Since the strongest ultraviolet
Fe I lines are not detectable in this spectral range, the
star furthermore acts as an effective Fe I-filter helping
to separate Fe I from Fe II lines in the laboratory spectra
where the strengths of neutral and ionized lines are comparable.

The majority of the new doubly-excited Fe II lines
(which have excitation potentials between 4.8 and 9.8 eV)
coincide with lines in 21 Peg. Due to different excitation
mechanisms in the laboratory source and the star we do not
expect a clear correlation between the laboratory intensities
and the stellar equivalent widths. In order to confirm the
identifications, we therefore apply model atmosphere calculations based upon quantum-mechanically calculated oscillator strengths for the new Fe II lines. We adopt Sadakane's
(1981) model parameters and his iron abundance for 21 Peg
to calculate an LTE model in hydrostatic and radiative
equilibrium with Kurucz' (1979) ATLAS 6 program and line
opacity distribution functions.

We note that even the faintest measurable UV lines in
21 Peg with central depression of about 0.1 to 0.2 in units
of the continuum are saturated, i.e. are located on the flat
part of the curve of growth because the instrumentally
determined line width is about 10 times larger than the thermal plus microturbulent width.

In the region between 1750 and 1900 Å, the model atmosphere calculations yield a remarkably good agreement between
the theoretical and the observed line strengths within the
uncertainties of the stellar equivalent widths of about 20
to 30 mÅ so that our investigation of the Fe II spectrum
contributes several tens of new identifications to the
stellar spectrum. On the other hand, in the region below
1600 Å the theoretical line strengths for the new lines -
as well as those for comparison lines of similar excitation
condition from Johansson's (1978) list with Kurucz' (1980)
f-values - come out too weak so that fortuitous coincidences
due to blending by as yet unknown contributors seem to
dominate here.

REFERENCES

Adam. J., Baschek, B., Johansson, S., Nilsson, A.E., Brage,
    T.: 1986, Astrophys. J. (Dec 15), in press
Baschek, B., Johansson, S.: 1986, contribution to this
    Colloquium
Brage, T., Nilsson, A.E., Johansson, S., Baschek, B., Adam,
    J.: 1986, contribution to this Colloquium

Johansson, S.: 1978, Physica Scripta 18, 217

Kurucz, R.L.: 1979, Astrophys. J. Suppl. 40, 1

Kurucz, R.L.: 1980, Smithsonian Astrophys. Obs. Spec. Rep.
    No. 390

Sadakane, K.: 1981, Publ. Astron. Soc. Pacific 93, 587

THE FORMATION OF FeII EMISSION LINES

H. Netzer
School of Physics and Astronomy, Tel Aviv University
Tel Aviv 69978, Israel

ABSTRACT. The excitation of FeII emission lines is reviewed, with
special emphasis on photoionized nebulae. The physical conditions, the
geometry and the techniques required to calculate FeII models are explained and the following mechanisms discussed: a: Excitation by absorption of continuum radiation. b: Excitation by inelastic collisions,
and c: Excitation by line fluorescence. Possibility a has been overlooked in the past and is probably not important in objects showing
strong optical FeII lines. Collisional excitation dominates in nature
but cannot, in itself, explain the strong observed lines from energy
levels of 7 to 10 eV. Line fluorescence, with L_α, CIV and most important among FeII lines, is probably the key to the understanding of the
high energy lines. A recipe for calculating FeII models is given,
showing the processes to include and the techniques to use. Major
recent findings about the FeII spectrum of AGN are discussed in detail,
in particular "the FeII problem" in these objects. Despite big improvements in the understanding of the 2000-4000A emission bump and other
FeII features, there are still several open questions like the role of
dust, the iron abundance, the geometry and density of the broad line
region and the general validity of photoionization models.

1.   INTRODUCTION

This review discusses the excitation of FeII lines in gaseous nebulae.
Two of the possible excitation mechanisms (absorption of continuum
radiation and collisions) have been known for years and there have been
big recent improvements in their treatment. A third process, (FeII
line fluorescence) has only been discovered recently and is, no doubt,
of great importance. All these will be discussed in Section 2 and a
recipe for calculating an FeII model will be given in Section 3.
    Most of the following applies to emission line nebulae where the
line emitting gas is removed from the photoionizing source. This is
similar to the situation in planetary nebulae, the Broad Line Region
(BLR) of Active Galactic Nuclei (AGN), novae at later stages of evolution and at least some symbiotic stars. Stellar photospheres, super-

R. Viotti et al. (eds.), Physics of Formation of FeII Lines Outside LTE, 247–257.

novae shells and shock excited gas may be different in some respects. The spectrum of AGN will be discussed, in more detail, in Section 4.

The bibliography prepared by Viotti and distributed at the meeting is a most important and useful collection of references. I shall heavily rely on it and will not refer, specifically, to many papers. My apologies to those whose excellent work on FeII has not been forgotten, but omitted from the present paper for the sake of brevity.

2.   EXCITATION MECHANISMS

Consider a gas cloud situated at a distance R from a photoionizing source whose dimension is much smaller than R. The degree of ionization, the gas temperature and the emitted spectrum, are all controlled by the ionization parameter (the incident flux of ionizing photons per hydrogen atom), the gas density and abundances. This is shown schematically in Fig. 1.

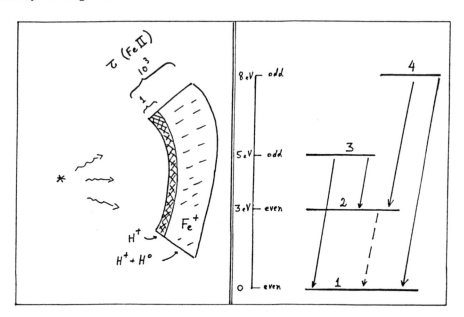

Fig. 1. Basic assumptions for photo-ionized gas. The ionizing radiation from the central source creates an H^+ zone, where most iron is at least doubly ionized, and a partly neutral zone where most iron is Fe^+. The layer where $\tau(FeII) \sim 1$ is very thin compared with the overall thickness of the Fe^+ zone.

Fig. 2. A four level Fe^+ atom showing all possible permitted (solid line) and forbidden (dashed line) transitions. Note that emission of a 3-1 line photon can result in absorption from 1 to 3 or 2 to 4 (line fluorescence).

Some iron is singly ionized and can be excited to high energy
levels in several different ways. The energy level diagram of Fe$^+$ is
extremely complex but it is possible to explain most processes using
a simple 4 level atom, as shown in Fig. 2. It consists of an even
parity ground level (level 1), a meta-stable even parity level at 3 eV
(level 2) and odd parity levels at 5 eV (level 3) and 8 eV (level 4).
The strong optical FeII lines (M37, M38, M42 ...) are due to 3-2 transitions. The UV lines (UV1, UV3, UV33 ...) are due to 3-1 and 4-1
transitions and the optical forbidden lines due to 2-1 transitions.
The spontaneous emission coefficient A_{32} is two to three orders of magnitude smaller than A_{31} so strong observed 3-2 lines must indicate
$\tau(3-1) \gtrsim 100$.
    Three excitation mechanisms will be considered.

2.1. Excitation by Absorption of Continuum Radiation

This process, sometimes called "continuum fluorescence" (a confusing
name, in view of the new line fluorescence process to be discussed),
involves the absorption of continuum radiation in the 1-3, 2-3, 1-4
and 2-4 transitions. Fig. 3 shows the expected absorption probability
of a gas with N $\sim 10^{10} cm^{-3}$ and large optical depth ($\sim 10^4$) in the 1-3
lines.

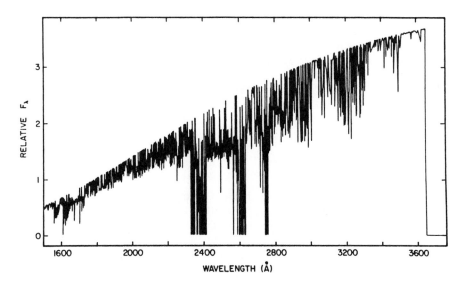

Fig. 3. The wavelength dependent absorption of an arbitrary continuum
(smooth upper envelope) by FeII lines in a high density gas where the
optical depth for the strong FeII resonance lines is 10^4. Only a small
part of the continuum (e.g. near 2400Å) is totally absorbed.

There are three requirements to produce strong emission lines by this process: <u>a</u> Broad absorption profiles: Continuum absorption is not efficient if lines have narrow thermal profiles, even at extreme optical depth (see Fig. 3). Large turbulent motion or differentially expanding clouds increase the absorption efficiency but such clouds may be unstable. <u>b</u> Large covering factor: The solid angle sustended by all clouds, Ω, may be small, enabling only a small fraction of the available continuum to be absorbed. This, for example, is the case in AGN, where the covering factor, $\Omega/4\pi$, is about 0.1. <u>c</u> Large optical depths: $\tau(FeII) > 1$ is required to absorb continuum photons. $\tau(FeII, 1-3) > 100$ is needed to produce strong optical (3-2) lines. The last requirement is in contradiction with the basic assumptions of the mechanism under discussion since most absorption takes place where $\tau(FeII) \sim 1$ and almost no radiation reaches the region where $\tau(FeII) > 10$. This has been overlooked in the past. It provides a strong argument against continuum absorption as the main excitation mechanism for object with strong optical FeII lines.

2.2. Collisional Excitation

Inelastic collisions with thermal electrons is important at high enough temperatures. $T_e > 7000K$ is required to produce strong 3-1 lines. Such a high temperature must be maintained throughout the cloud, in parts where $\tau(1-3) >> 10$, if strong 3-2 lines are to be explained.

The complexity of the Fe^+ atom, the poorly known atomic data and the complicated line transfer, made past modeling of this process rather uncertain. The introduction of the escape probability method has improved the situation considerably. This method involves a location dependent function, $\epsilon(\tau)$, that gives the escape probability of a line photon from a region where the line center optical depth to the surface of the cloud is τ. $\epsilon(\tau)$ is known, from analytical and numerical solutions, to a factor of ~ 3 and is given approximately by

$$\epsilon_{ij}(\tau_{ij}) = \frac{1 - e^{-\tau_{ij}}}{\tau_{ij}} \tag{1}$$

where i and j are the lower and upper levels, respectively. It is the only way to calculate, simultaneously, the statistical equilibrium equations where hundreds of optically thick transitions are involved. This is done by replacing, everywhere, A_{ji} by $A_{ji} \epsilon_{ij}$ and calculating the emitted line flux from $N_j A_{ji} \epsilon_{ij} h\nu_{ij}$. The largest number of levels attempted so far is by Wills, Netzer and Wills (1985, hereafter WNW) who have considered about 70 terms (not individual levels) and approximately 1000 FeII multiplets. I am now extending the calculations to include ~ 250 individual levels and more than 3000 lines.

For thermal line profiles:

$$\tau_{ij} \propto N_i f_{ij} \lambda_{ij} \propto \lambda_{ij}^3 A_{ji} N_i \frac{g_j}{g_i} \tag{2}$$

which can be used to show that the relative intensity of transitions
with a common upper level (j) and different lower levels (i) that are
in thermal equilibrium among themselves, is proportional to

$$\frac{1 - e^{-\tau_{ij}}}{\lambda^4_{ij}} \; e^{E_i/kT} \tag{3}$$

where E_i is the excitation energy of level i. The relative intensity
of such lines are in their LTE values, for large τ_{ij}, even if the upper
level is not thermally populated. This is relevant to AGN and other
high density nebulae and can be used to estimate the relative strength
of the optical (3-2) and UV (3-1) lines with a common upper level,
without the need to completely solve the multi-level atom. It is clear
that levels such as 4 in Fig. 2, 7-10 eV above the ground level, cannot
be efficiently excited if T_e < 12,000K. This has been realised as a
major difficulty since the discovery of many, strong 4-1 and 4-2 type
transitions in stars and AGN. Either T_e is much larger, which is in
conflict with the whole idea of photoionization, or we have neglected
another excitation process. I believe we now have found at least part
of the answer to this question, which is the third mechanism.

2.3. Line Fluorescence

Fluorescence of FeII lines with lines of other elements has been suggested in the past (e.g. Grahm 1974, Van der Hucht et al. 1978, Brown
et al. 1979, Penston et al. 1983). It has not been followed up by
accurate calculations until recently (see below and C. Jordan review)
and has been confused by inaccurate wavelengths and the treatment of
the observed line width rather than the one expected from individual
clouds.
      Netzer and Wills (1983) and WNW have considered the obvious
alternative of fluorescence among FeII lines. They found more than
300 (!) line pairs close enough in wavelength and suggested that this
is the main route for exciting the high energy levels. Examination of
Fig. 2 helps to clarify this point: Consider collisional excitation of
level 3 followed by the emission of a 5 eV (3-1) line photon. In a
large optical depth medium this photon can: a escape, b be absorbed by
1-3, and c be absorbed by 2-4. Route c is the one populating level 4.
The process is important if there are large enough numbers of wavelength coincidences like (1-3)↔(2-4) and if the relative absorption
probability in 2-4 is large enough compared with 1-3.
      Netzer and Wills (1983) give a simple way to calculate the process.
Elitzur and Netzer (1985) and Netzer et al. (1985) show a more general
scheme, based on the escape probability formalism, that enable the
simultaneous solution of a large number of levels with hundreds of line
fluorescences.
      Fluorescence with L_α is a second important example. L_α, which is
a strong and broad line, can pump some 10 eV Fe^+ levels, via wavelength
coincidence with several FeII lines. One should note, however, the
extremely large $\tau(L_\alpha)$ required for this. An estimate of the half width

(in units of Doppler width) X_* , of a resonance line, is given by
Adams (1972)

$$X_* = (a \tau_o / \sqrt{\pi})^{1/3} \tag{4}$$

where τ_o is the line center optical depth and $\underline{a}$ the damping constant.
The FeII lines that are most suitable for the process are 45 and
140 km/sec from the L_α line center. Using (4) we find a minimal optical depth of 1.5×10^5 for the first and 5×10^6 for the second. Such
large values of $\tau(L_\alpha)$ are unlikely to be present in nebulae excited by
a central thermal source since $\tau(L_\alpha) < 10^5$ is typical of their H^+ zone
and the neutral H^o zone is too cool for efficient excitation .

The situation is different in AGN where the hard non-stellar con-
tinuum produces a warm, partly neutral zone of $T_e > 7000$ and $\tau(L_\alpha) \sim 10^8$
and even larger. AGN high density clouds are also more favorable for
the process since its efficiency is proportional to the n_2/n_1 hydrogen
level population ratio.

Fig. 4 shows new calculations of high density $\tau(L_\alpha) = 10^8$ gas where
L_α fluorescence is included. A few FeII lines are indeed excited by
L_α but the energy in all of them is very small.

A few words of caution. L_α can be much broader than indicated in
equation 4, which gives only the part of the profile where the source
function is constant. The radiation field further away from the line
center is much weaker but can perhaps excite more FeII transitions.
In addition, the situation described here does not apply to stellar
atmospheres where conditions are rather different. There is a definite
need for more elaborated calculations of this process.

3.   HOW TO CALCULATE FeII MODELS

A complete model must incorporate the FeII solution into a self consistent calculation of the ionization and thermal structure. For photoionized gas the following steps should be followed:

   a Calculate a photoionization model by dividing the cloud into
small radial zones. Find the ionization equilibrium of all elements
in each zone.

   b Solve for the thermal structure to obtain the run of temperature across the cloud.

   c Solve the FeII equations in each zone, given N_e, T_e and the
Fe^+ abundance.

These steps are not independent since the total cooling may be a strong
function of the FeII level population. Best is to solve $\underline{b}$ and $\underline{c}$ simultaneously, or else use an iteration procedure. Regarding $\underline{c}$, the larger
the number of levels, the better. Schemes including less than 20
levels, are not too bad in working out the total cooling but fail badly
in producing the observed FeII spectrum (e.g. Grandi 1981). Since calculating a 250 levels atom is more complicated than the whole photoionization model it pays to use an iteration procedure with a simplified scheme for estimating the cooling due to FeII in each step (e.g.

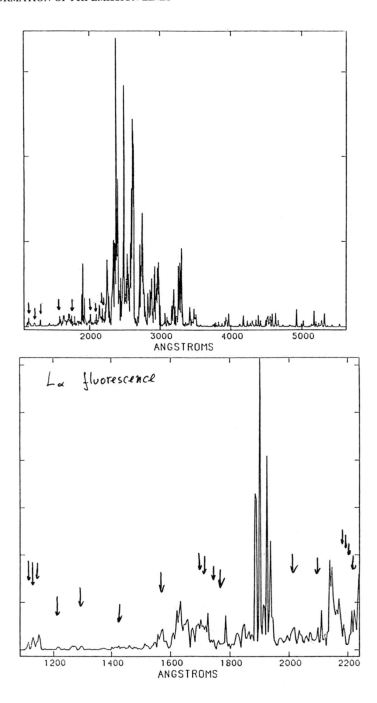

Fig. 4. The effect of L_α fluorescence in a high density $\tau(L_\alpha) = 10^8$ gas. FeII lines excited by this process are marked. A theoretical spectrum is shown on two different scales, to demonstrate the total energy in excited transitions.

WNW).

Several other processes must be taken into account. Most important for AGN clouds is the absorption of FeII lines by hydrogen atoms
in n=2. This is important where τ(Bal. continuum) approaches unity
and can change the ultraviolet FeII spectrum as well as the overall
ionization of hydrogen. Other, less important processes, are L_α and
CIV fluorescence and the absorption of the central and local (Balmer)
continua by FeII lines.

Several parameters in such calculations must be considered unknown;
most important – the iron abundance. FeII lines are so intense that
their cooling role in the Fe^+ zone must be significant. It is likely
that Fe/H is enhanced in some objects but this has only marginal
influence on the emergent FeII flux if most cooling is due to FeII.
The result is a large uncertainty in the calculated T_e in the Fe^+ zone
and thus an uncertainty in other lines that are formed there. FeII
line width is another unknown. Microturbulences may cause line broadening and change the efficiency of continuum absorption and line
fluorescence processes. There is no direct observational way to obtain
this parameter.

4.   FeII LINES IN AGN

The study of FeII lines in AGN has been an active and fast developing
subject since the early work of Wampler and Oke (1967) (Bahcall and
Kozlovsky 1969, Adams 1975, Oke and Shields 1976 , Osterbrock 1977,
Boksenberg and Netzer 1977, Phillips 1978a, 1978b, Collin-Souffrin et
al. 1979, Netzer 1980, Joly 1981, Kwan and Krolik 1981, Grandi 1981,
Wills et al. 1980, Netzer and Wills 1983, WNW 1985 and others). The
observational situation is reviewed by B. Wills in this volume.

Attempts to explain the FeII spectrum followed two parallel lines.
Collin-Souffrin (1979), Joly (1981) and collaborators (see review by
M. Joly) proposed that much of the FeII emission comes from gas which
is not photoionized but heated to about 10^4K by other mechanisms. Most
other workers have treated the FeII lines within the general framework
of photoionization models.

The central part of AGN contains a massive object, presumably a
black hole, with a very small region ($\sim 10^{16}$cm) around it where the
optical – UV and X-ray continuum originate. A likely, yet not the
only possibility is a massive accretion disk whose thermal emission is
the origin of the noticable continuum flattening at $\lambda < 7000$Å. The BLR
is the innermost line emitting region, with dimensions from $\sim 10^{17}$cm
(for faint Seyfert 1s) to few x 10^{18} (for bright QSO's). The gas is in
small clouds, or filaments, with typical column density of 10^{22-23}
atoms/cm^2 and small covering factor. Typical densities are $\sim 10^{10}cm^{-3}$.
The motion of the clouds, with relative velocity of up to 10,000 km/sec,
produces the broad emission lines but a single-cloud line has probably
a thermal profile. The Narrow Line Region (NLR) with lower densities
and velocities, is at several hundred parsecs from the center and does
not produce strong permitted FeII lines.

Conditions in the high density clouds are most favorable for the formation of FeII lines. The hard ionizing continuum produces enough photons that can penetrate deep into the gas and create an extended zone where $N(H^+)/N(H) \sim 0.05$ and $T_e \sim 7000-10,000K$. Most iron in this region is singly ionized and the optical depth in the resonance lines can reach 10^4 or 10^5.

Large improvement in the observations, over the last 3-4 years, show that <u>all</u> well studied broad lined AGN have strong ultraviolet FeII lines, although the optical counterparts may be weak. (Netzer et al. 1985 and Fig. 5 here). Improved modeling of those features (WNW) show that the total FeII emission is much larger than previously thought. A complete FeII model demonstrates the large contribution of hundreds of individually weak lines and show evidence that the prominent 2300-2600Å bump is only "the tip of the iceberg".

Fig. 5 shows how much energy is produced by the gas between 2000 and 4000Å. This wavelength region has been called "the small bump"

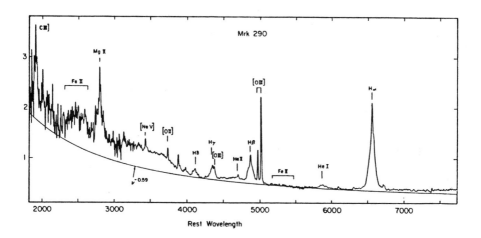

Fig. 5. Spectrum of the Seyfert 1 galaxy Mrkn 290. Note the very weak FeII (opt) and the strong FeII (UV). All the flux in the "small bump" between 2000 and 4000Å can be explained by FeII and Balmer continuum emission.

and has been attributed, in the past, to different continuum processes including a single 27,000K blackbody source. We now think that it can be explained by a combination of FeII and Balmer continuum emission.

The FeII spectrum of AGN is important in showing that <u>no reliable measurement of the FeII spectrum is possible without a good model for</u>

the line emission, i.e. the spectrum is so complex and line blending
so severe that   theoretical models are needed to enable the measurement of the lines. This is somewhat similar to the modeling of absorption lines in stellar atmospheres.

The work of WNW, based on the assumption that all FeII is from a
photoionized gas, can be summarized as follows:

a There are strong ultraviolet FeII lines in all broad-lined AGN.
The observed range in FeII(UV)/Fe(Opt) is large, from about 3 to 13.

b The total observed FeII emission is very large:
FeII(total)/L_α $\sim$ 1 and FeII(total)/H_β $\sim$ 10.

c Strong Balmer continuum emission is associated with strong
FeII.

d Models with enhanced iron abundance fit the observations best.

Photoionization theory cannot, yet, explain the FeII spectrum of
AGN. Several obvious improvements can be made. Firstly, the role of
dust and reddening must be further investigated. The number quoted in
b above is the observed ratio and may not be intrinsic to the source,
if some reddening is present. This will also affect the total energy
budget. Secondly, in most BLR models τ(Bal. cont) $\sim$ 1, which together
with τ(FeII) determines the relative intensity of FeII(UV)/FeII(opt).
There must be a clue of how to separate these two factors. Thirdly,
FeII emission may be predominantly from cloud that are different from
the one producing the high excitation lines (higher density, lower
ionization parameter, etc.). This possibility has not been given
enough attention, yet. Last, but not least, the possibility of a non-
photoionized gas, suggested by Collin-Souffrin and collaborators, must
be studied in more detail in connection with the overall spectrum of
AGN.

To summarize, presently there is an "FeII Problem" in AGN. There
has been much progress in the understanding of many spectral features
but the global energy budget is still not understood. The future
resolution of this problem may be, in my opinion, an important turning
point in the understanding of activity in nuclei of galaxies.

REFERENCES
Adams, T.F., 1972, Ap. J. 174, 439
Adams, T.F., 1975, Ap. J. 196, 675
Bahcall, J.N. and Kozlovsky, B.Z., 1969, Ap. J. 155, 1077
Boksenberg, A. and Netzer, H., 1977, Ap. J. 212, 37
Brown, A., Jordan, C. and Wilson, R., 1979, ("The First Year of IUE,"
      p. 232)
Collin-Souffrin, S., Joly, M., Heidmann, N. and Dumont, S., 1979,
      Astron. Ap. 72, 293
Elitzur, M. and Netzer, H., 1985, Ap. J. 291, 464
Gahm, G.F., 1974, Astron. Ap. Supp. 18, 259
Grandi, S.A., 1981, Ap. J. 251, 451
Joly, M., 1981, Astron. Ap. 102, 321
Kwan, J., and Krolik, J.H., 1981, Ap. J. 250, 478

Netzer, H., 1980, Ap. J. 236, 406
Netzer, H. and Wills, B.J., 1983, Ap. J., 275, 445
Netzer, H., Elitzur, M. and Ferland, G.J., 1985, Ap. J., 299, 752
Oke, J.B. and Shields, G.A., 1976, Ap. J., 207, 713
Osterbrock, D.E., 1977, Ap. J., 215, 733
Phillips, M.M., 1978a, Ap. J. Supp., 38, 187
                   1978b, Ap. J. 226, 736
Penston, M.V. et al., 1983, MNRAS 202, 833
Van der Hucht, K.A., Stencel, R.E., Haisch, B.M. and Kondo, Y.,
     1978, Astron. Ap. 36, 377
Wampler, E.J. and Oke, J.B., 1967, Ap. J., 148, 695
Wills, B.J., Netzer, H., Uomoto, A.K. and Wills, D., 1980, Ap. J. 237,
     319
Wills, B.J., Netzer, H. and Wills, D., 1980, Ap. J. Lett., 242, L1
Wills, B.J., Netzer, H. and Wills, D., 1985, Ap. J., 288, 94, (WNW)

FORMATION OF FEII LINES IN AGN

Monique Joly
Observatoire de Paris - Section Meudon
92195-Meudon Principal Cedex
France

ABSTRACT. Up to now photoionization models of the Broad Line Region of quasars fail to explain the strength of low ionization lines, in particular, FeII lines are too weak compared to Hβ. Using a code mainly designed to study low ionized and optically thick medium, I have calculated the radiation transfer in a non-photoionized homogeneous cloud in order to outline the physical conditions required to explain the observed FeII intensities. A strong constraint is set by the continuum radiation emitted by the cloud. It is a very efficient ionization factor of HI as well as of the heavy elements and the emission strength of the Balmer continuum gives limits to the temperature, the density and the column density. The results of the computations are compared to the published line ratios of about 30 well-observed AGN. I show that the "FeII/Hβ problem" is not easily solved as regards the strong UV FeII emitters, while low temperature clouds provide FeII opt/Hβ in agreement with observations.

1. INTRODUCTION

Wills et al. (1985) have stressed the existence of a "FeII problem" in the Broad Line Region (BLR) of Active Galactic Nuclei (AGN). Indeed FeII lines emit about half as much energy as the whole other lines : the observed ratios FeII tot/Lα and FeII tot/Hβ are respectively of the order of 2. and 12., while current photoionization models of the BLR provide much lower values of these ratios. The standard model of Kwan and Krolik (1981) provides FeII tot/Hβ ∿ 2. and the most favourable model (without any overabundance) computed by Wills et al. (1985) gives FeII tot/Lα ∿ 0.25 and FeII tot/Hβ ∿ 5. Therefore models which are more or less successful to account for the whole line spectrum fail to explain the FeII lines : theoretical FeII intensity is too weak compared to hydrogen lines.

In this paper, two attempts to improve the fit are presented.

First, Collin-Souffrin et al. (1986) have computed photoionization models under physical conditions much more favourable to the low ionization line emission than previous photoionization models :
- the low and the high ionization lines are emitted in two different kinds of clouds,

259

- the clouds responsible for the low ionization lines have a large neutral (HI*) zone produced by an intense X radiation, a high density and
possibly an overabundance of heavy elements.
    Second, Joly (1986) has computed the emission produced by homogene-
ous clouds in <u>collisional</u> equilibrium in order to outline the physical
conditions required to explain the observed FeII intensities. The emission region <u>does not</u> receive any external radiation.

2. THE MODELS

The code is described in detail in a paper by Collin-Souffrin and Dumont
(1986). It is mainly designed to study low ionized and optically thick
medium.
    The HI and MgII atoms are approximated by a 4-levels atom plus a
continuum and FeII by a 14-levels atom plus a continuum. Six photoionization models have been computed, numbered from 0 to 5 (0 stands for a
model similar to the Standard model of Kwan and Krolik 1981). The physical parameters and the results of the computations are given in the paper
by Collin-Souffrin et al. (1986).
    About 40 collisionnal models have been computed, with constant T_e
and n_H. Note that photoionization models exhibit also a region of more or
less constant T_e and n_H in the low excited zone, but the ionization equilibrium is quite different. The electron temperature ranges from 6000 to
15000K, the density from 10^{10} to 10^{12} cm^{-3} and the column density from
10^{21} to 10^{24} cm^{-2}.

3. INFLUENCE OF THE PHYSICAL PARAMETERS

3.1. The column density $\mathcal{N}_H$

At small column density, Hβ and the UV FeII lines (FeII UV = Σ lines
λ < 3000 Å) increase with n_eh (h = geometrical thickness). FeII opt (Σ
lines λ > 3000 Å) varies more rapidly because its lower level population
- and therefore optical thickness - is also increasing with n_e. Therefore
FeII opt increases compared to Hβ but the Balmer continuum (Bac) which is
proportional to n_e^2h increases much more rapidly.
    At large column density, the Local Thermodynamic Equilibrium (LTE)
is reached between the levels. Hβ ∝ √τ and FeII opt is constant while
FeII UV is absorbed by the Balmer opacity and converted into optical
photons (FeII UV/Hβ decreases dramatically). Balmer continuum increases
strongly compared to Hβ and FeII until it is optically thick and begins to
be self absorbed.

3.2. The density n_H

At low density, line emission increases with the number of particules.
Balmer lines reach first the LTE (collisions from level 2 populate efficiently the higher levels because level 2 is at thermodynamical equilibrium with level 1 thanks to the large optical thickness of Lα in most of

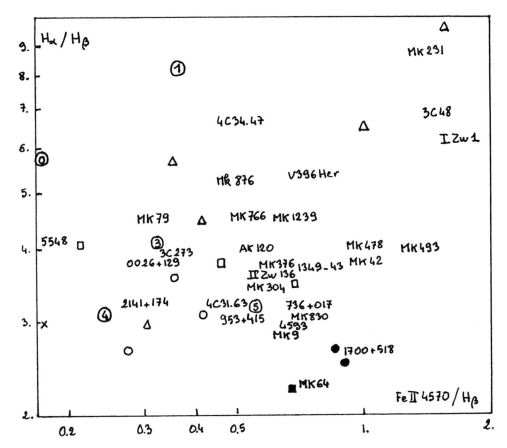

Figure 1. Hα/Hβ versus FeII 4570/Hβ observed in 30 AGN and computed under photoionization assumption (⓪ to ⑤) or collisional assumption (X : 12500K, 0 : 10000K, ◻ : 8000K, Δ : < 7000K). Full dots or squares : see text.

the present models) : Balmer lines saturate from 10^{11} cm^{-3} while FeII is still increasing. Balmer continuum increases continuously with n_e^2.

3.3. The electron temperature T_e

At T_e < 7000K, the population of the fourth level of hydrogen is negligible, the emissivity of Hβ is negligible, line ratios are very large. Increasing the temperature induces tremendous enhancement of the HI level population up to LTE which is reached at 8000K (if $n_H = 10^{12}$ cm^{-3}, $\mathcal{N}_H = 10^{23}$ cm^{-2}).

4. COMPARISON TO THE OBSERVATIONS

The results of the computations are compared to the published line ratios of about 30 well observed AGN.

On figure 1 is plotted Hα/Hß versus FeII 4570/Hß. FeII 4570 is an intense FeII feature, it is mainly the blend of multiplets 37 and 38. Symbols are according to the temperature of the collisional models, numbers are for the photoionization models (cf.§2 and Collin-Souffrin et al. 1986). A full symbol notes a collisional model which exhibits some unacceptably strong feature which excludes the possibility of those physical conditions : the model must be rejected. Generally the constraint is set by the Balmer continuum (cf.§3 and Joly 1986).

Weak FeII emitters are easily explained by either photoionization or collisional models and various physical conditions are possible For exemple 3C273 is well accounted for by :
- a collisional model with $T_e \sim 8000K$, $n_H \sim 10^{11}$ cm^{-3} and $\mathcal{N}_H \sim 10^{22}$ cm^{-2}
- a photoionized model of low ionization parameter with in the HI* region $T_e \sim 6500K$, $n_H \sim 10^{11}$ cm^{-3}, $\mathcal{N}_H = 10^{24}$ cm^{-2}.

Concerning the strong FeII emitters, it is not so easy : none of the photoionized models account for the intensity of FeII 4570 and only few collisional models are satisfactory. The physical conditions are tightly delimited : the temperature is low ($T_e \lesssim 7000K$), the density is high ($n_H \gtrsim 10^{12}$ cm^{-3}) and the column density is moderate ($10^{22} \lesssim \mathcal{N}_H \lesssim 10^{23}$ cm^{-2}). These three conditions are upperbounded by the strength of the Balmer continuum.

On figure 2 are plotted the Balmer continuum over Hß intensity ratio versus FeII 4570/Hß. Note first that the collisional models are more or less distributed within parallel areas following the temperature, the strong FeII emitter, corresponding to the low temperature and second that the photoionized models fall within the "10000K" range" in spite of the lower temperature of the HI* zone because of the influence of the radiation on the ionization degree. The lack of model in the range of low temperature is due to the huge variations of Hß in this range (cf§3.3). Along one of these "temperature bands" the Balmer continuum intensity is increasing strongly with the density. All the models with Bac/Hß > 10 are those indicated by full symbols in the other diagrams.

On figure 3 are the total optical FeII intensity (FeII opt) over Hß ratio versus FeII 4570/Hß. FeII opt in the sum of all the FeII multiplets arising from the 5eV levels down to the metastable levels, the wavelengths are larger than 3000 Å. The correlation between the theoretical ratios is very good owing to the similar behaviour of all optical multiplets to the physical conditions, once the optical thickness is large. Therefore, measuring the FeII blend around 4570 Å one can confidently infer the total optical intensity which is about 10 times multiplets 37 and 38. Comparing to the observed ratios we see that the correlation is indeed present but shifted down by a factor 3 showing that both the near UV lines (3000 Å < λ < 4000 Å) and many weak lines escape detection. For some intermediate redshift objects like 3C273, the 3000-4000 Å range has been measured and the agreement with the computations is good.

On figure 4 is plotted FeII opt/Hß versus FeII UV/Hß. FeII UV in the sum of all the FeII multiplets arising from the 5eV levels down to the four lower levels (< 2eV), their wavelengths are smaller than 3000 Å. Very few objects are gathered here because few observations have been performed simultaneously in the UV and the optical range. This is not the

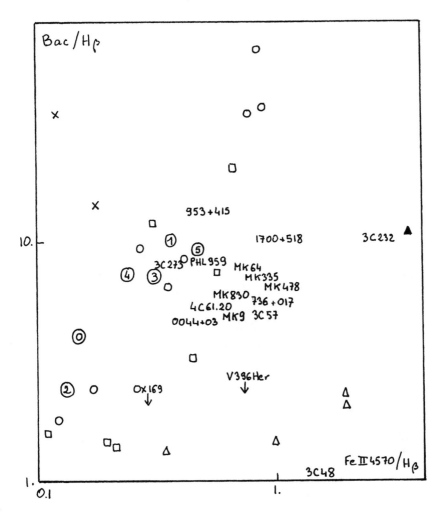

Figure 2. Bac/Hβ versus FeII 4570/Hβ. Same symbols as in figure 1.

case for IZw1, Mk 304, Mk 231 and Ak 120, and since these objects as many Seyfert 1 nuclei are variable, large error can occur on the UV over optical line ratios. Comparing to the models shows that it is more difficult to account for the FeII UV intensities than for the optical ones. Indeed we have seen previously (§3) that an increase of density or temperature which enhance the FeII UV emission increases dramatically the Balmer continuum, and an increase of optical thickness turns the FeII UV photons into optical ones. However FeII UV can be emitted partly by the transition region in the photoionized clouds emitting the high ionization lines.

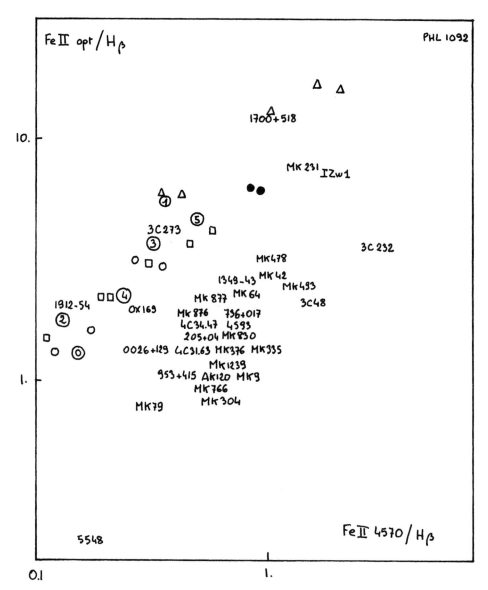

Figure 3. FeII opt/Hβ versus FeII 4570/Hβ. Same symbols as in figure 1.

5. DISCUSSION

We have seen in the previous section that if the observed optical FeII intensities are easily accounted for by collisional models, it is not the case for the UV ones. If we admit that the UV and the optical lines are emitted by the same region, how can we enhance preferentially the FeII UV/Hβ ratio computed in the present models ? We have to remind that some atomic data are poorly known, in particular the charge exchange

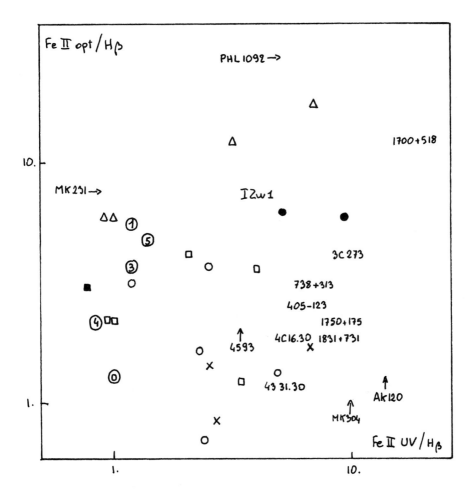

Figure 4. FeII opt/Hβ versus FeII UV/Hβ. Same symbols as in figure 1.

rates and the collisional strengths.

5.1. Charge exchange effect

The ionizing charge exchange rate between Fe^+ and Fe^{++} is supposed to be high (Baliunas and Buttler, 1980) but it is not known. It has been assumed equal to 1.5 10^9 cm/s. Thus, at $T_e \sim 12000K$, as the ionization degree of hydrogen is high, the ionizing charge exchange is efficient, $Fe^{++}/Fe^+ \sim 10$, the number of Fe^+ atoms and the FeII emission are weak. If the ionizing charge exchange is suppressed the FeII emission increases by a factor 5 and the ratio FeII UV/FeII opt ~ 2. The spectrum of 3C273 can be well accounted for by such a model ($n_H \sim 10^{11}$ cm^{-3}, $\mathcal{N}_H \sim 10^{22}$ cm^{-2}).

5.2 Collisional strengths

Very few ($\sim$ 10) collisional strengths have been computed (Nussbaumer et al., 1981). Some others were approximated according to a semi classical method established by Seaton (1962) and extended to positive ions by Burgess (1964). For the other transitions it is assumed $\Omega/g_u = 0.5$ (Ω is the collisional strength, g_u the statistical weight) which is much smaller than the few known values for UV lines ($\Omega/g_u \sim 4$). One computation has been performed with $T_e = 10000K$, $n_H = 10^{12} cm^{-3}$, $\mathcal{N}_H = 10^{21}$ cm^{-2} and $\Omega/g_u = 4$ for all UV lines. The FeII emission increases by $\sim$ 50% and FeII UV/FeII opt $\sim$ 4. The three AGN : 4C16.30, 1750 + 175 and 1831 + 731 are well accounted for by this model.

6. CONCLUSION

Optical FeII lines are likely to be emitted in a cold region ($T_e \sim 6500$-7000K) of high density ($n_H > 10^{12}$ cm^{-3}) and moderate column density. This region must be shielded from the observed power law continuum in order to have a low ionization degree which reduces Hβ. At low electron density and temperature, excited levels of HI are far from LTE, FeII opt/Hβ is high, Bac/FeII is low.

However this does not hold for FeII UV lines which require a lower column density and a more excited region, like a transition region in a photoionized cloud. Alternatively, some of the atomic data used in these computations could have to be revised and it is shown that low ionizing charge exchange rate or higher collisional strengths can adequatly increase the FeII UV/Hβ ratio. We urgently need good data for these two parameters in order to get the final answer.

REFERENCES

Baliunas, S.L., Buttler, S.E., 1980, Astrophys. J., 235, L45.
Burgess, A., 1964, Culham Conference on Atomic Collisions AERE, 63.
Collin-Souffrin, S., Dumont, S., 1986, Astron. Astrophys., in press.
Collin-Souffrin, S., Joly, M., Péquignot, D., Dumont, S., 1986, Astron. Astrophys., in press.
Joly, M., in preparation.
Kwan, J., Krolik, J.H., 1981, Astrophys. J., 250, 478.
Nussbaumer, H., Pettini, M., Storey, P.J., 1981, Astron. Astrophys., 102, 351.
Seaton, M., 1962, Proc. Phys. Soc., 79, 1105.
Wills, B.J., Netzer, H., Wills, D., 1985, Astrophys. J., 288, 94.

FeII and Balmer continuum emission in AGNs

R. Gilmozzi, W. Wamsteker, J. Clavel, A. Cassatella,
C. Gry, A. Talavera, C. Lloyd
IUE Observatory, ESA, Aptdo 54065, Madrid, Spain

H. Netzer
Wise Observatory, Tel Aviv, Israel

B.J. Wills and D. Wills
McDonald Observatory, Austin, Texas

We address in this short note a different aspect of the FeII problem in AGNs, one connected with treating the total FeII emission as a form of "continuum" rather than emission lines. The justification for this approach is quite clearly illustrated by Wills (this workshop) and the obvious consequence is that it is much easier to try and fit the FeII emission as a whole than to try and isolate the (thousands) of single lines (Wills, Netzer and Wills, 1985; WNW).

The presence of the large number of strong, blended and extremely broadened lines, combined with the Balmer continuum emission (which itself possibly suffers absorption in the FeII zone), gives rise to a continuum of extremely complex shape. This makes it very difficult to estimate the local continuum (as well as the ionizing continuum - a power law?) especially when attempting to analyze the strong emission lines in the spectrum of AGNs. This is particularly true in the case of "key" lines like MgII2800 or Hbeta, which are very much distorted by the underlying FeII continuum (see e.g. Wamsteker et al, 1985). An improper compensation of the FeII multiplets under the [OIII]4959/5007 lines, for example, is the most likely explanation for the presence of a broad component sometimes claimed for these lines.

The large UV variability of AGNs and the usually relatively poor signal to noise ratio in IUE spectra add some complexity to this picture. As a result, it is quite difficult to achieve a reliable estimate of the energy balance in the Broad emission Line Region (BLR) of AGNs.

R. Viotti et al. (eds.), Physics of Formation of FeII Lines Outside LTE, 267–271.

To overcome these problems to some extent, a program was started to obtain <u>simultaneous</u> spectra of a number of selected AGNs both in the UV and in the optical. The use of such larger wavelength interval helps in fact to put firmer contraints on the model parameters for both the FeII and the Balmer emissions. Figures 1 to 3 show some samples of the data obtained, together with the Balmer continuum and FeII models used in the continuum spectral decomposition.

The necessary ingredients of the fit are the following:

1) A relatively flat power law spectrum ($F_{\nu} \propto \nu^{-\alpha}$)
   NGC5548: $\alpha = 0.5$
   3C382 : $\alpha = 0.9$
   F-9   : $\alpha = 0.4$

2) Balmer continuum emission from a grid of models from WNW, adjusted to its observed level as determined in the FeII-free region at the BaC jump (usually unambiguously determined in such data), and convolved with the profile of the Balmer lines.

3) Also from WNW, a model which represents the UV and optical FeII emission line spectrum. The models cover a range of optical depths, temperatures and densities. The selected model was adjusted for the line widths to obtain a good fit at the 5190A blend. After this, the relative strenghts in the various wavelength domains was adjusted to optimize the fit. There are some indications that the FeII lines are somewhat less broad than the Balmer lines.

Table I shows the parameters derived from the fits to the spectra shown in the figures. A complete analysis of the physical conditions that can be inferred from these fits will be presented in a forthcoming publication. It is already clear, however, that the total FeII flux over H Ly-alpha intensity ratio is close to or larger than one in all three objects (and even larger than two for NGC5548 in 1980), a result which is difficult to accomodate within the framework of the photoionization models for the BLR.

As pointed out by Netzer (1985), this result would require that the very flat ionizing spectrum found in the optical and UV extend with the same index up to the KeV range, in clear contradiction with the observations.

References

Netzer,H, 1985, M.N.R.A.S. 216,63
Wamsteker,W, Alloin,D, Pelat,D, Gilmozzi,R, 1985, Ap J 295,
L33
Wills,BJ, Netzer,H, Wills,D, 1985, Ap J 288,94
Wills,BJ, 1987, These Proceedings

TABLE I

Lyman-alpha, Balmer continuum and Fe II fluxes for some AGN's.
(units are 10^{-14} erg/cm^2/s/A)

OBJECT	Ly-a	Ba-C	FeII Tot.	FeII(UV)	FeII(Int)	FeII(Opt)
NGC 5548						
1981	478	593	1010	852	57	101
1984	979	1335	1281	1111	74	96
3C382	300	261	319	254	19	46
F-9	1310	883	1085	780	120	185

FeII(UV) = 2000 - 3000 A
FeII(Int) = 3000 - 3500 A
FeII(Opt) = 3500 - 6800 A

Note: In the Lyman-α fluxes given above the narrow line component
is included in the flux. However, from independent estimates
of its strength through line decomposition its contribution
is judged to less than 20% of the flux given in the table.

Fig 1. (Next page) (a) Composite UV-optical spectra of NGC5548
in 1981 and 1984. Note the huge brightness variation.
(b) Top: The 1981 spectrum with the fit superimposed.
Bottom: The BaC and FeII models used in the fit.

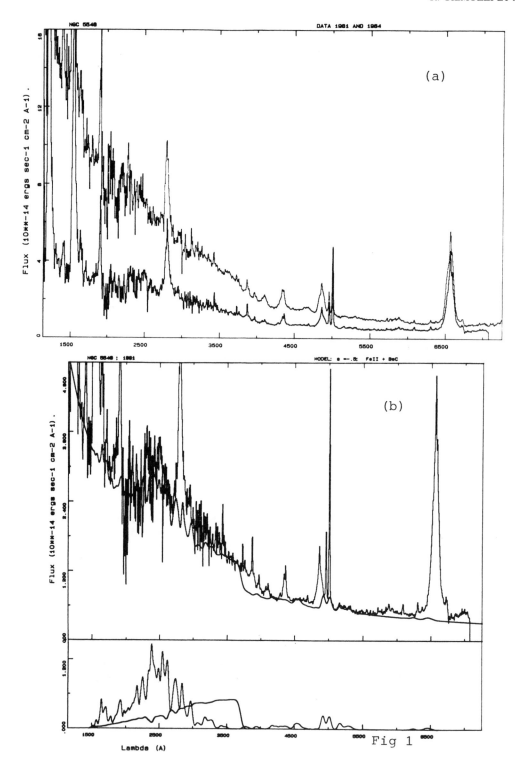

Fig 1

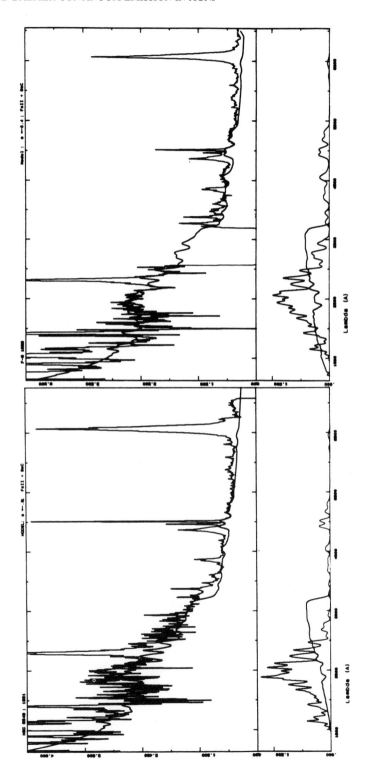

Fig 2. Top: The UV-optical spectrum of 3C382
and relative fit.
Bottom: The BaC and FeII models used
in the fit.

Fig 3. Same as fig 2 for Fairall-9 (1983)

A COMPUTER PROGRAM FOR SOLVING MULTI-LEVEL NON-LTE RADIATIVE TRANSFER PROBLEMS IN MOVING OR STATIC ATMOSPHERES

M. Carlsson
Institute of Theoretical Astrophysics
P.O.Box 1029 Blindern
N-0315 OSLO 3
Norway

ABSTRACT A summary of the characteristics of a non-LTE computer program is presented. The code can be used to solve non-LTE radiative transfer problems in semi-infinite, plane-parallel one-dimensional atmospheres with a prescribed velocity field. The model atom can contain many atomic levels and several ionization stages. The lines are assumed to be formed with complete redistribution over the line profile which is assumed to be a Voigt function.

1. INTRODUCTION

In many astrophysical objects, the simplification of local thermodynamic equilibrium (LTE) is not valid. For ions with a complicated structure, like Fe II, it is necessary to include many energy levels to obtain realistic results. The non-LTE calculations necessary tend to become numerically prohibitive and more efficient but approximate methods, like escape probability methods, have been used instead.

The present paper gives a summary of the characteristics of a computer code based on the method developed by Scharmer (1981) and Scharmer & Carlsson (1985). The method combines high efficiency with the accuracy obtained with standard methods like the complete linearization method (Auer & Mihalas 1969, 1970).

A detailed description of the code, a discussion of the numerical properties and a listing of the code is given in a paper by Carlsson (1986).

2. SUMMARY OF CHARACTERISTICS

The computer program has been written to be general and portable. Its main characteristics are:

273

R. Viotti et al. (eds.), Physics of Formation of FeII Lines Outside LTE, 273–275.

1. The code can be used to solve non-LTE radiative transfer problems in semi-infinite, plane-parallel one-dimensional atmospheres with a prescribed velocity field.

2. The model atom can contain many atomic levels and several ionization stages. Both line transitions and bound-free continua can be treated in detail simultaneously. The lines are assumed to be formed with complete redistribution over the profile function which is assumed to be a Voigt function.

3. The equations of radiative transfer and statistical equilibrium are solved simultaneously and for all lines and levels at the same time thus assuring first order consistency.

4. The basic concept of the method is to take advantage of approximations made possible by the physical nature of radiative transfer in spectral lines. A consequence is that the study of the numerical properties of the method illuminates the physical nature of nonlocal radiative interactions in spectral lines.

5. Single-precision arithmetics on computers with 32-bit word length provide enough numerical precision for convergence.

6. The code is relatively efficient, often ten times faster than the code LINEAR.

7. The CPU-time usage scales, at most, linearly with the number of frequencies and angles making the method well suited for problems involving velocity fields and many transitions.

8. The most time-consuming routines vectorize well which makes the code suitable also for vectorizing machines. These dominating routines also show a high degree of parallelism making the code suitable for computers with parallel architecture.

9. The code is in standard FORTRAN-77 making it highly portable.

10. The code is extensively commented and structured to facilitate changes and additions.

A calculation with a model atom with 18 atomic levels took six minutes on a VAX-11/750.

The limitations to one-dimensional media, plane-parallel geometry, complete redistribution and no blends are limitations of the present code, not of the basic methods. A partial redistribution formulation has been implemented and found to work well, although the convergence was slower than for a similar complete redistribution problem. It is straightforward to implement spherical geometry and to include blends.

The code is available on magnetic tape from the author.

3. REFERENCES

Auer, L.H., and Mihalas, D.: 1969, *Astrophys. J.* **158**,641.

Auer, L.H., and Mihalas, D.: 1970, *Mon. Not. R. Astron. Soc.* **149**,65.

Scharmer, G.B.: 1981, *Astrophys. J.* **249**,720.

Scharmer, G.B., and Carlsson, M.: 1985, *J. Comput. Phys* **59**,56.

SEMI-EMPIRICAL METHODS OF ANALYSIS OF FeII EMISSION AND ABSORPTION LINES FORMED IN DIFFUSE MEDIA

M. Friedjung
Institut d'Astrophysique, CNRS
98 bis, Boulevard Arago, 75014 Paris, France

ABSTRACT. Methods of directly analysing fluxes of FeII lines observed outside LTE are described, paying particular attention to the study of emission lines. Emission line curve of growth methods are described, but it is shown that emission lines are better studied by what are called 'self absorption curve' (SAC) methods. The latter are described in detail. Relative populations of terms, column densities and characteristic radii of line emitting regions can be found. When emission line analysis is combined with that of absorption lines, spectral synthesis can be performed. Application of such methods to observations of certain hot luminous Magellanic Cloud stars suggests the simultaneous presence of winds and disks.

1. INTRODUCTION

The full theory of the formation of FeII lines outside LTE in a non homogeneous medium is not simple because of the complexity of the Grotrian diagram. We are often concerned with the formation of FeII lines in media whose exact physical nature in uncertain, and for which it is not justified at the present stage of knowledge, to make detailed complex calculations, whose significance would be rather doubtful. The relevance of a synthetic spectrum calculated using rigorous theory is difficult to evaluate in such situations; and an approach is needed, through which one can obtain information about the general nature of the line formation region. I shall describe semi-empirical methods, which though less rigorous than detailed theoretical calculations, can give information about this general nature. Neglect of semi-empirical methods can, in my opinion, often lead to research continuing to develop in a wrong direction for a long time .....

After a historical description, I shall talk about recent work. I must apoligize for doing something which no

277

R. Viotti et al. (eds.), Physics of Formation of FeII Lines Outside LTE, 277–293.
© *1988 by D. Reidel Publishing Company.*

reviewer should do; I shall talk a lot about work in which I have personally been involved. Unfortunately the methods to be described have not been sufficiently studied by colleagues, and hopefully this talk will help to remedy the situation in the future.

2. HISTORY OF SEMI-EMPIRICAL METHODS FOR EMISSION LINES

Wellmann (1951a,b) developed a curve of growth method for the study of emission lines. From a consideration of line emission he obtained an expression for the integrated intensity over a line:

$$I = \frac{2h\nu_o^3}{c^2} \left(1 + \frac{w}{e^{h\nu_o/kT_s} - 1}\right) \frac{b_j}{b_i} exp\left(-\frac{h\nu_o}{kT_e}\right) A_\nu \qquad (1)$$

with:
$$A_\nu = \int_0^\infty \left(1 - exp(-\chi_\nu N_i H)\right) d\nu \qquad (2)$$

In these equations T_s is a radiation temperature and T_e an electron temperature, w is the dilution factor, b_j and b_i are factors of deviation of the upper and lower level populations from LTE, ν a frequency in the line and ν_o the central frequency of the line, χ_ν is the line absorption coefficient at ν, and N_iH the column density of atoms in the lower level. h, c and k have their usual meanings. The use of a geometric factor for the radiation dilution w supposes that the line is optically thin, which can in some circumstancies (T_s large) lead to errors. When lines are only broadened by the Doppler effect, and putting the observed emission line intensity equal to E, Wellmann obtained two quantities, one of which varies as a function of the other for b_j and b_i constant. These quantities are:

$$log\, A''' = log\, E + 4\, log\, (\nu_i/\nu) + 0.4343 \frac{h(\nu-\nu_i)}{kT_e} \qquad (3)$$

and:
$$log\, N''' = log\, (J_{th}\, exp((\chi_i - \chi_n)/kT_e) \qquad (4)$$

Deviations from LTE are supposed to be the same for lines of the same multiplet and supermultiplet and J_{th} is proportional to a prediction for relative line intensities according to Russel-Saunders coupling of atomic levels. ν_i is a standard frequency corresponding to 2.51 eV, while χ_i is the potential of ionization from the lower level of the line and χ_n a constant potential equal to 13.34 eV. Plots of logA"' against logN"' assuming different values for T_e give emission line curves of growth. When the method was applied by Wellmann to the FeII lines of the Be star γ Cas, no sign of line self absorption was seen.

Thackeray (1967) used forbidden lines to obtain an excitation temperature for η Car. . [FeII] lines should normally be optically thin. So a determination of the value of upper level statistical weight times transition probability of lines having the same visual estimate of intensity in multiplets with different upper term excitation potentials, gave an excitation temperature of 8500 ± 1500 K.

Viotti (1969) further developed such methods and also applied them to η Car. He plotted values of $\log(\lambda I/g_{\ell}A)$ for forbidden lines and $\log(\lambda^3 I/gf)$ for permitted lines against upper level excitation potential χ, where g_{ℓ} is the upper level and g the lower level statistical weight, A the transition probability and f the oscillator strength. When lines are optically thin the ordinates of such graphs are the logarithms of a quantity proportional to the total number of ions in the upper level divided by g_{ℓ}. Viotti found that the graph for both permitted and forbidden lines, indicated a linear relation between the abscissa and the excitation potential. This suggested not only that line self absorption was relatively small, but also that the levels were populated according to the Boltzmann's law. The permitted line upper levels gave a first value for the excitation temperature of 8700 K, while the lower lying levels of the forbidden lines gave one of 4500 K, the error being of the order of 1000 K. Correcting for interstellar reddening and line self absorption, and assuming the ordinate a linear function of $E(B-V)/\lambda$ (reddening), g.A or gf (self absorption), as well as of $\Theta\chi_{\ell}$ (Θ is $5040/T_{ex}$), excitation temperatures of T_{ex}=7800 and 11000 K were found for [FeII] and FeII respectively. Values of E(B-V) of 1.07 and 0.32 were determined from the forbidden and permitted lines; a 'self-absorption' effect also found for the forbidden lines could, according to Viotti, have been due to systematic errors. Measurements of equivalent widths as well as the previously used ones of total line intensity also enabled Viotti to determine the form of the continuous spectrum. FeII was discussed in detail and excitation temperatures were also given for η Car, XX Oph, V1016 Cyg, CH Cyg, VV Cep, WY Gem, NGC 1976, AG Car and AG Peg by Viotti (1970). It must be noted that Pagel (1969a) previously also used [FeII] emission line intensities to study the continuous spectrum of η Car. He found the lines optically thin and E(B-V) of 1.20.

In the short lived explosion of papers on this subject from 1969 to 1971, a lot of attention was directed to the determination of forms of the continuous spectrum and reddenings (Lambert 1969, Pagel 1969b, Ade and Pagel 1970, Caputo et al. 1970, Caputo 1971a,b). This was often done using forbidden lines, but permitted lines were also used, for which self absorption was usually found to be small.

Arguments about the importance of non-LTE level populations on the results of these studies also occurred, and it was considered more rigorous to compare lines having a common upper level, when looking for wavelength dependent effects. Excitation temperatures were sometimes also determined. A number of different stars were studied, including η Car (Pagel 1969a,b, Lambert 1969, Ade and Pagel 1970), the P Cygni star AG Car (Caputo and Viotti 1970, Caputo 1971a), the symbiotic stars V1016 Cyg (Caputo et al. 1970), AG Peg (Caputo 1971a), Z And (Caputo 1971b) and somewhat later RR Tel (Cassatella and Viotti 1974), the VV Cep binary Boss 1985 (=KQ Pup, Caputo 1971a), as well as the luminous Magellanic Cloud star S22 (Viotti and Ricciardi 1971, Cassatella and Viotti 1974). It should be noted that these early studies were not done with very good permitted line oscillator strengths, and this clearly influenced the results.

In work in which I was involved the self absorption of permitted lines was emphasized from the beginning. The methods applied to the narrow FeII emission lines of nova HR Del 1967, are described in Friedjung and Malakpur (1971). A somewhat different formalism to that of Wellmann (1951a) was used. A temperature T is taken which defines the ratio of the populations of the upper and lower levels of a line, so the emission in the line corresponds to the black body case with this temperature. Then

$$F_\lambda H = R^2/r^2 \left(1 - e^{-\tau}\right) \frac{2\pi h c^2}{\lambda^5} \frac{1}{\exp\left(1.44/\lambda T\right) - 1} \qquad (5)$$

Here F_λ is the continuum flux at wavelength λ, H is the ratio of the line central intensity to that of the neighbouring continuous spectrum, R the radius of the line emitting region (in fact the radius of a disk perpendicular to the line of sight with the same surface area in this direction as the line emitting region), r the distance of the object studied, and τ the optical thickness of the spectral line center. Expressing τ in terms of ϕ the population of the upper level divided by its statistical weight

$$\tau = constant \cdot (gf\lambda) \, \phi \left(\exp\left(1.44/\lambda T\right) - 1\right) \qquad (6)$$

so

$$\log \left(F_\lambda H\right) - \log(gf) + 4 \log \lambda = constant + \\ 2 \log(R/r) + \log \frac{1 - e^{-\tau}}{\tau} + \log \phi \qquad (7)$$

In Friedjung and Malakpur (1971) τ was found from a correlation between τ and $\Delta\lambda/\lambda$, the line width at half central
intensity divided by λ Assuming the population divided by
the statistical weight is constant for all levels inside a
term, one can plot a graph of $\log(\Delta\lambda/\lambda)$ against the left
end side of eq.(7) for lines having the same upper term. In
the paper the different graphs were horizontally shifted so
as to coincide. Graphs of $\log(\Delta\lambda/\lambda)$ could also be plotted,
and also horizontally shifted to coincide. The first type
of graph gave $\log(\Delta\lambda/\lambda)$ as a function of $\log(\frac{1-e^{-\tau}}{\tau})$, and
the second as a function of $\log\tau$. (In fact the exact fun-
ctional form of the second type is hard to determine
because the slope of each graph is small, so it was determined from that of the first type of graph). The relative
shifts of the first type of graph gave relative popula-
tion/statistical weight for the upper terms, while those of
the second type of graph gave this for lower terms. Examination of the graphs showed no evidence for non constancy
of population divided by statistical weight in each term,
and plots of relative population of both upper and lower
terms were consistent with a Boltzmann population distribution at a temperature of 4000 K. A lower limit to T in
eq.(5) could be found from the absence of forbidden lines
which have upper terms which are the same as lower terms of
permitted lines. A maximum R of 3x10+13 cm was then found,
supposing r equal to 500 pc, and correcting line fluxes for
interstellar extinction.

The method of determination of the size of the line
emitting region was applied to other objects including the
recurrent nova RS Oph (Friedjung 1972), a supernova (Friedjung 1974a), the symbiotic star RR Tel during its nova-like
outburst (Friedjung 1974b) and the R CrB type star MV Sgr
(Friedjung and Viotti 1976). In these studies the quality
of the data used was not always as good as it should have
been.

A detailed study of the luminous Magellanic Cloud star
S22 was performed (Friedjung and Muratorio 1980). The methods of this and more recent work will next be described.

3. ANALYSIS OF SELF ABSORPTION CURVES

A rather better method of studying emission lines than the
curve of growth method was defined by Friedjung and Muratorio (1980), Muratorio (1985), and Friedjung and Muratorio
(1987). It enables relative populations of upper and lower
terms to be simply determined, as well as other information
about the line formation region.

The present approach is based on an integration of
eq.(7) over the whole line profile, for a line which may be
emitted in an inhomogeneous medium. If the line is only

Doppler broadened, and exp$(1.44/\lambda T) \gg 1$, one obtains for lines emitted in a given medium

$$\log(F_\lambda W_\lambda) + 3\log\lambda - \log(gf\lambda) =$$
$$= \log(2k'\pi hc) + 2\log(R_c/r) + \log(\phi_c V_c) + Q(\tau_c) \quad (8)$$

Here W_λ is the equivalent width of the line, k' a constant equal to 0.02654 cm2/s, while R_c, ϕ_c, V_c and τ_c are characteristic values of the radius, population column density/ statistical weight per unit velocity range of the upper level, velocity and optical thickness for the medium emitting the line. $Q(\tau_c)$ is a function of τ which depends on the nature of this line formation region. At constant ϕ_c, the right end side of eq.(8) only varies with τ_c, which for Doppler broadening is a multiple of k'$(gf\lambda)$, and the lower level unit column population/statistical weight per unit velocity range characteristic of the medium. If the population/statistical weight has a Boltzmann distribution for the levels inside the same term, or what is the same is almost constant, a graph of $\log(F_\lambda W_\lambda)$ + $3\log\lambda$ - $\log(gf)$ against $\log(gf\lambda)$ for the lines of a multiplet, should give points lying on a curve. This curve is the Self Absorption Curve (SAC).

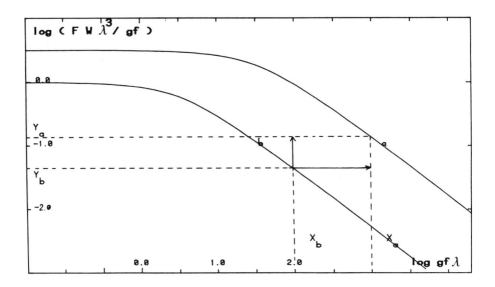

Figure 1. Self Absorption Curve: horizontal and vertical relative shifts of two multiplets a and b whose level populations O differ:
$$X_a - X_b = \log(\phi_{lb}) - \log(\phi_{la}); \quad Y_a - Y_b = \log(\phi_{ua}) - \log(\phi_{nb})$$

Curves for lines of different multiplets will be shifted with respect to each other. If two multiplets have the same upper term, the self absorption curve will be vertically shifted, and the shift gives the difference in log population of the upper term/statistical weight. This is seen in Fig.1. When both parts of the curve of large and also of intermediate or small optical thickness are clearly detected for different multiplets, relative populations of both upper and lower terms can be found at the same time.

It should be emphasized that classical emission line curves of growth cannot be used in this way. If one plots $\log(F_\lambda W_\lambda \lambda^4)$ against $\log(gf\lambda)$ to make a curve of growth, and none of the lines studied is optically very thin, one cannot independently determine the relative populations of the upper and of the lower terms. A change of the former would produce a vertical shift of the curve of growth, while one of the latter would produce a shift inclined $45°$. For a similar reason the SAC method of relative population determination only works when $1.44/\lambda T \gg 1$, i.e. when stimulated emission is small, as is generally the case for objects studied. The possibility of superposing vertically and horizontally shifted SAC's of different multiplets in addition enables the shape of the SAC to be better determined.

In the case of spectra which are rich in emission lines from many different multiplets, the shape of self absorption curves can be determined empirically, as can the validity of the assumption of a Boltzmann distribution of the population inside each term. If the different curves for each multiplet are shifted so as to be superposed, the shape of the curve can be better defined, while the shifts give relative term populations indipendent of an assumption of LTE. This method was that used to study the star S22 in the work already mentioned (Friedjung and Muratorio 1980). The metastable term population distribution was found to be much closer to Boltzmann's law, than that of the upper terms.

In the cases of other stars, self absorption curves calculated using simple models, are best used. In recent work by Muratorio and me (Muratorio 1985, Friedjung and Muratorio 1987), the form $Q(\tau_c)$ of the self absorption curve was found for a number of simple situations, including

(a) lines with a rectangular profile formed in a homogeneous slab (population, optical thickness, velocity dispersion constant)

(b) lines with a Gaussian profile formed in a homogeneous slab

(c) line formation in a high velocity wind

(d) line formation in a low velocity wind

(e) line formation in a wind confined to a disk with

constant angle.
The wind calculations were performed with certain simpli-
fying assumptions, based on the theory of Viotti (1976),
for optically thin lines formed by a three levels ion.
These assumptions made in the absence of more precise
theoretical calculations for a wind are, it must be emphasized, used for empirical fitting, and in any case can give
an indication of what may be expected. A wind velocity
power law is assumed of the form

$$V = V_c \, (R_c/R)^{2-\alpha}$$

(9)

with R a distance from the centre of the wind, and α constant. Then in a steady state the density of ions in the
ground state or in one of very low excitation potential n_o,
and the density in a metastable state n_m are in both
Viotti's radiative and collisional cases, given by:

$$n_o = n_o^c \, (R_c/R)^\alpha$$

(10)

$$n_m = n_m^c \, (R_c/R)^\alpha$$

(11)

n_o^c and n_m^c are constants in these expressions and represent
densities at an inner radius R_c. The ionization is supposed
constant or most iron is supposed to be in the form of Fe+.
The density of ions in an upper odd state is however:

$$n_u = n_u^c \, (R_c/R)^{2\alpha}$$

(12)

in Viotti's collisional case, and

$$n_u = n_u^c \, (R_c/R)^{2+\alpha}$$

(13)

in his radiative case, n_u^c being a constant. In the high
velocity wind situation (c), the wind velocity is much
larger than local thermal and turbulent velocities, so
observed emission and absorption at a given wavelength come
from a surface of constant radial velocity. V_c in eq.(9)
then is the same as V_c in eq.(8), R_c is the inner radius of
the wind, and if g_u is the upper level statistical weight

$$\tau_c = \frac{k' g f \lambda \, n_m^c R_c}{g \, V_c} \quad , \quad \phi_c = \frac{n_u^c R_c}{g_u V_c}$$

(14)

for a metastable lower level, n_m^c being replaced by n_o^c when
the lower level is not far from zero electron volts. The
physically less realistic low velocity case is one with a

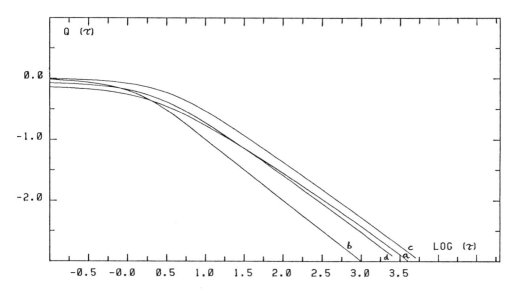

Figure 2. SAC'as for some of the situations described in review: (a) Rectangular line profile in uniform medium. (b) Gaussian line profile in uniform medium. (c) Line formation in high velocity wind - radiative constant velocity. (d) Line formation in low velocity wind - radiative case, constant velocity.

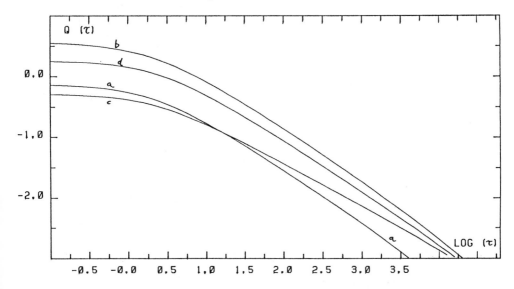

Figure 3. SAC's for high velocity winds with different velocity laws: (a) Constant velocity, radiative. (b) Constant velocity, collisional. (c) α = 2.5, radiative. (d) α=2.5, collisional.

wind velocity much less than the local thermal and turbulent velocities, and V_c in eqs.(8) and (14) now replaced by
V_t the local random velocity. In the disk situation (e)
optical thicknesses perpendicular to the disk are considered, and eq.(14) is replaced by

$$\tau_c = \frac{k'\theta f\lambda}{V_t} \int_0^\infty n_m^c/g \, dz \, , \quad \phi_c = \frac{1}{V_t} \int_0^\infty n_u^c/g_u \, dz \quad (15)$$

integration being performed in the direction z perpendicular to the disk, while R_c is still the inner radius of the
wind.

Different predicted $Q(\tau_e)$ are shown in Figures 2 and
3, and the differences between the different situations can
be seen. That between situation (a) and the others for
which the medium has a range of optical thickness is clear,
as is the effect of acceleration on a high velocity wind.
For a given model a comparison of a curve with observations
enables one to find the τ of a spectrum line in a multiplet.

Upper and lower limits to R_c can be found from the
self absorption curve method. An upper limit to the R.
where permitted lines are formed can be found, when as is
often the case, the lower terms of these lines (the metastable even terms) are the same as the upper terms of
forbidden lines. The latter are normally optically thin;
$Q(\tau_e)$ in eq.(8) is zero, so their self absorption curve is
a horizontal line. Supposing forbidden and permitted lines
formed in the same region, the difference between the
values of the ϕ_c's gives ϕ_c/τ_e for the permitted lines, and
R_c is then given by eq.(8). In general however, particularly for line formation in a wind, forbidden lines are not
formed in the same region as permitted lines, or no forbidden line is seen at all; ϕ_c/τ_e determined by this method is
then a lower limit, and R_c an upper limit. Conversely, one
can estimate ϕ_c/τ_e assuming LTE, from the relative populations of either the lower or the upper terms; deviations
from LTE are likely to decrease ϕ_c/τ_e so the R_c determined
in this way is a lower limit. It may also be noted that if
permitted and forbidden lines are formed in the same region, one can in addition obtain R_c for a forbidden line
from its upper term ϕ_c, since ϕ_c is by eq.(14) related to
the τ_c of a permitted line having as lower term this forbidden line's upper term. This forbidden line R_c is probably a lower limit when permitted and forbidden lines are
not formed in the same region.

When these methods are practically applied, one needs
to know V_c (or V_t) which can be found from the line width
when the spectral resolution is good enough to resolve the
line. In addition it must be pointed out that it is often

difficult to find the flat part of self absorption curves, as this is indicated by faint optically thin lines. This can lead to underestimates of ϕ_c and hence overestimates of R_c for permitted lines. It must in any case be emphasized that the assumption concerning the level population inside each term, which supposes that selective excitation is unimportant, can be tested. If such processes occur for certain lines they should give points which deviate from the self absorption curve.

Finally when lines are formed in a wind, one can in principle obtain information about the mass loss rate. If metastable and levels near 0 eV are in LTE, metastable or zero electron volt column densities can be used to derive that of Fe+, when the excitation temperature has been determined from relative term populations. For a high velocity wind the column density ϕ_c multiplied by V_c^2/R_c then leads to a mass loss rate of Fe+, which can be converted to a total mass loss rate if one has information about the abundance of Fe+.

4. ANALYSIS OF ABSORPTION LINES

The study of absorption lines, especially by curve of growth methods, is very classical in astrophysics, so I do not need to say very much. However winds can produce P Cyg absorption components, while certain stars showing many FeII emission lines in the optical, have ultraviolet spectra dominated by FeII absorption lines. The calculation of the equivalent width of such components is therefore very important in the synthesis of low and high resolution IUE spectra.

In the already mentioned work by Muratorio and me (Muratorio 1985, Friedjung and Muratorio 1987), the curve of growth of lines formed in a high velocity wind was calculated, assuming only Doppler broadening. Material at a certain negative velocity could cover part or the whole central star; and so absorption by it could contribute to the total line absorption found by integrating over different radial velocities.

5. EXAMPLES OF RESULTS OBTAINED USING THESE METHODS

I shall show in Figs. 4 and 5 examples of self absorption curves found by the method described, those for different multiplets being superposed after vertical and/or horizontal shifts. They are for one of a group of luminous stars of the Magellanic Clouds whose optical spectra are very rich in FeII emission lines, and for AG Car, a galactic P Cygni star. The relative populations of terms of

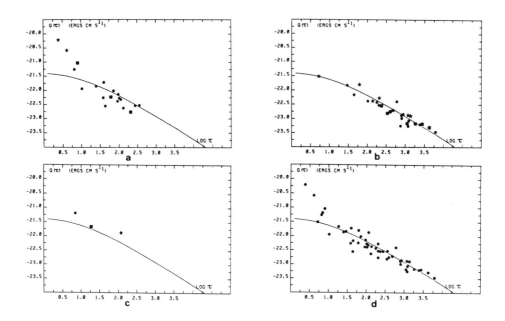

Figure 4. Wind confined to a disk self-absorption curve for R126. Each point represent for a line $\log(F_\lambda W_\lambda \lambda^3/gf)$ versus $gf\lambda$, horizontally and vertically shifted according to its upper and lower level population relative to that of multiplet 38. The relative level populations used for shifting the curves correspond to an excitation temperature of 5600 K.
a. Multiplets 3 ●, 14 ●, 15 ★, 21 ●, 25 ●, 26 ●, 32 ●, 43 ■ 44 ● (upper term excitation potential $< \sim 5.5$ eV).
b. Multiplets 27 ●, 28 ●, 37 ★, 33 ■ (upper term exc. potential ~ 5.5 eV)
c. Multiplets 23 ●, 154 ★, 173 ■ (upper term exc. potential $> \sim 5.5$ eV).
d. All the multiplets.

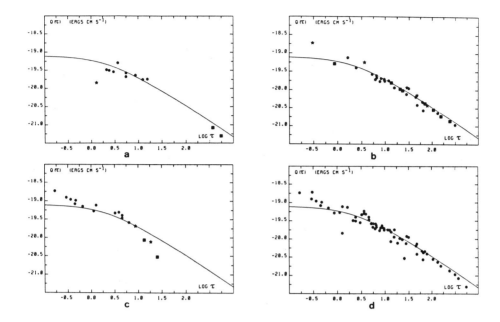

Figure 5. High velocity wind self-absorption curve for AG Car. Each point represent for a line log($F_\lambda W_\lambda \lambda^3$/gf) versus gf$\lambda$, horizontally and vertically shifted according to its upper and lower level population relative to that of multiplet 38. The relative level populations used for shifting the curves correspond to an excitation temperature of 7500 K. Same symbols as in Figure 4.
a. Multiplets 3 ●, 14 ●, 15 ★, 32 ●, 40 ●, 42 ■, 43 ●, 46 ● (upper term excitation potential < ∼5.5 eV)>
b. Multiplets 27 ●, 28 ●, 29 ●, 37 ★, 38 ■, 74 ● (upper term exc. pot. ∼ 5.5 eV).
c. Multiplets 127 ●, 144 ■, 153 ●, 154 ★, 173 ●, 186 ● (up- (upper term exc. pot. >∼ 5.5 eV).
d. All the multiplets.

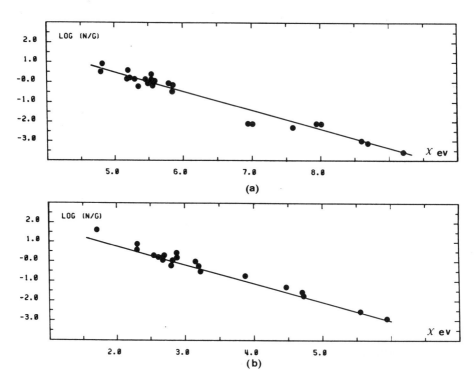

Figure 6. Level population of upper and lower levels of
FeII transitions in the LMC star S22, relative to that of
multiplet 38.

these stars, suggest that the relative population of
metastable terms on the one hand, and of upper odd terms on
the other hand are not far from Boltzmann distributions for
the lines and stars considered at least. The excitation
temperatures of both of these types of term found from the
permitted line intensities is similar. In the case of the
Large Magellanic Cloud star S22 the excitation temperature
of the upper terms of the forbidden lines appears to be
greater than that of the lower terms of the permitted
lines; the line formation regions are presumably
different.
      Synthesis of ultraviolet spectra in the range of IUE
only taking into account of the FeII lines, have been
performed by Muratorio for these stars (Muratorio 1985,
Muratorio and Friedjung 1987; see also Muratorio et al.
this volume). Usually only low resolution spectra are available, and spectra were sythesized using the parameters
found from optical FeII emission line self absorption curves. Results of various calculations for S22 are shown in
Figures 6 (level populations) and 7 (ultraviolet synthe-

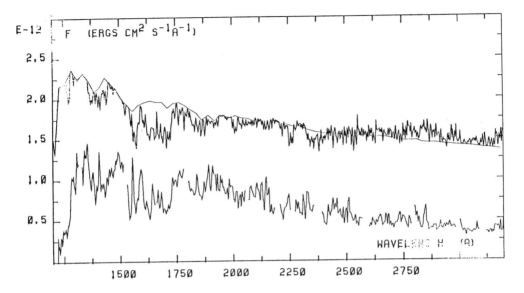

Figure 7. Comparison of the observed and theoretical ultra-violet spectrum of S22. (a) FeII synthetic spectrum (disk plus high velocity wind). The synthesis is superimposed on the 15000K Kurucz model atmosphere (smooth curve) to which absorption and emission FeII lines have been added. (b) Observed low resolution IUE spectrum of S22.

sis). The calculations indicate that, except for AG Car, it is difficult to explain the observed spectrum if both emission and absorption lines are formed in the same medium. Models with formation of all lines in a wind lead to excessive minimum mass loss rates, physical inconsistencies, or both. It appears that these difficulties can be overcome if the emission lines seen in the optical are formed in a region not in front of the photosphere. This region is most easily understood as a disk; its presence in addition to a wind can also explain the continuous energy distribution of these stars. The Heidelberg group has also independently found that these stars are probably surrounded by disks (Zickgraf et al. 1985, 1986; see also Zickgraf this volume), and call them B[e] stars. It is encouraging that no disk is needed to explain AG Car, which is another kind of star; if the method suggested disks around all stars, doubts can be cast on its validity!

Finally, examples of theoretical spectra of hot luminous stars having intense 'FeII winds' are presented in this volume in the paper by Muratorio et al., where the large amount of line absorption in the UV is emphasized.

6. THE FUTURE

The methods described here can be applied to many other sorts of stars, and interesting results can be expected. In particular when complex profiles are seen (see for example Carpenter this volume), it may be interesting to analyze different components separately by the self absorption curve method.

The theory of the form of the self absorption curves clearly needs to be refined. What is needed is theory which can be compared with the results of semi-empirical methods. A theory which just synthesizes the whole spectrum is difficult to judge; it is by a detailed analysis of different aspects such as self absorption curves, that such a judgement can be made. The analysis of observations and theory need to come closer together. If this happens semi-empirical methods have a promising future.

REFERENCES

Ade, P., Pagel, B.E.J.: 1970, Observatory 90, 6.
Caputo, F.: 1971a, Astrophys. Space Sci. 10, 93.
Caputo, F.: 1971b, Publ.astr. Soc. Pacific 83, 62.
Caputo, F., Viotti, R.: 1970, Astron. Astrophys. 7, 266.
Caputo, F., Gerola, H, Panagia, N.: 1970, Astrophys. Letters 5, 275.
Cassatella, A., Viotti, R.: 1974, Mem.Soc.astr.It. 45, 741.
Friedjung, M.: 1972, Astrophys. Space Sci. 19, 501.
Friedjung, M.: 1974a, Supernovae and Supernova Remnants, C. Batalli Cosmovici ed., Reidel, Dordrecht, 143.
Friedjung, M.: 1974b, Astrophys. Space Sci. 29, L5.
Friedjung, M., Malakpur, I.: 1971, Astrophys. Lett. 7, 171.
Friedjung, M., Muratorio, G.: 1980, Astr. Astroph. 85, 233.
Friedjung, M., Muratorio, G.: 1987, Astron. Astrophys. in press.
Friedjung, M., Viotti, R.: 1976, Astron. Astrophys. 53, 23.
Lambert, D.L.: 1969, Nature 223, 726.
Muratorio, G.: 1985, These de Doctorat d'Etat, Universite' de Marseille, Marseille.
Muratorio, G., Friedjung, M.: 1987, Astron. Astrophys. in press.
Pagel, B.E.J.: 1969a, Nature 221, 325.
Pagel, B.E.J.: 1969b, Nature 223, 727.
Thackeray, A.D.: 1967, Mon. Not. R. astr. Soc. 135, 51.
Viotti, R.: 1969, Astrophys. Space Sci. 5, 323.
Viotti, R.: 1970, Mem. Soc. astr. It. 41, 513.
Viotti, R.: 1976, Astrophys. J. 204, 293.
Viotti, R., Ricciardi, O.: 1974, in Colloquium on Supergiant Stars, M. Hack ed., Osservatorio Astronomico, Trieste, p.233.

Wellmann, P.: 1951a, Zeit. Astrophys. 30, 71.
Wellmann, P.: 1951b, Zeit. Astrophys. 30, 88.
Wellmann, P.: 1951c, Zeit. Astrophys. 30, 96.
Zickgraf, F.-J., Wolf, B., Stahl, O., Leitherer, C., Klare, G.: 1985, Astron. Astrophys. 143, 421.
Zickgraf, F.-J., Wolf, B., Stahl, O., Leitherer, C., Appenzeller, I.: 1986, Astron. Astrophys. 163, 119.

THE UV FE II EMISSIONS OF THE SYMBIOTIC STAR CH CYG[°]

C.Marsi, P.L.Selvelli
Astronomical Observatory of Trieste
Via Tiepolo 11 -Trieste - Italy.

ABSTRACT. Fe II is the principal component in recent UV spectra of CH Cygni. The total UV Fe II luminosity is of about
10^{34} erg s^{-1} if d=330 pc. The weakness of UV radiation since
Jan. 1985 rules out continuum fluorescence as excitation mechanism. The anomalous intensity of some multiplets whose
upper terms lie around 11 eV, is probably due to a selective
fluorescence mechanism with Lyman alpha.The SAC method has
been applied in order to determine the physical parameters
of the emitting region. The results indicate $T_{exc} = 7850 °K$,
$R \sim 10^{12}$ cm, and $n_e \sim 10^{10}$ el.cm^{-3} .

1-INTRODUCTION

The symbiotic star CH Cyg (M6III+?) during active phases shows emission
lines of Fe II both in the optical and in the UV spectral
range (Hack and Selvelli (1982), Marsi and Selvelli (1986)).
Starting in January 1985, in correspondence with a strong
decrease (by a factor of about 10) in the UV continuum intensity (Mikolajewska,Selvelli,and Hack,1986), the number of
Fe II emissions has dramatically increased, especially in the
UV range, where about 400 lines have been clearly detected.
        The IUE high resolution spectra that have been used
in this study of CH Cygni have been calibrated absolutely,
thus providing line flux measures (erg cm^{-2} s^{-1}) for all FeII
emissions. This fact has made possible both a quantitative
analysis and a detailed comparison with the expectations of
the semi-empirical method based on the self-absorption curve.

[°]Based on observations by the International Ultraviolet Explorer (IUE) collected at the Villafranca Satellite Tracking
Station (VILSPA) of the European Space Agency.

R. Viotti et al. (eds.), Physics of Formation of FeII Lines Outside LTE, 295–300.

2-THE DATA

The instrumentation and performances of the IUE satellite
are described by Boggess et al.(1978).
IUE images of CH Cyg have recently been taken at high resolution (SWP 24955, SWP 25833, SWP 28011 for the λ 1175-2050
range, LWP 5256, LWP 5880, LWP 7553, for the λ 1900-3200
range). Since the y-axis of the IUE high resolution spectrum
is not calibrated, we have followed the method described by
Cassatella et al. (1982,1983) in order to obtain the total
flux (erg cm^{-2} s^{-1}) in the line.

3-THE FE II SPECTRUM

Fe II is the principal component of the line spectrum of CH
Cyg. It is worth recalling that up to, but not including,
January 1985 the UV continuum was quite strong and the Fe II
spectrum generally appeared in absorption in the far UV range
(UV 8,9,37-43,96-102), with the sole exception of UV 191,
and generally in emission in the near UV (UV 1-3,35,36,60-64
78,391). This rather peculiar behavior has been interpreted
(Hack and Selvelli,1982) in terms of continuum fluorescence.
Starting from Jan. 1985, the UV continuum decreased strongly
and all Fe II lines appeared in emission, together with several lines of higher and lower excitation, e.g. OI,NI,SiII,
SiIII,CIII,NIII,CIV,NV, and possibly HeII.(Selvelli and Hack,
1985). Most FeII emissions show a triangular profile , with
a mean value for FWHM of about 55 km s^{-1}.The few exceptions
are: UV 1, in which the width is larger but the blue wing
is weakened by an absorption component; UV 62,63,and 64 which
show an absorption dip whose exact position changes with time;
UV 60 and 78, whose lines have larger wings than the other
lines (the FWHM is ~60 km s^{-1}, while FWOI is ~300 km s^{-1}).
The emission profiles for the emissions other than FeII
range from a quite sharp one (FWHM~0.3-0.4 Å) for the neutral
species like OI and NI,to a rather broad one (FWHM~400 km s^{-1})
for the more ionized species like CIII ,SiIII ,SiIV, and CIV.
It is notable that in January 1985 a very broad and strong
Lyman-alpha emission, previously completely absent, has appeared (Fig.1). Its half-width at zero intensity is about 2200
km s^{-1}.Its center appears red-shifted by 300 km s^{-1} and is cut
by a strong absorption (3.8 Å wide) centered at the nominal λ .
The Lyman-alpha emission flux is of the order of 1.3 10^{-10}
erg cm^{-2}s^{-1} , but the calibration curve is quite uncertain at
this wavelength. The ratio of the sum of all the UV FeII
emissions to Lyman alpha is larger than 5, but this value is
affected by the severe re-absorption in Lyman-alpha.
      Prominent UV multiplets are UV 60,62,63,64,78,158, and
especially UV 191 (Fig. 2), whose lines have intensities of
the order of 10^{-11} erg cm^{-2}s^{-1} . Spectrograms taken after Jan.
1985 have shown, by and large, the same FeII features as tho-

se of Jan. 1985, the main difference being found in the line flux; the FeII emissions observed in May 1985 appeared weaker by roughly a factor of about 3 in flux.

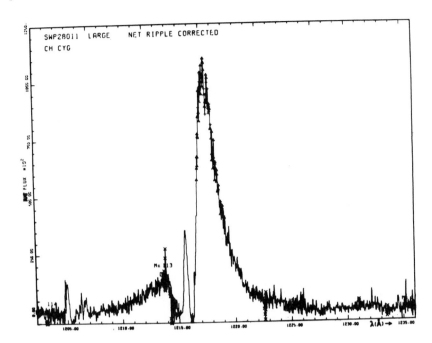

Figure 1. The Lyman-alpha emission.

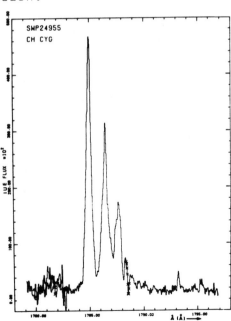

Figure 2. The region near multiplet UV 191.

Some lines and multiplets, i.e. λ 1869.53,2406.66,2503.87,
2506.74,2508.30, UV 363,373,380,391,399, are abnormally in-
tense(Fig. 3). These emissions come from high-lying levels,
odd quartets and sextets higher than 11 eV, which, as suggested by Johansson and Jordan (1984), can be populated by
Lyman-alpha absorption from a ^{4}D. The decay can be direct, as
in the case of λ 2406.66 and 2506.74, or there may be a cascade emission trough IR lines which gives rise to UV 373,
UV 391, UV 399, etc.

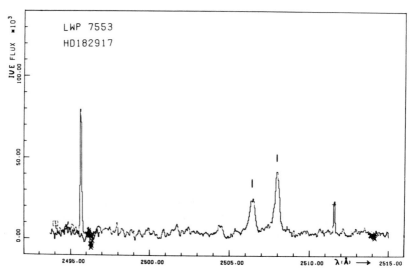

Figure 3. The emissions at λ 2506.7 and 2508.3 whose anomalous
strength is due to Lyman-alpha pumping.

        The anomalous intensity of UV 191 has not yet been ex-
plained satisfactorely.Viotti et al.(1980) have suggested
fluorescence with Si II UV 4. In CH Cyg, however, in all
available spectra, the UV 191 emissions are much stronger
than those of Si II UV 4. Alternative explanations have been
proposed by Nussbaumer (1981), in terms of a selective collisional excitation from a ^{6}S to x ^{6}P, by Jordan (these proceedings) in terms of Lyman-alpha pumping if collisional transfer of populations between nearby levels occurs, and by
Johansson (these proceedings) in terms of dielectronic recombination. All these suggestions are attractive but require
a detailed study of the physical conditions in the emission
region of CH Cyg where UV 191 is produced.

4-THE SELF-ABSORPTION CURVE (SAC)

Muratorio and Friedjung (1986) have developed the SAC method
for studying FeII lines observed in emission and absorption.

We refer to this paper for the several details which cannot
be described in this contributed paper. Essentially, the SAC
is a curve of growth for emission lines obtained in the case
of an optically thick medium. Physical information is derived
from plots of the quantity $\log (F\lambda^3/gf)$, related to the line
flux F (in erg $cm^{-2} s^{-1}$) contained in an emission line, versus
the quantity $\log gf\lambda$, related to the optical thickness of
that line. A comparison and contrast of different plots made
for different multiplets, which are then compared with a "theoretical SAC", whose slope depends on the geometrical model
(spherical wind, disk, etc.) of the emitting region, provides
information about the excitation temperature and size of
the volume where the Fe II lines are produced. The empirical
SAC for a generic multiplet i can be superimposed on that
for muliplet 1 (reference multiplet) with a shift in both
axes: the orizontal shift is related to the relative populations of the lower (L) terms n_i^L/n_1^L, while the vertical shift
is related to the relative populations of the upper (U) terms
n_i^U/n_1^U. Plots of $\log (n_i^L/n_1^L)$ versus E_i^L-0 eV and $\log(n_i^U/n_1^U)$
versus E_i^U-4.80 eV give two straight lines whose slope gives
the excitation temperature T_{exc}. From the upper terms we have
obtained T_{exc}=7600$\pm$200 °K, while the lower terms give T_{exc}=
8100$\pm$250 °K.
       The construction of a unique empirical SAC and its
comparison with a theoretical one provides information on
the size and structure of the emitting region. In the theoretical curve we have assumed for the line-broadening velocity the value v_o=55 km s^{-1} and for the distance the value
d=330 parsecs.

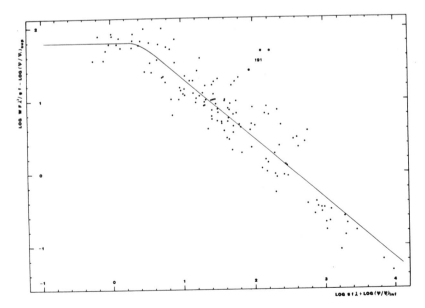

Figure 4. The self-absorption curve.

Figure 4 represents the observed SAC and its theoretical fit.
The slope is -0.87 ± 0.03 , close to that (-0.84) of a model
of a spherical wind with a velocity greater than the turbulent velocity. The horizontal shift which makes the theoretical and the observed SAC_U to overlap gives log $n_1^L \sim 16.0$.
The vertical shift gives n_1^U and R_o, the radius of the emitting region. n_1 can be estimated in two ways: 1) assuming a
Boltzmann distribution with $T_{exc} = 7850 \,^\circ K$, which sets an upper
limit for log n_1 of about 12.93^c, 2) using a 3-levels atom
model populated by collisions and decays (this is justified
by the lack of UV radiation), which gives log $n_1 \sim 10.8$.
The two corresponding values for R_o are $R_o \gtrless 2.1 \times 10^{11}$ cm and
$R_o \sim 2.2 \times 10^{12}$ cm ,respectively. If R_o is known, it is possible
to estimate the electron density N_e assuming standard Fe
abundance and that all Fe is Fe^+ . The N_e values which
correspond to the radii reported above are $N_e \lesssim 5 \times 10^{10}$ el.
cm^{-3} and $N_e \sim 4.4 \times 10^9$ el. cm^{-3} ,respectively. It is notable
that the ratio SiIII 1892/CIII 1908 gives $Ne \sim 9 \times 10^9$ el. cm^{-3}.
     The slope of the SAC indicates a wind model. A rough
estimate of the mass-loss rate can be obtained from the
above reported values of v_o and R_o with the assumptions
$Fe \sim Fe^+$ and $N(H) \sim N_e$. The result is $\dot{M} \sim 4 \times 10^{-8}$ $M_\odot$ yr^{-1} .

REFERENCES

Boggess A. et al. 1978, Nature Vol. 275 p.2.
Cassatella A.,Ponz D., Selvelli P.L. 1982, ESA IUE Newsletter
     No. 10, p. 31.
Cassatella A.,Ponz D., Selvelli P.L. 1983, ESA IUE Newsletter
     No. 15, p. 43.
Hack M.,Selvelli P.L., 1982 Astron.Astrophys.107,200.
Johansson S., Jordan C.,1984 MNRAS 210,239.
Mikolajewska J.,Selvelli P.L. , Hack M., IAU Coll. 93,in press
Muratorio G., Friedjung M, 1986, MNRAS ,in press.
Nussbaumer H.,Pettini M., Storey P.J., 1981 Astron. Astrophys.
     102,351.
Selvelli P.L., Hack M., 1985 ,Astronomy Express 1,115.

ANALYSIS OF THE FeII LINES IN THE ULTRAVIOLET SPECTRUM OF THE VV CEP STAR KQ PUPPIS

A. Altamore[1], G.B. Baratta[2], M. Friedjung[3],
G. Muratorio[4], R. Viotti[5]

1. Istituto Astronomico, Universita la Sapienza, Roma, Italy
2. Osservatorio Astronomico, Roma, Italy
3. Institut d'Astrophysique, CNRS, Paris, France
4. Observatoire de Marseille, Marseille, France
5. Istituto Astrofisica Soaziale, CNR, Frascati, Italy

ABSTRACT. The ultraviolet spectrum of the VV Cep variable KQ Pup is very rich in FeII lines with a P Cygni profile. We have measured the FeII absorption and emission lines from 1250 A to 3214 A in the IUE high resolution spectra of the star taken in 1979 and 1980. Emission lines were analyzed using the Self Absorption Curve method of Friedjung and Muartorio. The level population is close to a Boltzmann-type distribution with T=6700 K and 6000 K for the lower and upper terms respectively. The FeII column density is about 2x10+20 cm-2. Possible models of the emission and absorption line regions are discussed.

The VV Cep variable KQ Pup (HR 2902, Boss 1985) is known for having a very rich FeII spectrum in the near-UV (Swings 1969) and in the space ultraviolet (Altamore et al. 1982). Altamore et al. have identified some 600 spectral lines, the large majority of which belong to FeII transitions. This ion is represented by narrow absorption and/or emission lines with a mean radial velocity difference emission minus absorption of +40 km/s. It was also found that the FeII line flux varies according to the excitation potential, oscillator strength and wavelength. Figure 1 shows a portion of the ultraviolet spectrum of KQ Pup where lines of both low and high excitation potential and of different oscillator strength are present. In particular, Altamore et al. (1982) identified several high excitation emission lines not present in the Moore's (1952) Ultraviolet Multiplet Table, but included in the extensive line list of Kurucz (1981). Thus this star represents an ideal target

301

R. Viotti et al. (eds.), Physics of Formation of FeII Lines Outside LTE, 301–305.

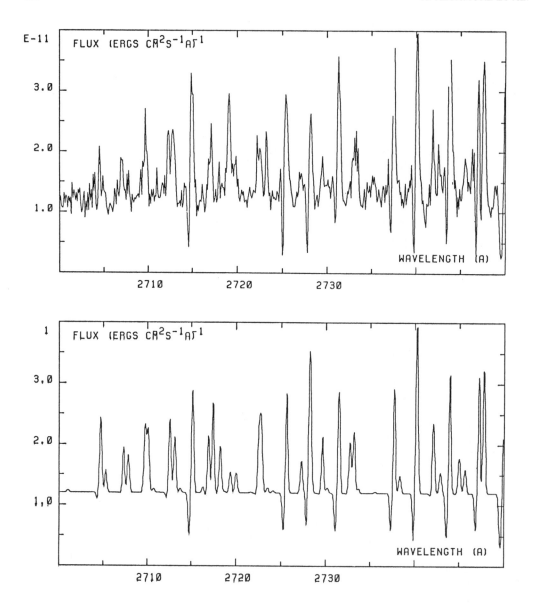

Figure 1. Above: the high resolution IUE spectrum of KQ Pup near 2700-2750 A. Several narrow FeII lines with a P Cyg profile are present belonging to both low and high excitation levels. Below: the synthetic spectrum for the same region (see text).

the preliminary results of an analysis of the FeII lines in the ultraviolet spectrum of KQ Pup based on IUE high reso- lution data obtained in Febrary 1979 and February 1980.

The IUE spectra were reduced at the Centro d'Informa- tica of the Istituto di Astrofisica Spaziale, Frascati and the line intensities were measured at the Centro di Calcolo of the Istituto Astronomico, Universita La Sapienza, Roma, using the CRAS interactive code developed by A. Altamore and C. Rossi. Both equivalent widths and line intensities were obtained for all the absorption and emission lines. For each spectral range (SWP and LWR) of IUE, long and short exposure images were used for the faintest spectral features and to avoid saturation of the strongest emission lines.

The line intensities were analyzed at the Marseille Observatory using the Self Absorption Curve method deve- loped by Friedjung and Muratorio (1986) and described by Muratorio (1985) and by Friedjung in this volume. The interstellar extinction was assumed equal to zero. For the emission lines, self-absorption plots were drawn for diffe- rent multiplets, and the excitation temperature of the lower and upper levels derived from the relative shifts of the plots. In this study the Kurucz (1981) extensive list of oscillator strength was used. We find that the popula- tion of the lower (0-3.4 eV) and upper (4.8-8.6 eV) levels are not far from a Boltzamann-type distribution with mean excitation temperatures of 6700 K and 6000 K respectively. Using these values we have obtained for the FeII emission lines the Self Absorption Curve in Figure 2. The large majority of the lines fall near the sloping part of the curve indicating that most of the emission lines are strongly self-absorbed. The observational points have been fitted with a theoretical curve which has been computed assuming a disk-like structure of the emitting region and a FeII column density of 2x10+20 cm-2. The scatter of the points appears larger than expected on the basis of the observational errors. This could be explained in different ways: (i) uncertainty on the atomic data; (ii) line blends; (iii) deviation of the level population from a Boltzmann distribution as a result of the presence of selective mechanisms of level population. This last point is particu- larly evident in a number of cases, including the lines of multiplet 191 at 1785.26 and 1786.74 A (the third component of the multiplet is affected by an IUE reseau mark, see Figure 2), line 1362.77 of multiplet 152, and the 5p6F^o-c4F and 4p4G^o-c4F transitions at 2506.76 and 2508.34 A. The observed fluxes of these lines are much larger than the computed ones, and suggest the presence of a very effective excitation mechanism in the envelope of the KQ Pup system. We have also analyzed the violet-shifted absorption lines of FeII and found that they mostly fall in the saturated

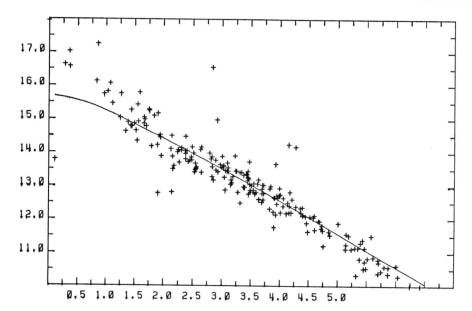

Figure 2. The Self Absorption Curve for the FeII emission lines in the UV spectrum of KQ Pup. Abscissae are logarithm of the normalized optical thickness, ordinates are log of normalized line intensity. The excitation temperature is 6700 and 6000 K for the lower and upper terms respectively. The line is the theoretical Self Absorption Curve computed as discussed in the text.

part of the curve of growth for an expanding atmosphere (see Friedjung this volume for more details).

The theoretical emission and absorption line curves were used to synthetize the ultraviolet spectrum of KQ Pup. Figure 1b shows the synthetic spectrum of KQ Pup computed for the same spectral region of Figure 1a. In general the computed spectrum well represents the observed one within the uncertainty represented by the scatter of the points in Figure 2, whose nature has been discussed above. In addition, neglecting lines of other ions does not significantly affect the goodness of our fit (with the obvious exception of few strong lines, such as the MgII doublet).

In conclusion, KQ Pup certainly represents a good laboratory for the study of the formation of FeII lines in dilute media and, especially to investigate the excitation mechanisms of this ion. Future high S/N ultraviolet observation of this and similar objects are important for a progress in our understanding the FeII problem in astrophysical objects.

REFERENCES

Altamore, A., Giangrande, A., Viotti, R.: 1982, Astron. Astrophys. Suppl. Ser. 49, 511.
Friedjung, M., Muratorio, G.: 1986, Astron. Astrophys. submitted.
Kurucz, R. L.: 1981, Smith. Astr. Obs. Spec. Rep. No. 390.
Moore, C. E.: 1952, Ultraviolet Multiplet Table, NBS Circular No.488, Section 2.
Muratorio, G.: 1985, These de Doctorat d'Etat, Universite de Marseille, Marseille.
Swings, J.P.: 1969, Astrophys. J. 155, 515.

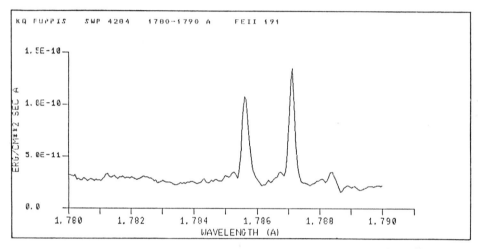

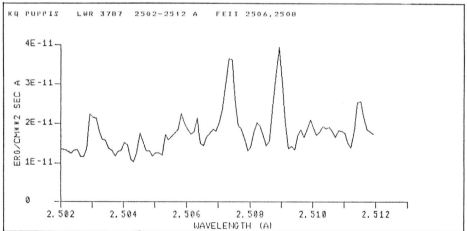

Figure 3. The spectrum of KQ Pup near 1780-1790 A and 2502-2512 A showing the high excitation FeII emission lines.

FeII SPECTRAL SYNTHESIS OF THE ULTRAVIOLET SPECTRUM OF HOT STARS WITH COOL WINDS

G. Muratorio
Observatoire de Marseille, Marseille, France

M. Friedjung
Institut d'Astrophysique, CNRS, Paris, France

R. Viotti
Istituto Astrofisica Spaziale, CNR, Frascati

FeII emission lines are frequently observed in the optical spectra of luminous blue stars. They provide evidence for the presence of cool extended envelopes, probably resulting from intense stellar winds. The ultraviolet spectra of these stars are generally strongly affected by absorption and emission lines of singly ionized iron formed in the expanding atmospheric envelope. A typical example is the LMC emission line hypergiant S22. Bensammar et al. (1983) and Muratorio and Friedjung (1986) have shown that the ultraviolet energy distribution of this star is strongly depressed by FeII envelope absorption, especially between about 1550 and 1750 A. The strength of these features is related to the FeII column density through the envelope and to the expansion velocity, hence to the mass loss rate. FeII line absorption dominates the ultraviolet spectrum of peculiar galactic stars such as P Cyg, Eta Car and AG Car, of the LMC variable S Dor and of the Hubble-Sandage variables. In most cases the study of the UV spectrum of these objects is difficult because of their faintness and no high resolution spectra are available which could allow a detailed analysis of the FeII line profiles. On the other hand, at low spectral resolution the many FeII lines present in the UV are strongly blended with each others, so that their study requires a different kind of approach involving the comparison of the observed UV spectra with theoretical spectra which include the effects of the envelope absorption and emission, as first suggested by Bensammar et al. (1983).

The technique of spectral synthesis has been developed by Muratorio et al. (1984), Muratorio (1985), Friedjung and Muratorio (1986) and applied to the study of the early type emission line stars (Muratorio, 1985; Muratorio and Fried-

R. Viotti et al. (eds.), Physics of Formation of FeII Lines Outside LTE, 307–310.

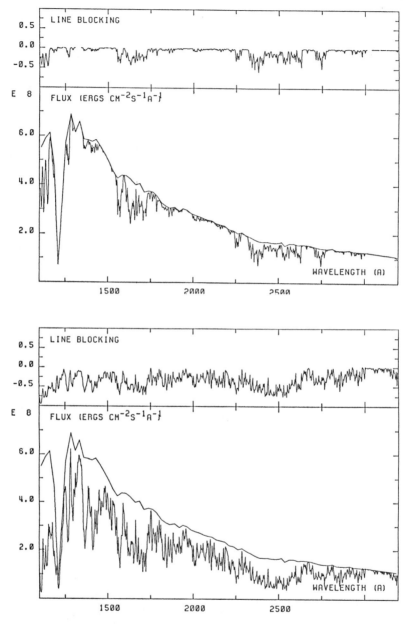

Figure 1. Synthetic UV spectrum of a luminous hot star with cool wind. The Kurucz' (1979) model atmosphere with T=15000 K and a bolometric luminosity of -9 mag was adopted for the central star. For the wind we assume T(FeII)=6000 K and Ṁ/Vexp = 10-9 (upper plots) and 10-7 (lower plots). For each case the FeII line blocking and the energy distribution are given.

jung 1986). Details concerning the technique are given in Friedjung talk in this volume. It has been found that these stars are surrounded by rather cool winds were UV FeII absorption takes place. In some cases FeII emission lines are most easily understood as formed in a disk-like region, rather than in an isotropic wind. But this result requires further studies. In the following we assume spherical symmetry of the winds.

Following the guidelines described in the above quoted works, we have computed a grid of synthetic spectra with the aim of investigating the dependence of the UV spectrum on the structure of the wind, namely density, expansion velocity and mass loss rate. For this study we have used the Kurucz (1981) FeII data and Kurucz (1979) model atmospheres. Computations were carried out for wind velocities (Vexp) from 100 to 500 km/s with relative mass loss rates (M/Vexp) from 10-7 to 10-10 Mo/yr.km/s. Iron was assumed to be all in the first stage of ionization with level excitation temperature below 10000 K. These conditions generally apply to all the early type stars with optical FeII emission and, in particular the Luminous Blue Variables (LBV's).

Figures 1 gives the results of the computations for two representative cases. The adopted energy distribution of the central star is that of a Kurucz' (1979) model atmosphere with stellar radius of 9x10+12 cm, corresponding to a bolometric luminosity of -9 mag, typical of a P Cyg star. We suppose the wind to start near the stellar photosphere with a constant expansion velocity (the results however do not critically depend on the velocity law). The excitation temperature of FeII is taken Tex=6000 K, in agreement with the values found by Muratorio and Friedjung (1986) in luminous blue stars. For the mass loss rate, normalized to the expansion velocity Vexp, we used two values, M/Vexp=10-9 and 10-7 Mo/yr/km/s which should encompass the typical values found in luminous early type stars. For each case we give the synthetic spectrum and the FeII line blocking, that is the computed spectrum normalized to the Kurucz model atmosphere. The blocking by the FeII absorption is particularly strong in a number of spectral regions near 1600-1800 and between 2200 and 2800 A. In these regions the energy distribution is depressed by about 30% and 60-70% in the two cases respectively. Thus the amount of energy substracted by the FeII blocking in the wind is quite large in the conditions which are expected to occur in several luminous stars in the MC's and in outer galaxies, and should play a fundamental role in any energy balance consideration.

Although these preliminary models are rather crude, they indicate the need for including the effects of a cool wind in the model atmospheres. Future developments of this method should take into account the asymmetry of the winds and the contribution of other atomic species such as NiII and CrII and doubly ionized metals, and the effects of different chemical composition.

REFERENCES

Bensammar, S., Friedjung, M., Muratorio, G., Viotti, R.:
    1983, Astron. Astrophys. 126, 427.
Friedjung, M., Muratorio, G.: 1986, Astron. Astrophys. sub-
    mitted.
Kurucz, R.L.: 1979, Astrophys. J. Supplem. Ser. 40, 1.
Kurucz, R.L.: 1981, Smithonian Astrophys. Obs. Special
    Report no.390.
Muratorio, G.: 1985, These d'Etat, Universite' de Provence.
Muratorio, G., Friedjung, M.: 1986, Astron. Astrophys. sub-
    mitted.
Muratorio, G., Friedjung, M., Viotti, R.: 1984, Proc.
    Fourth European IUE Conference, ESA SP-218, p.309.

SESSION 4

FUTURE PLANS AND CONCLUSIONS

FUTURE LABORATORY WORK ON FE II

Sveneric Johansson
Department of Physics, University of Lund
Sölvegatan 14, S-223 62 LUND, Sweden

I have been asked by the Chairman of the Conference to say something
about future laboratory work on Fe II. In a way it is impossible to make
a specification of the future experimental program at the laboratories,
where measurements of atomic parameters of Fe II are in progress.
Present activities concerning energy levels, transition wavelengths,
oscillator strengths and branching ratios have been reported at this
colloqium with promises about future work in these subfields.
Theoretical calculations of all kinds of quantities including ionization
parameters and collision strengths have been reviewed and new
computations of transition probabilities for forbidden lines have been
reported. We have also been informed about an extensive theoretical
treatment of the total Fe II spectrum with millions of lines and
gf-values.
    All atomic data on Fe II are of astrophysical significance.
Statements by the astrophysicists express the needs for more data. The
question for the future is how far the atomic physicists are willing to
go in order to meet the demands from the astrophysical community.
Several laboratories have funding problems doing e.g. classical
spectroscopy in the shadow of "Big Science" or applied physics. Joint
projects on detailed analyses of laboratory and stellar spectra may be a
possibility to illustrate the interplay and mutual dependence of two
closely related branches, which normally are ascribed to two different
scientific fields: Atomic physics and astrophysics. The composition of
the program and the participation list of this colloquium demonstrates
the strong interaction between the two. A look into the future has been
made by American scientists in the book "Physics through the 1990s.
Atomic and Molecular Physics" sponsored by the National Research
Council. From Chapter 7, dealing with scientific interfaces I want to
make a quotation from the section on Astrophysics. "Interpreting the
abundant data of astrophysics demands a deep understanding of atomic,
molecular and optical processes. In addition, it demands a broad data
base of atomic and molecular parameters such as transition energies,
oscillator strengths, and photon and particle cross sections. Providing

R. Viotti et al. (eds.), Physics of Formation of FeII Lines Outside LTE, 313–315.

these data is a major challenge for atomic and molecular physics.
Experimental data flow from all branches of the field, particularly from
the discipline that has come to be called laboratory astrophysics. These
experimental data are vital, but more data are required than the
experimental community can possibly provide us. Thus, theoretical data
are also vital. The need to generate theoretical data for
astrophysicists motivates a major portion of the theoretical effort in
the atomic and molecular community."

Current experimental work on Fe II, that I am aware of, concerns
atomic data for energy levels, transition wavelengths, life times and
f-values. Studies of atomic processes like charge-transfer and the
PAR-process (Photoexcitation by accidental resonance) are also in
progress. In general all these types of work are reported in journals of
Atomic Physics, but all published papers, which are of relevance for
astrophysics, are communicated to the "consumers" by the Chairmen of the
Working Groups in Commission 14 of IAU. Every three years a short review
and an exhaustive reference list are given on atomic transition
probabilities (WG 2), collision cross sections and line broadening (WG
3), structure of atomic spectra (WG 4) and molecular spectroscpopy (WG
5) in the "Reports on Astronomy" in the series of transactions of IAU.

The current work on energy levels and transition wavelengths in Fe
II by B. Baschek and myself has been reported at this meeting. New
spectroscopic data are now available in the wavelength region 850 -
50000 Å but our first goal is to analyse the region covered by IUE and
the Space Telescope. Eventually we hope to update the Ultraviolet
Multiplet Table. The laboratory data have been recorded at the NBS,
Washington D.C., and at Kitt Peak Solar Observatory, in the latter place
with a Fourier Transform Spectrometer (FTS). A similar instrument for
the VUV region has been built at Imperial College, London (ICL) by A. P.
Thorne and coworkers and nice spectra of Fe II down to 1850 Å have
recently been recorded and communicated to me by R.C.M Learner. The
uncertainty in wavelength is of the order 0.1-0.3 mÅ at 2000 Å. The
inclusion of the FTS data in our analysis will definitely improve the
accuracy, facilitate the analysis and - due to the wealth of new
information - delay the work. The project with C.R. Cowley on high-level
transitions of Fe II in optical spectra of stars will continue and
demands further analysis of highly excited states.

Laboratory work on oscillator strengths by means of a
wall-stabilized arc has been reported by Moity. His published data cover
the long-wavelength range of IUE, and new recordings in the optical
region are planned. S. Kroll and M. Kock in Hannover have combined hook
measurements with emission FTS data in their measurements of f-values
for 124 Fe II lines in the region 2200 - 5400 Å. The combination of new
measurements of branching ratios and the life-time data from Hannaford
and Lowe have permitted U. Pauls and coworkers in Zurich to get very
accurate f-values for faint Fe II lines around 7500 Å, suitable for
abundance determinations. Again the emission data refer to the Kitt Peak
FTS, now used with a narrow band filter in order to improve the signal
to noise ratio. All experimental efforts to obtain accurate oscillator
strengths and more energy levels will improve the conditions for such
theoretical calculations of gf-values, where a parameter fitting method

is used. In this context one could also mention the possibilities to
obtain relative gf-values from stellar spectra, which will be more
reliable with the use of highly-resolved spectra from HST.

Two technical achievements have definitely had a great influence on
the renewed interest in stellar and laboratory spectroscopy of iron
group elements in general and Fe II in particular, viz IUE and FTS. For
both facilities the wavelength accuracy is set by the observed radiating
plasma itself, i.e. the star and the laboratory source. IUE spectra of
metal-rich stars contain still a large number of unidentified lines, for
which a good wavelength determination increases the possibility to make
positive identifications. Differences in the plasmas that generate the
stellar and laboratory spectra can be systematically studied and used
for the term analysis. The next generation space telescopes will
hopefully improve the possibilities to do interactive analyses of
stellar and laboratory spectra. As regards the FTS it provides an
unparalleled accuracy for wavenumbers in spectra, which are feasible to
register with the instrument. Compared to photographic recordings the
number of lines is less at short wavelengths, but this shortage is well
compensated for by the improved accuracy. Thus, a combination of
highly-resolved spectra on photographic plates and the Fourier Transform
recordings gives the optimum conditions for a complete analysis with
accurately determined energy levels in neutral and singly ionized
elements. The latest contribution for achieving this optimum is the FTS
at Imperial College, London.

Atomic processes that play an important role for excitation of Fe
II in astrophysical plasmas have in recent years been investigated in
the laboratory. Charge-transfer collisions (in the laboratory source)
between iron atoms and noble gas ions in low-temperature plasmas have
been studied at Lund and a detailed study of this mechanism by means of
FTS data is in progress at Caltech by Whaling and coworkers. Population
of Fe II levels by recombination has been proposed as an important
mechanism both in laboratory and astrophysical plasmas. Future work is
planned at Lund trying to distinguish between the contributions from
charge-transfer and recombination in the population of highly-excited
states in Fe II. The PAR-process has proven to be prominent in different
astrophysical objects or systems, revealed by anomalous intensities in
specific emission lines. Experiments are planned to study this mechanism
in the laboratory using synchrotron radiation from the new facility MAX
at Lund. By irradiating a gas of iron ions with monochromatized
radiation, corresponding to wavelengths for resonance lines of
cosmically abundant elements, we hope to be able to detect fluorescence
from selectively photoexcited levels and thereby simulate the conditions
in stellar atmospeheres.

Finally, I wish to thank the initiators of this conference, R.
Viotti and M. Friedjung, for bringing together people with experiences
in theoretical and experimental/observational atomic physics and
astrophysics, who all have strong feelings for the spectrum of Fe II,
wherever it appears. Undoubtedly, the great success of the meeting is
partly due to the conference site, chosen with distinction by the Local
Organizing Committe.

Fe II PROSPECTS IN SOLAR PHYSICS

Robert J. Rutten
Sterrewacht "Sonnenborgh"
Zonnenburg 2
3512 NL Utrecht
The Netherlands

1. SOLAR PHYSICS OVERVIEW

Just like the other fields of astrophysics, solar physics changed with the advent of radio astronomy and space astronomy. The emphasis shifted from radiative transfer toward the nonthermal structures and phenomena exhibited by the solar atmosphere, most notably the flare and the magnetic structuring into photospheric tubes and coronal loops. Solar physics has become a field of MHD and plasma physics.

This is evident if you read the reviews collected in the NASA-CNRS volume "The Sun as a Star" (Jordan 1981; free copies are still available from S.D. Jordan, code 682, NASA-GSFC, Greenbelt MD 20771). That volume was intended to help re-establish the link between solar and stellar physics, traditionally strong in the era of radiative transfer research but weakened when the shift towards hydromagnetics made solar physics a somewhat isolated field. Now that the IUE and Einstein satellites have demonstrated that the other cool stars exhibit similar magnetic complexity, a new solar-stellar link called "magnetic activity" has formed, mainly populated by solar physicists turned stellar (e.g. A. Dupree, C. Jordan, J. Linsky, R. Noyes, R. Rosner and C. Zwaan with their coworkers).

(Another new solar-stellar link may be forged out of helioseismo-logy. Its diagnostics of the solar interior are of obvious importance to the theory of stellar structure and stellar evolution. However, I wonder whether resolving the magnetic fine structure of the solar interior and generating the concomitant interpretative complexity may not temporarily result into similar isolation as studying the magnetic fine structure of the solar atmosphere did in the past).

There is a second space revolution coming up that follows naturally on the opening-up of the full electromagnetic spectrum: acquiring sufficient spatial resolution to study the processes that underlie the phenomena. This requires resolving the latter on the scale of the former, often about 0.1 arcsec in the photosphere (photon mean free path, pressure scale height). This revolution will undoubt-edly come, the only question being when; by providing insights in dynamic, magnetic and plasma processes it will establish guidelines to

R. Viotti et al. (eds.), Physics of Formation of FeII Lines Outside LTE, 317–320.
© *1988 by D. Reidel Publishing Company.*

the rest of astrophysics much as resolving the solar spectrum did for radiative transfer in the past. Fe II lines will play their role in this coming revolution. They are bound to be important diagnostics of the processes in the upper photosphere and lower chromosphere, which is precisely the layer where the magnetic field replaces the gas pressure as the chief structuring agent of the atmosphere.

The Sun does not stop at the angular scale of 0.1 arcsec in providing astrophysical enigma. The space age should mature enough that scales much smaller than 100 km become observable, through space interferometry, short-wavelength imaging and near-Sun observation. Far-future prospects have already been formulated ("Prospects for the 21st Century", report of a National Solar Observatory Workshop, Jan. 1986).

2. SOLAR PHYSICS FACILITIES

The spatial-resolution revolution requires instrumentation that is now being realised, though much slower than anticipated. First and foremost, a large solar telescope must be put into orbit, able to resolve the photospheric flux tubes which constitute the basic ingredient of the activity phenomena. The decade-old Solar Optical Telescope project, originally planned for shuttle flights in the early eighties, was finally killed this spring just before the Challenger disaster. A scaled-down version called HRSO has taken its place, with its first flight planned for 1992. This is a facility as important for solar physics as Space Telescope will be for non-solar astrophysics.

In addition to a solar space telescope, there must be adequate ground-based instrumentation to pursue the new insights that space-resolved flux tubes, granules, sunspots etc. will generate. Ground-based observation can not compete with space observation in spatial resolution, but it provides the flexibility and the extended time coverage needed for evolutionary and follow-up studies.

The current situation is not good. The only facility where high spatial resolution is regularly obtained is the Sacramento Peak Observatory. Although an attempt to kill it last year was aborted, it appears now to be bled to death slowly. Taken together with the demise of solar physics at Mount Wilson and the scarcity of university programs in solar physics in the USA, this may imply jeopardizing the next generation of American solar physicists.

Indeed, the forefront in ground-based high-resolution solar physics seems to be shifting to Europe. The German solar physicists are installing a large observatory on Tenerife, including a major vacuum tower telescope (Schröter et al. 1985). The French are going to build a large polarization-free telescope at the same site (Mein and Rayrole 1985). The Swedes have just completed a simple but superb vacuum telescope on La Palma (Scharmer et al. 1985) which may prove to be the best on earth at the best location on earth for spatial resolution. The Swedes and the Norwegians are the driving force behind the LEST Foundation which aims to build a 2.40 m polarization-free vacuum telescope in the nineties (Stenflo 1985). The LEST Foundation is already the most international of the telescope-building consortia: current members are from Australia, China, Germany, Israel, Italy, Norway, Sweden, Switzerland and the USA. It has moved into its first phase of realization, with O. Engvold (Oslo) as project director.

3. FUTURE Fe II OBSERVATIONS

In conclusion I want to point out a promising specific type of Fe II observation from space, which is to study the behaviour of the Fe II lines near the solar limb. Although the solar limb spectrum observed during eclipse was the original motivation and testing ground for developing solar NLTE radiative transfer theory (Thomas and Athay 1961), the hazards and problems of eclipse spectrometry combined with its difficulty of interpretation have resulted in its decline when the outer solar atmosphere became observable from space, seen in projection against the solar disk at short wavelengths.

With the space revolution the limb becomes resolvable and highly interesting again. Optically thick observation inside the limb yields valuable diagnostics to radiative transfer because it shows lines still in photospheric conditions but raised to the height of formation where NLTE effects are dominant rather than second-order refinements. Examples are Canfield's (1969) classic analysis of rare earth limb emission lines, emission wings caused by coherency in resonance lines (Rutten and Milkey 1979; Rutten and Stencel 1980) and the large spatial intensity variation of Fe II lines seen near the limb (Canfield and Stencel 1976; Canfield et al. 1978; Rutten and Stencel 1980; Cram et al. 1980).

Above the limb, in optical thin conditions at least for the continuum, the solar spectrum at high spatial resolution is the ideal testing ground for checking NLTE mechanisms such as the pumping of Fe II by Ly-α (Johansson and Jordan 1984). The inhomogeneous nature of the chromosphere enriches this diagnostic by providing information with large variation in density and other state parameters, seen radially against the disk and laterally from the limb. This richness is evident in the literature on solar prominences, which seem to exist specifically to satisfy NLTE-PRD radiative transfer specialists.

REFERENCES

Canfield, R.C.: 1969, Astrophys. J. _157_, 425

Canfield, R.C., Stencel, R.E.: 1976 Astrophys. J. _209_, 618

Canfield, R.C., Pasachoff, J.M., Stencel, R.E., Beckers, J.M.: 1978, Solar Phys. _58_, 263

Cram, L.E., Rutten, R.J., Lites, B.W.: 1980, Astrophys. J. _241_, 374

Johansson, S., Jordan, C.: 1984, Mon. Not. Roy. Astr. Soc. _210_, 229

Jordan, S.D. (ed): 1981, "The Sun as a Star", NASA-CNRS Monograph
        Series on Non-thermal Phenomena in Stellar Atmospheres", NASA
        Special Publication 450

Mein, P., Rayrole, J.: 1985, Vistas in Astronomy 28, 567

Rutten, R.J., Milkey, R.W.: 1979, Astrophys. J. 231, 277

Rutten, R.J., Stencel, R.E.: 1980, Astron. Astrophys. Suppl. 39, 415

Scharmer, G.B., Brown, D.S., Pettersson, L., Rehn, J.: 1985, Applied
        Optics 24, 2558

Schröter, E.H., Soltau, D., Wiehr, E.: 1985, Vistas in Astronomy 28,
        519

Stenflo, J.O.: 1985, Vistas in Astronomy 28, 571

Thomas, R.N., Athay, R.G., 1961: "Physics of the Solar Chromosphere",
        John Wiley Co., New York

PROSPECTIVES OF GROUND BASED OBSERVATIONS

E. Joseph Wampler
European Southern Observatory
Karl-Schwarzschild-Str. 2
D-8046 Garching bei München, FRG

The study of FeII emission phenomena in quasars and other extragalactic objects will be revolutionized by the next generation of instruments being planned by ESO. In ten years it will be possible to obtain, for 16-mag objects and in a reasonable observing period, spectra with resolution exceeding $\lambda/\Delta\lambda = 5 \times 10^3$ with signal-to-noise ratios better than 30 throughout the optical/near infrared wavelength region. These spectra, when combined with UV spectra obtained with the Hubble Space Telescope (HST) will permit the detailed examination of FeII emission in the rest frame of selected extragalactic objects from below the Lyman limit to wavelengths longer than 1 μ-meter. It will then be possible to extend the techniques pioneered by H. Netzer and B. and D. Wills (Netzer and Wills, 1983; Wills, Netzer and Wills, 1985) to constrain the acceptable models of the F_e^+ region in quasars and active galaxies. The next generation spectra will have sufficient signal-to-noise ratios to identify and measure the strength of key multiplets, such as multiplets 191, 188 and 167 (Jordan, 1986; Wampler, 1985). It will be possible to measure the relative strengths of density dependent multiplets, such as the ratio of multiplet 36 to multiplet 42 (Wampler, 1985). And using deconvolution techniques one will be able to compare the FeII line profiles with those of the Balmer lines, MgII λ 2800 and CIV λ 1548. We will then be in a position to identify the excitation and ionization mechanisms of F_e^+, accurately determine the abundance of F_e^+ and study the possible evolution of this abundance with redshift.

Some of these next-generation instruments exist or are in construction. Oliva et al. (1986) have reported in this conference the first study of infrared forbidden FeII lines in supernovae remnants using the new ESO infrared spectrophotometer, IRSPEC. A description of the instrument is given in the June 1986 ESO Messenger (Moorwood et al., 1986). IRSPEC opens up the 1-3 μ meter band for spectrophotometric studies of the brighter quasars, nearby galaxies and galactic objects. Its resolution $(\lambda/\Delta\lambda \approx 2000)$ is sufficient for isolating interesting features and resolving broad emission lines in active galaxies and quasars.

R. Viotti et al. (eds.), Physics of Formation of FeII Lines Outside LTE, 321–322.
© *1988 by D. Reidel Publishing Company.*

The New Technology Telescope (NTT) is now under construction and will be equipped with a high efficiency multi-purpose optical spectro-photometer. A description of this instrument (EMMI) is given by Dekker et al. (1986). It will give resolution up to $\lambda/\Delta\lambda = 40 \times 10^4$. The NNT/EMMI combination will give about a factor of 2 greater speed than can now be obtained using CASPEC on the 3.6-meter telescope.

The greatest step in sensitivity will be taken when the ESO Very Large Telescope (VLT) is commissioned about a decade from now. The VLT will have 20 times the light gathering power of the 3.6-meter tele-scope. Because the study of FeII objects is usually not sky background limited this increase in collecting power can be translated directly into an improvement of nearly a factor of 4.5 in the signal-to-noise ratio or an improvement in resolution by a factor of 20 at the same signal-to-noise ratio.

The instrumentation on the VLT will take full advantage of the experience ESO is acquiring in instrumenting the NTT and the 3.6-meter telescope. By the end of this century the VLT when teamed with the new space telescopes will represent the most powerful observing facility in the world.

REFERENCES

Dekker, H., Delabre, B., and D'Odorico, S.: 1986, Proc. SPIE 627, 339.
Jordan, C.: 1986, IAU Coll. No. 94, "Physics of Formation of FeII
      Lines Outside LTE" (Reidel: Dordrecht).
Moorwood, A.F.M., Biereichel, P., Finger, G., Lizon, J.-L., Meyer, M.,
Nees, W., and Paureau, J.: 1986, The Messenger 44, 19.
Netzer, H., and Wills, B.J.: 1983, Ap.J. 275, 445.
Oliva, E., Moorwood, A.F.M., and Danziger, I.J.: 1986, IAU Coll. No.
      94, "Physics of Formation of FeII Lines Outside LTE" (Reidel:
      Dordrecht).
Wampler, E.J.: 1985, Ap.J. 296, 416.
Wills, B.J., Netzer, H., and Wills, D.: 1985, Ap.J., 288, 94.

SPACE TELESCOPE AND THE FE II PROBLEM

Kenneth G. Carpenter
Center for Astrophysics and Space Astronomy
University of Colorado
Boulder, CO 80309-0391 USA

ABSTRACT. The capabilities of the Hubble Space Telescope, especially as they apply to the "Fe II problem" are reviewed. The major spectroscopic observing modes of both the Faint Object Spectrograph and the High Resolution Spectrograph are discussed and compared with those of the IUE satellite, with a detailed illustrative look at the options available for observations of Fe II (UV 1) near 2600Å. Finally, the major impacts to be expected on astrophysical investigations of Fe II from the Space Telescope and associated instrumentation are summarized.

1. INTRODUCTION

The Hubble Space Telescope (HST) is frequently thought of solely in terms of "edge-of-the-universe" astronomy and physics. However, HST will be extremely valuable to a very wide range of astronomical research topics, many of which will in fact not be directly related to the origin or final fate of the universe as a whole. In this category, we can certainly place the "Fe II problem" that we have discussed in great length at this meeting. Without a doubt, our attempts to "solve" this problem and understand the physics of Fe II and of the plasmas in which it is formed will benefit greatly from the capabilities of this first large-scale orbiting astronomical observatory.

The HST has an f/24 optical system with a 2.4 meter diameter primary. The telescope design provides a clear collecting area of nearly 40000 square centimeters and an angular resolution on the order of 0.1 second of arc. The large size of the primary mirror, in combination with five science instruments built to take advantage of the telescope's light gathering power and spatial resolution, will provide a quantum leap in our abilities to obtain high signal-to-noise UV spectra, high spatial resolution two-dimensional images, and high-speed photometry, of faint astronomical targets. In addition, it will permit very high resolution (R=100000) spectroscopic observations of brighter targets. The five science instruments on HST consist of the High Resolution Spectrograph (HRS), the Faint Object Spectrograph (FOS), the Wide-Field/Planetary Camera (WF/PC), the Faint Object Camera (FOC), and the High Speed Photometer (HSP).

R. Viotti et al. (eds.), Physics of Formation of FeII Lines Outside LTE, 323–328.

   In this paper I shall summarize the characteristics and
capabilities of HST with frequent reference to those of the extremely
successful International Ultraviolet Explorer (IUE) satellite, which
has provided for more than eight years a reliable facility for the
acquisition of ultraviolet spectral data. I will concentrate
primarily on the spectrographic instruments on-board HST, since they
will be of the greatest interest to those interested in the Fe II
problem; however, the possible usefulness of other instruments will be
briefly noted as well.

2. SPECTROSCOPIC MODES AVAILABLE WITH IUE AND HST

2.1 Overview

There are a large number of possible spectroscopic observing modes on
HST. Without counting the numerous additional combinations possible
when the various apertures are paired with each of the spectrograph
gratings, the HST has 22 grating modes available to the user between
the FOS and HRS. The FOS has 12 available apertures while the HRS has
two science apertures. In addition, the two cameras have objective
prism/grism capabilities and the FOC a long-slit spectrographic mode.
The capabilities of all the instruments, both spectroscopic and otherwise, are described in detail in a series of handbooks prepared by the
Instrument Support Branch of the Space Telescope Science Institute.
These handbooks were distributed with the call for guest observer
proposals distributed in the fall of 1985. The reader is referred to
these handbooks for details beyond those presented in this very brief
summary.
   The major grating configurations available with the FOS and HRS
are listed in Table 1, along with the IUE grating modes. The first
column gives the mnemonic for the instrument and observing mode, the
second column the wavelength range in A over which the mode can be
utilized, the third column the resolution of the mode in A, and the
final column the wavelength coverage (in A) in one frame of data
(i.e., one observation) for each mode. The FOS entries in the table
are for those modes available with the blue tube, except for the G780H
mode which is only available with the red tube. All of the listed FOS
configurations have very similar counterparts with the red tube,
except for mode G130H, which is centered at 1300Å where the red tube
has no sensitivity.
   The FOS provides in most of its modes a resolution intermediate
between the low and high resolution IUE modes, with a comparably wide
wavelength coverage. The HRS provides, in all but one mode, a
superior resolution to IUE, but at the expense of very limited wavelength coverage per obervation. However, perhaps the most important
distinctions between the IUE spectrographs and the two HST spectrographs are the increases in sensitivity, dynamic range, and achievable
signal/noise ratio available with the latter two instruments. Both
the FOS and HRS use photon counting linear digicons instead of the SEC
vidicons of IUE. The use of these digicons provides a tremendous

Table 1. Summary of Spectrographic Modes on IUE and HST

Instrument/ mode	Wavelengths accessible	Resolution (Å)	Angstroms per single observation
IUE			
Low	1150–2000	6	850
Low	2000–3230	6	1230
High	1150–2000	.16	850
High	2000–3230	.26	1230
FOS/BL			
G160L	1150–2523	6.5	1373
G650L	3530–5500	25.	1970
G130H	1150–1608	1.0	458
G190H	1575–2332	1.5	757
G270H	2227–3306	2.1	1079
G400H	3244–4827	3.1	1583
G570H	4583–6885	4.5	2302
RD G780H	6274–9259	5.8	2985
HRS			
G140L	1050–1800	.57	286
G140M	1050–1700	.054	27
G160M	1200–2000	.069	34.5
G200M	1600–2400	.078	39.5
G270M	2200–3200	.091	46
ECH A	1060–1700	.011–.018	5.4–9.2
ECH B	1700–3200	.017–.035	8.4–17.6

increase in dynamic range over that possible with IUE, with the HRS and FOS providing a dynamic range in excess of 10^{+7} versus the IUE dynamic range of 255. At the same time, with photon statistics limiting the achievable signal/noise (at least up to a S/N near 60), the S/N achievable with the HST spectrographs is much greater than the 20 or so available with IUE. In fact, with some care, a S/N on the order of 100 should be possible with either the FOS or the HRS. In addition, the much smaller apertures of the HRS and FOS will provide a much enhanced spatial resolution over that of IUE, if desired. However, there are penalties to be paid in going from IUE to HST. The FOS cannot match the high resolution capability of IUE, while the HRS provides greater resolution only at the expense of losing the comprehensive wavelength coverage available in a single observation with IUE.

In order to illustrate some of these points, let us examine in more detail the options available for observing the Fe II resonance multiplet UV 1 near 2600Å with the various UV spectrographs.

2.2 Detailed Example: 2600Å (Fe II UV 1)

The lines of UV multiplet 1 are among the most interesting of the Fe II transitions in the mid-UV spectral range since they appear in a wide variety of objects and are usually quite strong. It may thus be most instructive to make a detailed comparison of the instruments for the region around 2600Å. Table II summarizes the instruments and modes available with IUE and HST for use in this spectral region. The first column defines the instrument and mode, the second column the resolution in A, the third column the resolution in velocity units (km/sec), the fourth column the sensitivity in units of counts/diode/sec per ergs/cm^2/sec/Å, and the last column the wavelength range covered by a single observation centered at 2600Å. The IUE sensitivity has been cast into these units by assuming that the inverse sensitivity curves published in the IUE guides (for use in determining optimal exposure times) yield a signal-to-noise ratio of 10 and thus are comparable to exposures with photon-counting devices yielding a total of 100 counts/diode. The entries are arranged in order of increasing resolution.

Table II. Characteristics of UV spectrographs at 2600Å

Instrument Mode	Resolution Angstroms	km/sec	Sensitivity (cds)/(ergs/cm^2/s/Å)	Angstroms per observation
IUELO	6	692	$5 \times 10^{+11}$	1230
FOS/BL G270H	2.1	241	$5 \times 10^{+14}$	1079
IUEHI	0.26	30	$8 \times 10^{+10}$	1230
HRS G270M	0.091	10.5	$8 \times 10^{+12}$	46
ECH B	0.030	3.5	$2 \times 10^{+12}$	13

The various entries in this table represent a wide range of possible resolutions, sensitivities, and wavelength coverage per observation. IUE and FOS provide a wide wavelength coverage per single observation and are thus most appropriate for studying a large number of lines at moderate resolution. The FOS has the greatest sensitivity of any of the instruments at 2600Å, while providing a resolution of about 2Å (intermediate between IUEHI and IUELO) and will be the best instrument for very faint sources. However, if high resolution is the major scientific requirement, then clearly the HRS is the instrument of choice. The HRS provides both superior resolution and sensitivity to that of the IUE high resolution mode (IUEHI). The only disadvantage, relative to IUEHI, is the very small wavelength range available in a single frame of data. It will take a large number of exposures with the HRS to scan the full range obtainable with IUE in a single observation. This will of course not be a problem if the observer is interested only in the UV 1 lines which occur in a fairly small range of wavelengths. Table 3 summarizes the advantages and disadvantages of each instrument for observations centered on 2600Å.

Table III.   Relative advantages/disadvantages of UV spectrographs at
             2600Å

Instrument	Advantages	Disadvantages
FOS	sensitivity, coverage S/N	resolution
IUEHI	resolution, coverage	sensitivity, S/N
HRS	resolution, sensitivity, S/N	coverage
IUELO	coverage	resolution, sensitivity, S/N

One advantage of IUE not pointed out in the above table is that it is actually in orbit and currently available for use. The same may not be the case for the HRS and FOS for quite some time yet.

3. MAJOR IMPACTS OF HST ON FE II RESEARCH

The Hubble Space Telescope will provide us with many enhanced capabilities that should be of great utility in the pursuit of the Fe II problem.
     The HST with the FOS and HRS provide us with a much higher sensitivity and a much larger dynamic range than IUE. This will place a much wider range of both stellar and non-stellar objects into the

"observable universe", and allow us to better judge the generality of conclusions based, in many cases, only on observations of the nearest or brightest members of a particular class.

The higher spectral resolution, in combination with the higher signal-to-noise, possible with HRS will allow us to obtain a much better velocity resolution and more detailed line profiles. This will allow us to make significantly more precise determinations of the velocities of Fe II plasmas both in absolute terms and relative to other ions in the same or nearby regions. The higher signal-to-noise profiles will further enhance our ability to study the velocity fields in the plasmas in which the Fe II emission originates.

A higher spatial resolution will also be possible with both the FOS and HRS. The IUE has just two, relatively large apertures: a 3 arcsec circle and 10x20 arcsec oval. In contrast, the FOS has a dozen possible apertures ranging in size from 0.3 arcsec round to 4.3 arcsec square, while the HST has two small, square apertures 0.25 arcsec and 2.0 arcsec on a side, respectively. I should point out that these smaller apertures, while useful for most observations of point sources, may actually be a disadvantage for faint extended sources, where all that matters is the number of photons to be captured.

Finally, there is a possibility of high spatial resolution imaging with the FOC and WF/PC. This presents the possibility of obtaining two-dimensional information on the source regions. Furthermore, these images can be obtained through filters to study the distribution of emission from specific ions (such as Fe II), which should be very helpful in our attempt to understand the Fe II line formation mechanisms and the plasmas in which they operate.

CAN WE BRIDGE THE GAP BETWEEN OBSERVATION AND THEORY?

M. Friedjung
Institut d'Astrophysique, CNRS
Paris, France

It is important to improve methods for interpreting observations of FeII lines produced by real astronomical objects. Detailed complex calculations including a very large
number of energy level (e.g. one million) are not necessarily very useful, as many assumptions about the nature of
the emitting regions and their geometry are included. It is
not clear what role the assumptions play, and those which
might be valid for one kind of object (e.g. Active Galactic
Nuclei) are not necessarily valid for others (e.g. stellar
winds).

What I believe to be a better approach, is to improve
semi-empirical methods. Theory and observation need to come
closer together. The shapes of emission line self-absorption curves need to be calculated in a more rigorous way
than done up to now, as a function of physical parameters,
taking radiative transfer properly into account. We also
need to know about what is sensitive to what, so different
geometries and physical models can be tested using self-
absorption-curve methods. Spectral synthesis may also be
improvable in the light of such considerations, both for
emission line self-absorption-curves and absorption line
curves-of-growth.

It may be interesting in the future to analyze FeII
emission line profiles using self-absorption-curve methods.
Emission at different radial velocities, when most line
broadening is due to the Doppler effect, may have different
self-absorption-curves, and the properties of these curves
could give useful clues about the physics. One might be
able to apply thsi to the results presented by Capenter at
this meeting. The same considerations are also true for
absorption line curves-of-growth.

Better quality observations of faint objects will
become available in future years, and more refined methods
will then be usable and justifiable.

R. Viotti et al. (eds.), Physics of Formation of FeII Lines Outside LTE, 329.
© *1988 by D. Reidel Publishing Company.*

CONCLUDING REMARKS

R. Viotti
Istituto Astrofisica Spaziale, CNR
Frascati, Italy

Why a conference on FeII? The spectrum of any astronomical object includes lines of so large a variety of atomic species in many different ionization stages that it is hard to understand the reason for organizing a scientific meeting purely devoted to a single ion, which certainly is not the most abundant one in nature. To give a clear answer, let me put the question in another way. Is it possible to do nowadays good Physics in Astronomy, or, in other words, is it possible to investigate quantitavely the physical processes that cause the observed astronomical phenomena? Is on the other hand Modern Physics able to explain all the celestial phenomena, or do we need more atomic data, and a better knowledge of the physical processes in the extreme conditions which characterize astrophysical plasmas?

To give an answer to these questions we need to study in detail a single, well representative case. In this regard, FeII is probably the best choice because of its very rich spectrum, and, as discussed during this meeting, for being so frequently observed and identified as 'peculiar' in very many stars and galaxies.

This conference was an attempt to gather experts from many different fields in order to discuss the FeII problem. In this regard, one major 'problem' of the FeII problem was to avoid that each expert would speak a language not understandable for persons working in other fields. That is that the conference become a kind of Tower of Babel.

My impression is that, in spite of the many lively discussions (or because of them!), and thanks to the wanderful environment of the Capri island, this conference has been a fruitful occasion for exchange of ideas, and to better understand the role of FeII in Astrophysics.

In this talk, instead of making a summary of the conference, I wish to discuss some examples, mostly based on my personal experience, which can better illustrate the early development of the FeII problem. I shall also discuss some methodological aspects of the problem and summarize the main ideas about present and future work.

R. Viotti et al. (eds.), Physics of Formation of FeII Lines Outside LTE, 331–340.

1. AGAIN ABOUT THE HISTORY OF THE FeII PROBLEM

When was the 'FeII Problem' raised for the first time? Professor Gratton discussed the first impact with the problem on him when he studied with Prof. Cecchini the spectrum of nova DQ Herculis. Indeed the optical spectrum of a nova near maximum is very rich in strong FeII emissions with a broad P Cygni profile. In the following decline phase, [FeII] lines appear and gradually strengthen with respect to the permitted ones. A similar behaviour was also shown by the well known southern variable η Car during the small outburst which occurred in 1889 and in the following years. It was fortunate that at that time the Harvard Observatory had a telescope (a 13-inch Boyden telescope) in the southern emisphere which allowed recording of stellar spectra. η Car was in fact observed since May 1892. Actually, the first report on the spectrum of this peculiar star is a short notice by Le Sueur which appeared in 1869. Charles Witney (1952) identified in the first spectra of the star the earlier members of the Balmer series in emission, and more than 100 absorption lines of neutral and singly ionized metals (obviously including FeII) displaced by -180 km/s with respect to the radial velocity of the hydrogen emission lines. Since 1895 the spectrum has consisted almost entirely of emission lines, and Dorrit Hoffleit (1933) described the spectral evolution of the star from 1892 to 1930 with the gradual increase of the intensity of the forbidden FeII lines. The large spectral change of η Car is illustrated in Fig.1. Actually, the earlier investigators did not identify [FeII]. It was Paul Merrill (1928) who has first identified [FeII] in η Car, and this was the

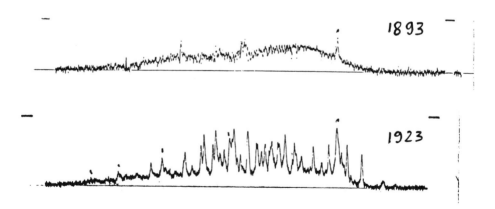

Figure 1. Tracings of the old spectropgrams of η Car of the Harvard Observatory obtained in 1893 during a minor "outburst", and at minimum in 1923.

first identification of these lines in an astrophysical (and laboratory) object. From this point of view, η Car may be considered as a good laboratory for investigating the FeII atomic structure. This is for instance illustrated by the work of Garstang (1962) and Johansson (1977). Presently η Car exhibits one of the most spectacular FeII spectrum both at optical (e.g. Gaviola 1953, Thackeray 1953) and ultraviolet wavelengths (Cassatella et al. 1979, Altamore et al. 1986, see Fig.2).

About 20 years ago I started my work on this puzzling object using several old spectroscopic plates collected at the Bosque Alegre Observatory during 1944 to 1954 and kindly put at my disposal by Professor Gratton. One could imagine the shocking impact of such a hard problem on a young beginner in Astronomy what I was at that time. Why does this star present so many strong emission lines of ionized iron? why does this ion which is not one of the most abundant ones in the stellar atmospheres contribute so much to the total energy output in the visual? why do permitted and forbidden lines have nearly the same intensity? My first years of research were devoted to this problem (Viotti 1969, 1970, 1976a,b etc.), but many questions remained without an answer. After the successful launch of the IUE satellite it was hoped to have a better insight on the problem from the ultraviolet observations. But IUE observations have instead added new problems. For instance, if we look the low resolution spectrum shown in Figure 2, we find that two of the strongest emission features, at 1785 and 2508 A, belong to transitions for high excitation FeII levels. The second one in particular is even not included in the Charlotte Moore's multiplet tables! Multiplet UV 191 is shown in Fig.3 at high resolution. Emission from high lying FeII levels was one of the major arguments during this conference.

There are two other objects which have played an important role in the FeII problem, γ Cas and the Sun. γ Cas is the prototype of the Be stars, and it is also the first emission line star observed spectroscopically. It was father Angelo Secchi who first identified long time ago, in 1866, the presence of bright lines in its spectrum. 70 years later during 1932 to 1942, γ Cas undewent a phase of large photometric and spectroscopic variations. At that time, the star displayed a rich FeII emission spectrum which was the basis for the important work of Wellman (1951) which is the first attempt of a non LTE analysis of FeII emissions in an astrophysical object. FeII lines are frequently observed in the optical spectra of blue emission line stars and display a variety of profiles not yet well understood.

Our stellar neighbor, the Sun, has been as always a good laboratory for the study of the fundamental data on

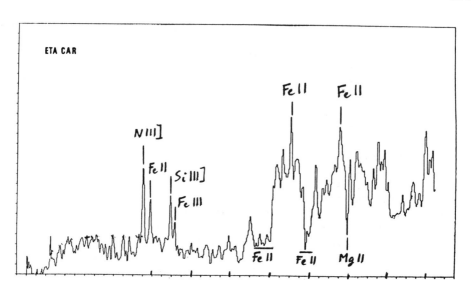

Figure 2. The low resolution ultraviolet spectrum of η Car
observed with the International Ultraviolet Explorer (IUE).
The strongest emissions are marked, including two FeII
features at 1785 and 2508 A which belong to transitions
from highly excited lelvels.

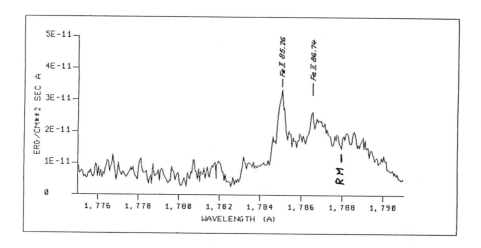

Figure 3. The high resolution ultraviolet spectrum of η Car
near the high excitation multiplet no.191 at 1785-88 A.

FeII, and for the study of LTE and non LTE formation of this ion. In particular, forbidden FeII lines were identified in absorption in the solar photospheric spectrum by Swings (1965). This provided an estimate of the solar iron abundance higher than that derived from the analysis of the permitted photospheric lines. A revision of the absolute scale of the FeII oscillator strengths finally solved this problem (e.g. Baschek et al. 1970). It should be recalled that FeII is present with a large amount of emission lines in the solar limb UV spectrum (Doschek et al. 1976), and the study of the FeII line intensity and profile may provide basic information about the structure of the solar and stellar chromospheres.

I think that still nowadays these three objects, η Car, γ Cas and the Sun, are among the best celestial laboratories for our work, both for suggesting what is wrong on the atomic data and for inspiring new researchs about the physical processes in stellar atmospheres and envelopes.

2. METHODOLOGICAL CONSIDERATIONS

Observation and modelling are the two opposite aspects of our work, sometimes in conflict with each other. But no model can be done without a clear knowledge of the processes of line formation and transfer. However, it is not easy to find which physical processes have to be included in the model, and whether all the important processes have been considered, or we have to introduce more physics. We therefore need some means to make a diagnosis, and in the case of FeII, this is helped by the presence of so many lines with different strength, excitation potential and wavelength.

Therefore, in many cases a kind of statistical approach appeared quite appropriate. This allowed for instance to separate the dependence of line intensity on the different parameters, such as the level population (or the level excitation temperature, Thackeray 1967), size of the emitting region (Friedjung and Malakpur 1971), continuum energy distribution (Viotti 1969) or interstellar extinction (Pagel 1969).

This approach was seriously humpered by the large uncertainty on the oscillator strengths and the lack of high quality data on emission line intensity which gives a large scatter of the observational points around the mean curves. Small deviations due to the departure of the level population from a Boltzmann-type distribution are therefore lost, and we cannot investigate in detail the mechanisms of level population. We hope to have in near future a new set of oscillator strengths with an accuracy comparable (or better) than that now possible for the line intensity.

In spite of the above problems concerning the accuracy
of the atomic and observational data, the statistical method appears quite stimulating and has already provided a
number of fundamental results which have been largely discussed during this conference. It is clear that in many
cases the observed deviations cannot be totally explained
by errors on the atomic data or on the measurements. This
is for instance true for the well quoted UV multiplet 191.
In figure 3 we show the shape of this multiplet in the high
resolution IUE spectrum of η Car. Like the other FeII
emissions, also this multiplet presents both broad and
narrow emission components. It should be also noted that in
Fig.3 we see only two narrow emission peaks, corresponding
to the lines 1785.26 and 1786.74, while the third one at
1788.00 A is missing. This may be due to an IUE Reseau Mark
near the position of that line. I think that this could be
taken as an example of how 'instrumental' effects could
produce misleading results.

Several works were devoted in the earliest senventy to
the determination of the excitation temperature of FeII
level from the emission line intensities. These temperatures should not considered as real gas temperatures. As
discussed by Viotti (1976a) and others, in low density
astrophysical media the FeII level population is a complex
function of the electron density and temperature and of the
diluted radiation field. The population of each level is
the result of the balance of many atomic processes which
are difficult to describe because of the very complex term
diagram. In other words, we should expect all the levels to
deviate from the mean curve, by an amount more or less
largely depending on the presence of selective excitation
mechanisms, like in the case of the upper term of multiplet
191.

The statistical approach appeared also fruitful in
those cases - novae, Seyfert galaxies - where the emission
lines are very broad and seriously blended with each other.
Viotti (1976b) has shown that there are characteristic
'signatures' of the FeII emission in the optical and
ultraviolet region which can be easily identified at low
resolution. Table 1 summarizes the main emission humps of
FeII and the relative intensity in the optically thin case
and assuming an excitation temperature of 10000 K. These
humps have been observed in many objects. For instance
Baratta et al. (1976) identified broad emission features of
FeII in objective prism spectra of Nova Cygni 1975, 2 to 7
days after maximum. The wavelength ranges of the FeII
features were 3480-3520, 4140-4190, 4210-4310 and 4450-4626
A. Their relative strength clearly indicate that FeII emission was optically thick. Wampler and Oke (1967) first
noted broad emission features in the optical spectrum of
3C273. They also noted a striking similarity to the

Table 1
Characteristic broad emission humps of FeII

Mean wavel.	wavel. range	intensity
2380 A	2320-2420 A	3.72
2610	2580-2640	2.15
2750	2730-2760	0.97
3210	3150-3350	1.77 (-2)
4210	4150-4250	3.54 (-4)
4370	4300-4450	2.94 (-4)
4550	4500-4650	7.8 (-4)
5290	5200-5350	2.42 (-4)
6190	6100-6250	4.2 (-5)
6460	6400-6500	3.52 (-5)

spectrum of Nova Herculis 1934 near maximum light. The
recent developments of this field have been largely discussed during this conference. The identification of the FeII
humps in the spectra of novae and Seyfert galaxies only
tells us about the presence of this ion. But in order to
determine its actual contribution to the energy spectrum,
in particular including the thousands of weak lines, it is
necessary to compute a theoretical FeII spectrum based on
some sorts of models and with a number of free parameters
to be adjusted in order to fit the observed spectrum. An
example of an FeII synthetic spectrum is shown in Fig.4.

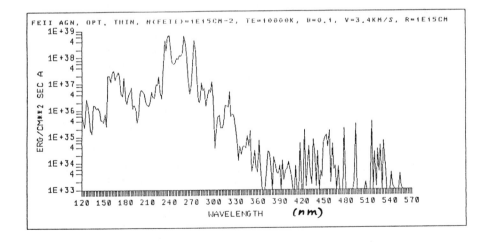

Figure 4. Example of a theoretical FeII emission spectrum.

FeII frequently is the dominant contributor to the absorption spectrum in the ultraviolet of many astrophysical objects, including novae, supernovae, cool and hot variable stars. Fig.2 is an example of an UV spectrum deeply 'shaped' by the FeII absorption. In most cases we are dealing with faint objects only observable at low resolution. For their analysis the computation of synthetic spectra is essential, but this is quite a hard job since one has to compute for a large amount of FeII transitions, line formation in a diluted atmosphere and in the presence of a velocity field.

At low resolution we lose many details and only obtain limited information about the physical structure of the emitting, or absorbing regions. A great progress will be done from the extensive study of the FeII line profiles at high resolution which is presently only at its beginning.

Finally one should consider that ions other than FeII could contribute to the emission and absorption spectrum of peculiar objects. For instance FeIII and NiIII are probably responsible for the broad humps at 1900 and 1640 A observed in the low resolution UV spectrum of β Lyrae (Viotti 1976b). The CoII 'signature' is present in the UV spectrum of Type I supernovae (Branch and Venkatakrishna 1986). Thus in many cases it appears necessary to include other ions in the theoretical models.

3. THE FUTURE

The FeII problem is so complex that it is hard to give a realistic prospect of its future. In any case Astronomy is beautiful because it is full of unexpected results! Let me discuss here only a few points. During this conference it has been agreed that FeII plays an important role in many aspects of Astrophysics. The FeII line emission gives a non-negligible contribution to the energy spectrum of novae and Seyfert galaxies. The FeII line opacity may deeply affect the UV spectrum of novae, supernovae and Be stars. In these cases FeII should have a major role in the energy balance, and this raises problems of inconsistency of the adopted models. In addition the 'cool' FeII is formed on a 'hot' continuum, what means that there is a temperature gradient. The geometry could also be far from spherical symmetry. In the next future the new generation of large collecting area ground telescopes, and the hopefully successful launch of the Hubble Space Telescope, will provide the optical and UV spectra of very distant objects, such as supernovae and Hubble Sandage variables in far away galaxies. New more refined model atmospheres and synthetic spectra should thus be developed to make a correct interpretation of their spectra, especially to derive an esti-

mate of the chemical abundance of their atmospheric envelopes.

Many interesting results are expected in the near
future from the infrared. FeII transitions between the
lower metastable levels produce near-IR lines which should
be visible in many astrophysical objects, such as emission
line stars, supernovae and SNR's, Seyfert galaxies, and
give precious information about the physical structure of
the cool parts of their envelopes. Fine structure transitions of the ground term should be poweful cooling agents
in diffuse media and nolecular clouds. Systematic study of
the 26 μm [FeII] line is thus very promising.

Concerning the FeII ion, sveral efforts have been made
to understand the mechanisms of level population and depo-
pulation in different physical environments. There is an
increasing interest on the upper Grotrian diagram of FeII
and on the doubly excited levels. It will be important to
further investigate the role of dielectronic recombination
on the level population and on the FeIII/FeII ionization
balance, which is also important in any abundance determination.

The above ones are only a few examples of the many
important results which have already been achieved on the
'FeII problem' in many different fields of Physics and
Astrophysics. In many cases FeII appeared as a real 'probe'
for the diagnostics of the non LTE phenomena. And an impul-
se for similar studies in other ions. It is clear that the
FeII problem is full of promises of fundamental future
developments. I hope that this conference has stimulated
enough atomic physicists to do more work, astronomers to
collect data, more systematically and with better accuracy,
and theoreticians to include in their models more and more
physics. This would be the best achievement of this conference.

REFERENCES

Altamore, A, Baratta, G.B., Cassatella, A., Rossi, L.,
    Viotti R.: 1986, New Insights in Astrophysics, Proceed.
    NASA/ESA/SERC IUE Conference, ESA SP-263, p.303.
Baratta, G.B, Smriglio, F., Viotti, R.: 1976, Astron.
    Astrophys. 53, 329.
Branch, D., Venkatakrishna, K.L.: 1986, Astrophys. J. 306,
    L21.
Cassatella, A., Giangrande, A., Viotti, R.: 1979, Astron.
    Astrophys. 71, L9.
Friedjung, M., Malakpur, I.: 1971, Astrophys. Lett. 7, 171.
Garstang,P.H.: 1962, Mon. Not. R. astr. Soc. 124, 321.
Gaviola, E.: 1953, Astrophys. J. 118, 234.
Hoffleit, D.: 1933, Harvard Bull. 893, p.11.

Johansson, S.: 1977, Mon. Not. R. astr. Soc. 178, 17P.

Le Sueur, A.: 1969, Proc. R. Soc. 18, 245.

Merrill, P.: 1928, Astrophys. J. 67, 391.

Muratorio, G.: 1985, These de Doctorat d'Etat, Universite'
    de Marseille, Marseille. See also Muratorio G. et
    al.,Fourth European IUE Conf., ESA SP-218, p. 309 (1984).

Pagel, B.E.J.: 1969, Nature, 221, 325.

Swings, J.P.: 1965, Ann. Astrophys. 28, 703.

Thackeray, A.D.: 1953, Mon. Not. R. astr. Soc. 113, 211.

Thackeray, A.D.: 1967, Mon. Not. R. astr. Soc. 135, 51.

Viotti, R.: 1969, Astrophys. Space Sci. 5, 323.

Viotti, R.: 1970, Investigations on the Emission Lines in
    Stellar Spectra, Thesis Scuola di Perfezionamento in
    Fisica, Roma University, July 1969, published in Mem.
    Soc. Astr. Ital. Vol.XLI, Fasc.4, pp. 513-542 (1970).

Viotti, R.: 1976a, Astrophys. J. 204, 293.

Viotti, R.: 1976b, Mon. Not. R. astr. Soc. 177, 617.

Wampler, E.J., Oke, J.B.: 1967, Astrophys. J. 148, 645.

Wellmann, P.: 1951, Zeit. Astrophys. 30, 96.

Whitney, H.B.: 1952, Harvard Bull. No.921, 8.

(*) Page numbers followed by "n" are those of the title
page of the article in which the subject is mentioned
several times.

ASTROPHYSICAL OBJECTS INDEX

STARS: